Low-temperature Activation and Catalytic Transformation of Methane to Non-CO_2 Products

Low-temperature Activation and Catalytic Transformation of Methane to Non-CO_2 Products

Franklin Tao

WILEY

This edition first published 2026

Registered Offices
John Wiley & Sons, Inc., 111 River Street, Hoboken, NJ 07030, USA
John Wiley & Sons Ltd. New Era House, 8 Oldlands Way, Bognor Regis, West Sussex, PO22 9NQ, UK

For details of our global editorial offices, customer services, and more information about Wiley products visit us at www.wiley.com.

The manufacturer's authorized representative according to the EU General Product Safety Regulation is Wiley-VCH GmbH, Boschstr. 12, 69469 Weinheim, Germany, e-mail: Product_Safety@wiley.com.

Wiley also publishes its books in a variety of electronic formats and by print-on-demand. Some content that appears in standard print versions of this book may not be available in other formats.

Library of Congress Cataloging-in-Publication Data
Names: Tao, Feng, 1971– author.
Title: Low-temperature activation and catalytic transformation of methane to non-CO_2 products / Franklin Tao.
Description: Hoboken, NJ: Wiley, 2026. | Includes index.
Identifiers: LCCN 2024016451 (print) | LCCN 2024016452 (ebook) | ISBN 9781394193257 (hardback) | ISBN 9781394193264 (adobe pdf) | ISBN 9781394193271 (epub)
Subjects: LCSH: Methane. | Activation (Chemistry) | Catalysis.
Classification: LCC TP359.M4 T36 2024 (print) | LCC TP359.M4 (ebook) | DDC 665.7/76–dc23/eng/20240515
LC record available at https://lccn.loc.gov/2024016451
LC ebook record available at https://lccn.loc.gov/2024016452

Cover Design: Wiley
Cover Image: Angew. Chem. Int. Ed. Volume 55, Issue 43, October 17, 2016

Set in 9.5/12.5pt STIXTwoText by Straive, Pondicherry, India

Contents

Preface

Methane, despite being the simplest organic molecule, holds immense significance in our daily lives. From household cooking and heating, chemical production, to electricity generation in power plants, methane plays a pivotal role. It serves as a raw material for producing essential intermediate compounds used in manufacturing a wide range of carbon-containing functional materials, including polymers, textiles, cosmetics, medicines, and fertilizers. The utilization of methane thus presents a double-edged sword, necessitating efficient utilization for chemical production and energy conversion while mitigating its effects on environment.

This book aims to act as a textbook for junior or senior majoring in chemistry or chemical engineering to provide them with insights into the formation, sources, emissions, and transformations of methane, particularly focusing on its activation and catalytic conversion to non-carbon dioxide products at room temperature or relatively low temperature (25–250 °C). In addition, it is intended to serve as a reference book for researchers and scientists in chemistry, chemical engineering, energy science, materials science, and environmental science as well as engineers and experts in these fields.

The journey through this book begins with an exploration of the significance of methane in Chapter 1, followed by an examination of its unique physical and chemical properties contributing to its chemical inertness in Chapter 2. Chapter 3 briefs how methane is formed in nature including various rocks and seabeds, through biomethanation, and from anthropogenic activities including industrial processes. How methane is extracted from nature is introduced in Chapter 4. Chapter 5 introduces how methane emission contributes to environmental issues. Chapter 6 discusses the high-temperature catalytic transformation of methane to syngas, ethane/ethylene, aromatics, hydrogen, and formaldehyde. The conversion of chemical energy in methane to electrical energy through intermediate temperature fuel cell processes are briefly discussed in Chapter 7. Chapter 8 provides the background information for understanding why chemical transformations of methane under mild conditions in terms of relatively low temperatures (<250 °C) are feasible and preferred.

Chapters 8–11 lay the groundwork for understanding various activation and catalytic transformation mechanisms of methane on metal atom-containing sites under mild conditions, paving the way for in-depth discussions in these chapters after them. These four chapters discuss fundamental studies on methane activation and catalytic conversion by single-atom cations or metal-oxygen cluster cations at low pressure and low temperature, providing insights into the development of catalytic sites consisting of metal atoms. The thermodynamics of low-temperature chemical transformations of methane are discussed in Chapter 8. The activation of methane by free-standing single-atom cations of transition metals in vacuum in terms of low pressure of methane is presented in Chapter 9. Activation of methane by free-standing MO^+ clusters at low pressure of methane is reviewed in Chapter 10. Oxidation of methane catalyzed by free-standing MO^+ clusters at low pressure of methane is reviewed in Chapter 11. These fundamental studies discussed in Chapters 8–11 provide insights on the origin of the activity of catalytic sites consisting of metal atoms to be discussed fully in Chapters 12–21. The catalytic oxidation of methane through single-atom metal oxide clusters M_1O_n anchored on an open support is reviewed in Chapter 12. The organometallic approach of transforming methane to value-added chemicals under mild oxidative catalytic conditions in liquid solutions is briefed in Chapter 13. Then, the selective low-temperature oxidation of methane by solid organic catalysts is discussed in Chapter 14.

Chapters 15–21 discuss the activation and catalytic transformation of methane by microporous materials at low temperatures (mostly ≤150 °C). Confinement effect of micropores of microporous aluminosilicate is briefed in Chapter 15 since the next several chapters deal with the activity and catalytic transformation of methane on the metal sites encapsulated in the micropores of various zeolites. Experimental methodologies employed in exploring low-temperature activation and catalytic conversion of methane are briefly introduced in Chapter 16, followed by detailed discussions on activation and catalytic oxidation mechanisms using N_2O, O_2, and H_2O_2 on metal cation-based sites encapsulated in zeolite in Chapters 17–21. These chapters delve into the selective oxidation of methane to form value-added non-carbon dioxide products such as acetic acid and methanol. In Chapter 17, activation of methane to form methanol through the oxygen atoms bonded to metal atoms anchored in zeolite activated with N_2O is discussed. The catalytic oxidation of methane on M–O sites encapsulated in zeolite and on the surface of AuPd nanoparticles are discussed in Chapter 18. The activation and catalytic oxidation of methane by molecular O_2 on M–O clusters immobilized in zeolite are reviewed in Chapter 19. Catalytic synthesis of acetic acid through carbonylation of methane with the participation of molecular oxygen on M–O clusters embedded in zeolites is discussed in Chapter 20. The challenges to be tackled toward efficient, green transformation of methane to value-added non-CO_2 chemicals at low temperatures are presented and potential solutions discussed in Chapter 21.

I highly appreciates the support from the Catalysis Science Program, Chemical Sciences, Geosciences, and Biosciences Division, Office of Basic Energy Sciences, Office of Sciences, the U.S. Department of Energy. The published results from my research group cited in this book were achieved in the project supported by the Catalysis Science Program, Chemical Sciences, Geosciences & Biosciences Division, Office of Basic Energy Sciences (BES), Office of Sciences, the U.S. Department of Energy, under grant no. DE-SC0014561.

2025

Franklin Tao

Acknowledgments

I sincerely express my gratitude to the community of scientists for their outstanding contributions to the field of methane activation and catalytic transformation. Their work has been invaluable in the preparation of this book. Although only a limited number of examples from them and my research group are discussed or cited here, readers are encouraged to explore the broader literature for more comprehensive information and details. As some readers may know, I endured a horrific ordeal stemming from false accusations and a legal prosecution for which I was, ultimately, fully exonerated. Nevertheless, the experience nearly cost me everything and was devastating emotionally and financially. What sustained me through those years were my family, my love for science, my faith, and my preservation in fighting for justice. From the outside, it may be hard to imagine how extraordinarily difficult it was for me and my family in these years, and how hard it continues to be. I deeply grieve for the tragic cases of professors who, under similar pressure and mistreatment, chose to end their lives. Few can understand their feelings more profoundly than I do. This is a very heavy page in the history of scientific community. I hope that my persistent pursuit of justice—and the ordeal I endured—has helped prevent similar wrongful allegations.

The tremendous support from my wife, my son, and my daughter enabled me to step away from despair and move forward. I am deeply grateful for their unwavering love, as well as for the help and encouragement I received from my children in the preparation of this book. My devotion to science gave me the strength to finish it. This book serves as my contribution to the scientific community, reflecting my perseverance through extreme hardship during the exceptionally challenging period of my life. I hope that this book provides you with valuable insights into the chemical transformation of methane. I wish you all the very best!

1

Why Do We Care About Methane?

1.1 Chemical Production

Methane (CH_4) plays a significant role in two key sectors: chemical production and energy supply. It is anticipated to emerge as a substitute for petroleum, serving as a foundational raw material in chemical industries. Over the last decades, the maturation of shale gas extraction techniques, such as horizontal drilling and hydraulic fracturing, has significantly lowered the cost of shale gas. Consequently, the usage of CH_4 as a raw material in the production of intermediate compounds has increased in the past decade, although presently, the primary raw material for chemical and manufacturing industries is still petroleum.

One of the major applications of CH_4 is the production of H_2 at an industrial scale through steam reforming for chemical production, food processing, and fertilizer synthesis. Other than these conventional applications, industries have increasingly turned to CH_4 to produce chemical intermediates, thereby creating value-added chemicals for manufacturing industries, although petroleum is still the dominant raw material for years or even decades. CH_4 can undergo chemical transformation through syngas route and direct transformation routes (Figure 1.1). In direct pathways, CH_4 is converted into synthesis gas (a mixture of CO and H_2) via steam reforming, dry reforming, or partial oxidation. Synthesis gas serves as a precursor to numerous vital intermediate compounds of chemical industries, including methanol, hydrocarbons, hydrogen, and many others. These intermediate can be used to produce a great number of value-added chemicals such as acetic acid, formaldehyde, vinyl acetate, allyl alcohol, alkylbenzene, propylene oxide, ethanol, acrolein, ethyl chloride, other alcohols, Hydrogen peroxide, alkylamine, aniline, nitrile, and ammonia. These value-added chemicals serve as building blocks for manufacturing various functional materials, including plastics, polymers, cloths, cosmetics, and medicines (Figure 1.1).

Particularly noteworthy is the routes of chemical production from CH_4 that avoids releasing environmentally harmful CO_2. Unfortunately, most CH_4 transformation reactions listed in Figure 1.1 occur at high temperatures, demanding substantial energy input, costly production facilities, and extensive maintenance. Furthermore, many of these transformations yield CO_2 as a byproduct. Particularly noteworthy are these routes of chemical

Low-temperature Activation and Catalytic Transformation of Methane to Non-CO_2 Products,
First Edition. Franklin Tao.

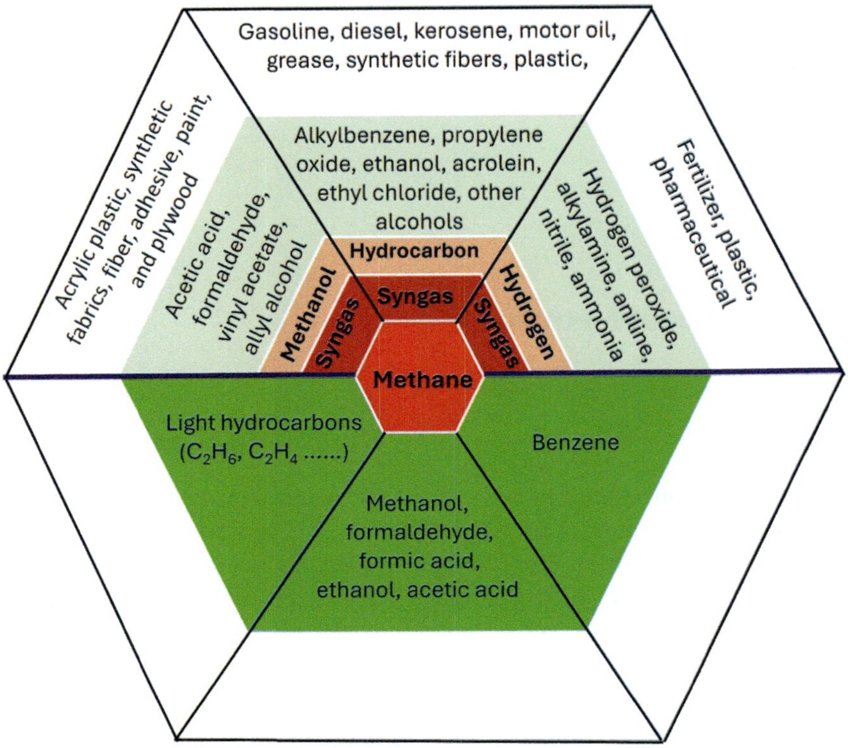

Figure 1.1 Expected chemical transformation map taking CH_4 as a substitute for petroleum in synthesizing crucial chemical intermediates to produce value-added chemicals and materials used in healthcare and daily household products.

production from CH_4 that do not release environmentally harmful CO_2. Therefore, the development of low-temperature reactions without releasing CO_2 and suitable catalysts for these reactions is vital to achieve a green transformation of CH_4 with low energy input and zero environmental impact. This pursuit underscores the significance of fundamental studies of activation and catalytic transformation of CH_4 under a mild condition in terms of low reaction temperatures.

1.2 Energy Supply

In the energy sector, CH_4 presently contributes predominantly to electricity generation [1]. The combustion of CH_4 at high temperatures provides heat to vaporize water, generating steam to drive electrical generators in CH_4 power plants. The advent of shale-gas-based electricity generation has considerably reduced electricity prices, leading to the retirement of numerous coal plants in America. According to the United States Energy Information Administration, 39.8% of electricity in the United States is generated through CH_4 combustion-based power plants (Table 1.1) [2]. It is the largest component in fossil-fuel-derived electricity, including 39.8% from CH_4, 19.5% from coal, and 0.6% from petroleum.

Table 1.1 The US utility-scale electricity generation by source, amount, and share of the total in 2022 (data as of October 2023).[a]

Energy source (Category level 1)	Energy source (Category level 2)	Energy source (Category level 3)	Amount (kWh)	Share of total
All sources			4231	**100%**
Fossil fuels (total)			**2553**	**60.4%**
	Natural gas		1687	39.9%
	Coal		832	19.7%
	Petroleum (total)		23	0.5%
		Petroleum liquids	16	0.4%
		Petroleum coke	7	0.2%
	Other gases[b]		12	0.3%
Nuclear	**Nuclear**		**772**	**18.2%**
Renewables (total)			**901**	**21.3%**
	Wind		434	10.3%
	Hydropower		255	6.0%
	Solar (total)		144	3.4%
		Photovoltaic[c]	141	3.3%
		Solar thermal	3	0.1%
	Biomass (total)		52	1.2%
		Wood	35	0.8%
		Landfill gas	9	0.2%
		Municipal solid waste	6	0.1%
		Other biomass waste	2	<0.1%
	Geothermal		16	0.4%
Pumped storage hydropower[d]			**−6**	**−0.1%**
Other sources[e]			**11**	**0.3%**

[a] Utility-scale electricity generation is electricity generation from power plants with at least 1 MW (or 1000 kW) of total electricity generating capacity. Data are for net electricity generation.
[b] Small-scale solar photovoltaic (PV) systems are electricity generators with less than 1 megawatt (MW) of electricity generating capacity, which are not connected to a power plant that has a combined capacity of one MW or larger. Most small-scale PV systems are at or near the location where the electricity is consumed, and many are net metered systems. Smaller PV systems are usually installed on building rooftops.
[c] Other gases include blast furnace gas and other manufactured and waste gases derived from fossil fuels.
[d] Pumped storage hydroelectricity generation is negative because most pumped storage electricity generation facilities use more electricity than they produce annually. Most pumped storage systems use fossil fuels or nuclear energy for pumping water to the storage component of the system.
[e] Other (utility-scale) sources include nonbiogenic municipal solid waste, batteries, hydrogen, purchased steam, sulfur, tire-derived fuel, and other miscellaneous energy sources.
Source: [2] / U.S. Energy Information Administration / Public Domain.

In addition, nuclear power plants and renewables contribute 18.2% and 21.5%, respectively. Essentially, CH_4, coal, nuclear, and renewables account for 39.8%, 19.5%, 18.2%, and 21.5% of the electricity generated in the United States (Table 1.1) [2].

Apart from electricity generation for industries and households, CH_4 serves as the primary energy source for three energy-end uses, including space heating, water heating, and cooking for households across the United States (Figure 1.2). As per the 2020 report from the United States Energy Information Administration, 61% of households in the United States utilize natural gas for at least one energy-end use [3]. As shown in Figure 1.2, the Midwest Census Region and West Census Region have the highest share of households at 74% in 2020. The share of households for spacing heating, water heating, and cooking in the South was lowest.

In certain European and Asian countries, CH_4 is allowed to directly be employed as fuel to combust in natural gas engines for vehicles, while it remains prohibited for combustion engines in North American. Consequently, CH_4 serves as the fuel for methane-powered vehicles in these regions. In addition, CH_4 is utilized in emergency hospital situations to generate electricity through high-temperature fuel cells [4]. As shown in Figure 1.1, the syngas derived from CH_4 can be employed in producing gasoline through the Fischer–Tropsch synthesis. However, this route is neither cost-competitive in contrast to gasoline production of petroleum refineries nor environmentally friendly [5].

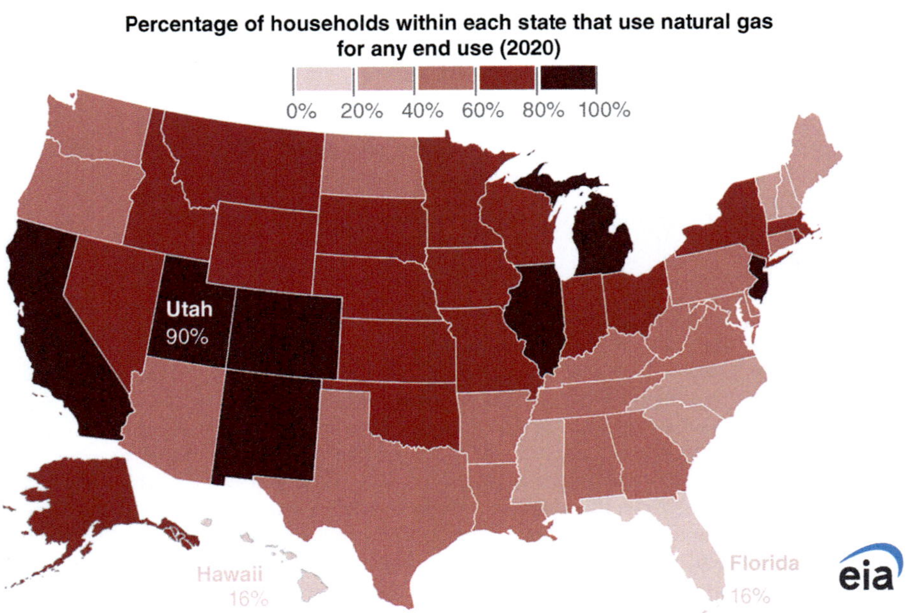

Figure 1.2 Percentage of households within each state that use natural gas for any end use in 2020. *Source:* Reproduced with permission from U.S. Energy Information Administration, Residential Energy Consumption Survey [3] / U.S. Energy Information Administration. All words and percentages in the figure were added by the source of this figure.

1.3 Climate Change

The utilization of CH_4 has resulted in a significant environmental challenge concerning the greenhouse gas effect [6, 7]. The 2016 Paris Agreement established a goal aimed at curbing climate change, specifically limiting the global average temperature increase to well below 2 °C above preindustrial levels. This legally binding international treaty, adopted by 196 countries worldwide, came into effect on 4 November 2016. The objective was described as restricting global warming to an increase of 1.5 °C by the end of this century. To achieve this goal, global emissions by 2030 need to be halved from the 2010 emission levels, with emissions reaching net zero by 2050.

CH_4, in terms of the greenhouse effect, holds a potency 26 times greater than that of carbon dioxide. As briefed above, the Global Methane Pledge was introduced during the 26th Conference of the Parties (COP-26) in Glasgow in November 2021 to address CH_4 emissions [8]. This pledge garnered endorsement from 111 countries, committing to collaborating to reduce CH_4 emissions by 30% of the 2020 levels by 2030 [8, 9]. The pledge sets forth an ambitious reduction within a mere nine years, expected to lead to a net cooling effect within a decade. It stands as the first international policy commitment solely addressing CH_4 emissions.

Unfortunately, CH_4 emission sources span a wide spectrum, ranging from coal mines, wetlands, petroleum industries, power plants, nonelectric vehicles, and biomass burning to ruminant digestion in livestock. Understanding the formation of CH_4 from anthropogenic activities and its subsequent emission into the atmosphere is crucial in minimizing CH_4 emissions and ensuring the environmentally friendly utilization of CH_4 derived from these activities. This underlines the critical importance placed on CH_4 within our environmental concerns. Chapters 3 and 5 discuss the formation of CH_4 and its emission to the atmosphere, respectively.

1.4 Reconciling Shale-Gas Utilization and Environmental Issue

The utilization of CH_4 in chemical and manufacturing industries revolves around its fundamental roles as a raw material for producing essential intermediate compounds of chemical industries for manufacturing functional materials for our daily life – such as plastics, rubber, textiles, cosmetics, fertilizers, and medicines – and as a source of hydrogen (H_2) for various chemical processes and material manufacturing. In the carbon cycle, CH_4 plays a vital role unless existing carbon-containing materials can be efficiently recycled to meet the demand for chemical and functional materials without requiring additional carbon sources. Achieving such an ideal carbon cycling world would necessitate the development of a comprehensive array of recycling techniques, with economic feasibility being a key determinant of their adoption. While it is conceivable that future industries may rely solely on recycling existing carbon-containing chemicals and materials without extracting additional raw materials from shale gas, it is not sure when such a transition could be accomplished.

In the interim, until sustainable recycling-based chemical and manufacturing industries are realized, the substitution of petroleum with natural gas remains an option. Biomass also presents an alternative source of carbon-based chemicals, but the portion of its application to consumable carbon-containing materials is still temporarily limited for the issue of costs and other reasons. Therefore, from the perspective of chemical production and material manufacturing, CH_4 is expected to maintain a certain share in an unknown period of time.

CH_4 emission is one of the major contributors to global warming. CH_4 emission primarily stems from leaks at shale gas production sites. Mitigating CH_4 emissions can be achieved by employing technologies such as green completions or reduced-emissions completions, which can reportedly reduce leaked CH_4 by up to 90% [10]. Addressing the impact of CH_4 emissions on climate change necessitates strategies such as isolating wells and implementing leak prevention measures. However, a long-term solution to minimize the contribution of CH_4 to global warming involves transitioning the current energy supply system towards a cleaner alternative, such as an optimized composite energy system consisting of nuclear energy and various forms of renewable energies such as wind, hydropower, solar, biomass, and geothermal energy.

Nevertheless, the transition to a cleaner energy system does not imply the elimination of shale gas from either industries or our daily lives. Compared to petroleum and coal for chemical and manufacturing industries, CH_4 is often a preferable raw material. This transition towards cleaner energy sources is expected to occur and last over decades, during which CH_4 is likely to continue contributing to the energy sector. Balancing the anticipated utilization of CH_4 in chemical and manufacturing industries with its contribution to global warming requires stringent policies to manage CH_4 emissions, technological advancements in shale gas drilling with minimal leak and zero environmental impact, and the development of new catalytic processes for transforming CH_4 into value-added chemicals with low-energy cost and zero CO_2 emission under mild conditions in terms of a reaction temperature of room temperature or low temperature (25–250 °C). Therefore, it is significant to develop efficient, green transformations of CH_4 into highly valued intermediate compounds for chemical and manufacturing industries, such as methanol and acetic acid, under mild conditions. The global community has made significant strides in this direction in the last decade, with ongoing efforts focusing on room temperature and low-temperature activation and catalytic transformation of CH_4 into value-added chemicals, as discussed in the following chapters.

References

1. U.S. Department of Energy (US DOE). 2009. "Modern shale gas development in the United States: A primer." Rep. DoE-FG26-04NT15455, Washington, DC. 116 pp.
2. U.S. Energy Information Administration. 2022. "What is U.S. electricity generation by energy source?" https://www.eia.gov/tools/faqs/faq.php?id=427&t=3
3. U.S. Energy Information Administration. 2023. "The majority of U.S. households used natural gas in 2020." https://www.eia.gov/todayinenergy/detail.php?id=55940#:~:text=Although%2061%25%20of%20U.S.%20households,%2C%20water%20heating%2C%20and%20cooking,%2C%20water%20heating%2C%20and%20cooking

4 Chen, Y., deGlee, B., Tang, Y., Wang, Z., Zhao, B., Wei, Y., Zhang, L., Yoo, S., Pei, K., Kim, J. H., Ding, Y., Hu, P., Tao, F., Liu, M. 2018. "A robust fuel cell operated on nearly dry methane at 500 °C enabled by synergistic thermal catalysis and electrocatalysis." *Nat. Energy* 3, 1042–1050.

5 United Nations Climate Change. "COP28 Agreement Signals "Beginning of the End" of the Fossil Fuel Era.", 2023. https://unfccc.int/news/cop28-agreement-signals-beginning-of-the-end-of-the-fossil-fuel-era

6 U.S. Energy Information Administration. "Technically recoverable shale oil and shale gas resources: An assessment of 137 shale formations in 41 countries outside of the united states.", 2013. https://www.eia.gov/analysis/studies/worldshalegas/pdf/overview.pdf

7 Weinhold, B. 2012. "The future of fracking: New rules target air emissions for cleaner natural gas production." *Environ. Health Perspect.* 120, a272–a279.

8 United Nations Climate Action. "COP26: Together for our plant.", 2021. https://www.un.org/en/climatechange/cop26

9 International Energy Agency (IEA). 2022. "The global methane pledge." https://www.iea.org/reports/global-methane-tracker-2022/the-global-methane-pledge

10 Jenner, S., Lamadrid, A. J. 2013. "Shale gas vs. coal: Policy implications from environmental impact comparisons of shale gas, conventional gas, and coal on air, water, and land in the United States." *Energy Policy* 53, 442–453.

2

Properties and Chemical Inertness of Methane

CH_4 is an odorless, colorless gas exhibiting various unique and chemical properties. Its normal boiling and melting points are −162 °C and −182.5 °C, respectively, with a triple point at the pressure of 0.115 atm and the temperature of −182.5 °C (Figure 2.1) [1]. At 0 °C and 1 atm, its density is 44.7 mol m^{-3} or 0.717 kg m^{-3}, which is lower than that of air at the same temperature and pressure. According to its phase diagram in Figure 2.1, CH_4 can exist as a gas at room temperature and high pressure or as a liquid under low-temperature and ambient-pressure conditions. The critical point of CH_4 occurs at −83 °C and 46 bar (Figure 2.1). The supercritical liquid density of CH_4 at −83 °C and 46 bar is 162.7 kg m^{-3}, while the densities of its liquid phase at −162 °C and gas phase at 0 °C and 1 atm are 422.6 and 0.7168 kg m^{-3}, respectively. Furthermore, its flash point is −188 °C.

The heat of vaporization for CH_4 is 8.19 kJ mol^{-1}, with specific heat capacities at constant pressure (C_p) and constant volume (C_v) at 35.8 and 27.4 J mol K^{-1}, respectively. In addition, its heat of formation and Gibbs free energy of formation are −75 and −51 kJ mol^{-1}, respectively, while its standard molar entropy (S^o) is 186 J mol·K^{-1}. Notably, the autoignition temperature of CH_4 is 537 °C, and its heat of combustion is −890.8 kJ mol^{-1}.

CH_4 is regarded as the simplest yet most stable hydrocarbon. Its carbon atom undergoes sp^3 hybridization, forming four equivalent C–H sp^3-s sigma bonds. These bonds exhibit a length of 1.087 Å and all H–C–H bond angles of 109.5°. With a bond strength of 438.8 kJ mol^{-1} [2], it is the strongest among hydrocarbons. Notably, CH_4 demonstrates a high ionization energy (1206.1 kJ mol^{-1}), [3] low proton affinity (424.5 kJ mol^{-1}), and low acidity (pK_a of 48) [4, 5]. The chemical inertness of CH_4 is due to the highly localized C–H sigma bond in terms of the head-to-head overlapping of electron densities of a sp^3 orbital of a C atom and the 1s orbital of a H atom. This highly localized electron density makes the C–H sigma bond lack a high-energy occupied orbital and a low-energy unoccupied orbital, which are frontier orbitals to couple with orbital of other molecules or to receive electron density from or donate to other species such as an atom, anion, or cation, resulting in the chemical inertness of a CH_4 molecule. In addition, its polarizability is 2.8×10^{-40} C^2 m^2J^{-1} [6]. Such quite low polarizability requests a high electric field to polarize the nonpolar CH_4

Low-temperature Activation and Catalytic Transformation of Methane to Non-CO_2 Products,
First Edition. Franklin Tao.

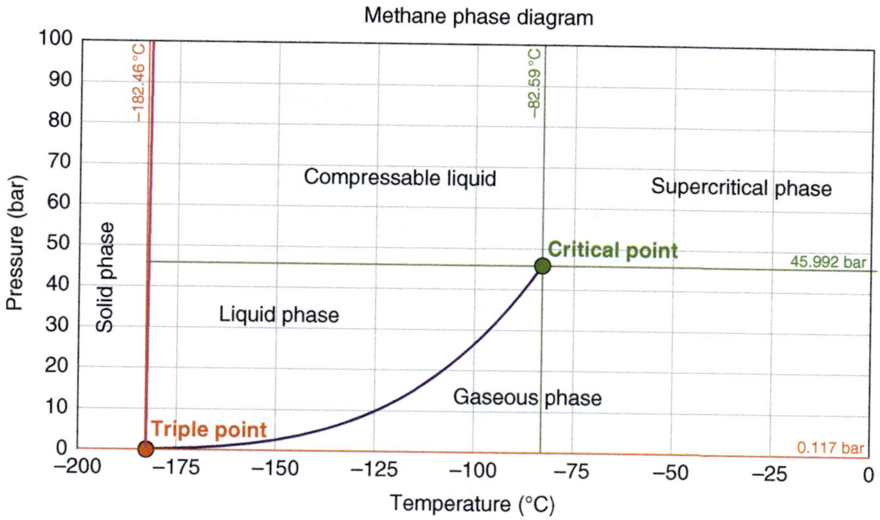

Figure 2.1 Phase diagram of pure CH_4. *Source:* https://www.engineeringtoolbox.com/methane-d_1420.html#phases. [1] / The Engineering Toolbox.

molecule before any nucleophilic or electrophilic attack could apply to CH_4. These structural properties suggest the chemical inertness of CH_4 as the routine redox or acid–base chemistry is inaccessible to any molecular orbital or electron state of this molecule. These features make the activation of CH_4 extremely challenging.

The bond strength of C–H of CH_4 is 438.8 kJ mol^{-1}. Once a H atom of CH_4 is lost, a $CH_3\cdot$ in terms of a methyl group is formed. The $CH_3\cdot$ is planar since the carbon atom takes a sp^2 hybridization instead of the sp^3 hybridization of the C atom of CH_4. The C–H bond length of $CH_3\cdot$ is slightly reduced, and thus, its C–H bond strength is slightly increased by about 20 kJ mol^{-1}. Notably, the ionization energy of CH_3 is obviously decreased by 265 kJ mol^{-1} in contrast to CH_4 [7]. The loss of one electron from $CH_3\cdot$ causes it to form a cation, CH_3^+, which is the simplest carbenium ion in organic chemistry. Based on the literature, the substitution of an H atom of $CH_3\cdot$ or CH_3^+ with an alkyl group (C_nH_{2n+1}–) can stabilize the methyl radical or carbenium since the larger alkyl group can aid in decreasing charge density on the carbon atom of $CH_3\cdot$ or CH_3^+. The reactivity of such a radical or carbenium decreases along the order of tertiary, secondary, and primary CH_4. From this point of view, CH_4 is the most inert organic molecule.

The inertness nature of CH_4 makes its activation quite challenging. As CH_4 is the most inert molecule in organic chemistry, any derivatives including molecules, radical, carbenium, and anions are more active and thus could be immediately activated once they are formed. These features make the selective transformation of CH_4 to a specific product challenging.

References

1 The Engineering ToolBox (2008). Methane - Thermophysical Properties. [online] Available at: https://www.engineeringtoolbox.com/methane-d_1420.html#phases. [Accessed July 18, 2022].

2 Luo, Y. R. 2007. "*Comprehensive Handbook of Chemical Bond Energies*." CRC Press. 19–145.

3 Berkowitz, J., Greene, J. P., Cho, H., Ruscic, B. 1987. "The ionization′ potentials of CH4 and CD4." *J. Chem. Phys.* 86, 674–676.

4 Shilov, A. E., Shul′pin, G. B. 1997. "Activation of C–H bonds by metal complexes." *Chem. Rev.* 97, 2879–2932.

5 Geerts, J. W. M. H., Hoebink, J. H. B. J., van der Wiele, K. 1990. "Methanol from natural gas. Proven and new technologies." *Catal. Today* 6, 613–620.

6 Amos, R. 1979. "An accurate ab initio study of the multipole moments and polarizabilities of methane." *Mol. Phys.* 38, 33–45.

7 Dixon, D. A., Feller, D., Peterson, K. A. 1997. "Accurate calculations of the electron affinity and ionization potential of the methyl radical." *J. Phys. Chem. A* 101, 9405–9409.

3

Formation of Methane in Nature and by Anthropogenic Activity

CH_4 is produced through both natural processes and human activities. The majority of CH_4 originates from natural sources, while human activities, including production from industrial processes, create a very small portion of CH_4. However, about 60% of CH_4 released to the aerosphere is contributed by anthropogenic activities due to its prevalence in human-related processes and its direct emissions into the ambient environment [1]. In many cases, CH_4 forms in various rocks and beneath the seabed becomes trapped or sealed, often remaining isolated from the atmosphere. Techniques such as hydraulic fracturing and horizontal drilling involving deep boreholes to be briefed in Chapter 4 are employed to extract CH_4. The CH_4 released into the atmosphere due to human activities has a significant impact on climate change. CH_4 formed on the Earth's surface readily escapes into the atmosphere without any delay, contributing significantly to its influence on climate change.

3.1 Methane Formed in Rocks

The carbon cycling within the deep biosphere of the Earth is a complex process. Both deep crystalline and sedimentary rocks play essential roles in the biosphere. Sedimentary rocks, such as shale, constitute a significant part of the biosphere. The formation of CH_4 in the deep biosphere involves intricate mechanisms. Generally, CH_4 originates from either relatively low-temperature abiotic reactions within the upper crust or microbial activity. Abiotic reactions, including the direct autotrophic synthesis of CH_4 through the Sabatier reaction, contribute to CH_4 formation in a rock. The hydrogen needed for this synthesis within bedrock can arise from water–rock interactions [2–4] or the dissociation of water driven by energy from radioactive decay [5–7]. In addition, microbial processes, particularly the involvement of fermentative bacteria, can produce H_2 [8, 9]. Both inorganic (such as carbonate minerals and graphite in rock bed) and organic sources provide the carbon needed for these abiotic reactions [10]. These hydrogen and carbon sources facilitate CH_4 production through hydrogenotrophic methanogens.

Microbes within sedimentary rocks can utilize stored organic materials to generate hydrogen and small organic molecules. For example, clostridial fermenters decompose

Low-temperature Activation and Catalytic Transformation of Methane to Non-CO_2 Products,
First Edition. Franklin Tao.

organic materials to produce carbon and H_2, enabling methanogens to convert them into CH_4 [11]. Fungi also contribute to CH_4 formation by breaking down recalcitrant organic materials present in black shale, creating a carbon source that fuels microbial methanogenesis, leading to CH_4 production [12]. Moreover, dead microbial biomass serves as a carbon source for methanogens, facilitating CH_4 formation.

3.2 Methane Hydrate Formed in Seabed

Methane hydrate forms as an ice-like material at low temperatures and under water pressures exceeding 30 atm. Its formula, $CH_4 \cdot 5.75H_2O$, represents a crystal structure of water with entrapped CH_4 molecules, as shown in Figure 3.1a. A single unit cell of methane clathrate comprises 8 CH_4 and 46 H_2O molecules, although its chemical formula is $(CH_4)_4(H_2O)_{23}$. This structure, known as structure-I methane hydrate (Figure 3.1b), includes 2 dodecahedral cages, each containing 1 CH_4 and 12 H_2O molecules, alongside 6 tetradecahedral cages, with 1 CH_4 and 14 H_2O molecules per cage. The shared vertices among these cages account for a total of 8 CH_4 and 46 H_2O molecules. Other than CH_4, the structure I can be filled with C_2H_6, H_2S, or CO_2. Figures 3.1c and d present a unit cell of the gas hydrate structure II and III, respectively [13]. There are 24 cages including 16 small and 8 large ones in a unit cell of structure II (Figure 3.1c). The relatively large cage cavity of structure II allows the C_2H_6, C_3H_8, and C_4H_{10} molecules to be kept. Structure III is also called the H structure (Figure 3.1d). Its unit cell consists of small and large cages. The large cages of the H structure are filled with large molecules.

The majority of methane hydrates exist within marine continental margin sediments, typically several hundred meters below sea level, as schematically shown in Figure 3.2 [14]. They form through precipitation and crystallization when gaseous CH_4 interacts with water at low temperatures and high pressures on the seabed. These CH_4 molecules, originating from geological faults under the ocean floor, likely migrated from these faults before combining with water molecules to form methane hydrates at the low-temperature, in high-pressure environment. The initial sources of these released CH_4 molecules are biosystems located on the seabed or in deeper rocks beneath the ocean floor. A conservative estimate of global methane hydrate is approximately 1.8×10^3 Gt carbon [15], equivalent to about 3.0×10^{15} m^3 at a density of 0.717 kg m^{-3}, although different scientific studies propose values as high as 1.3×10^4 Gt carbon [13, 16].

3.3 Bio-methanation

This and the following sections elaborate on the formation of CH_4 through human-induced or anthropogenic activities. These activities involve enteric fermentation, oil and gas extraction, coal mining, rice cultivation, and wastewater treatment, among others, as highlighted by the global methane initiative [1, 17, 18]. Figure 3.3 illustrates the proportions of CH_4 contributed by various anthropogenic activities globally, with contributions from enteric fermentation (27%), oil and gas (24%), municipal solid waste (11%), coal mining (9%), wastewater (7%), rice cultivation (7%), stationary and mobile sources (4%), biomass (4%), and other sources (3%).

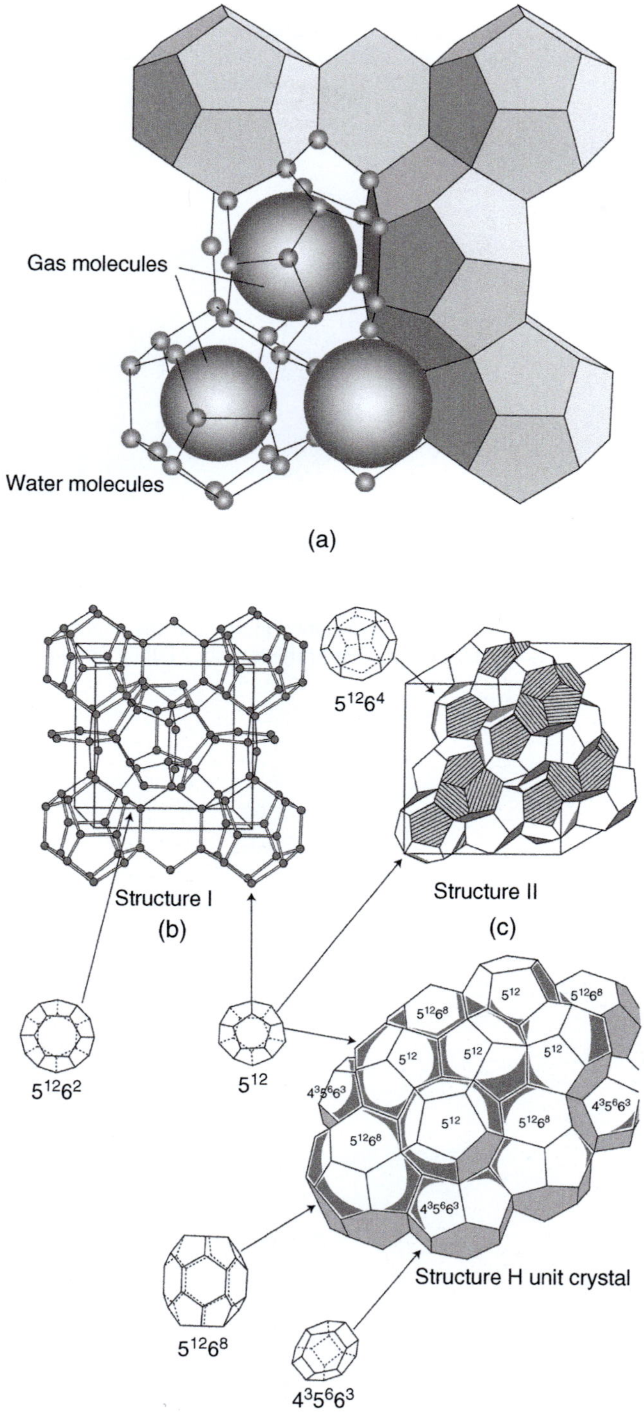

Figure 3.1 Cage structure and unit cell of three different types of gas hydrates. (a) The typical structure of a gas hydrate with water molecules is linked together to form a cage trapping a gas molecule such as methane within. (b–c) Illustration of the three different gas hydrate cage structures including structure I in (b), structure II in (c), and structure III or called H structure in (d). *Source:* Reproduced with permission from [13] / The Royal Society.

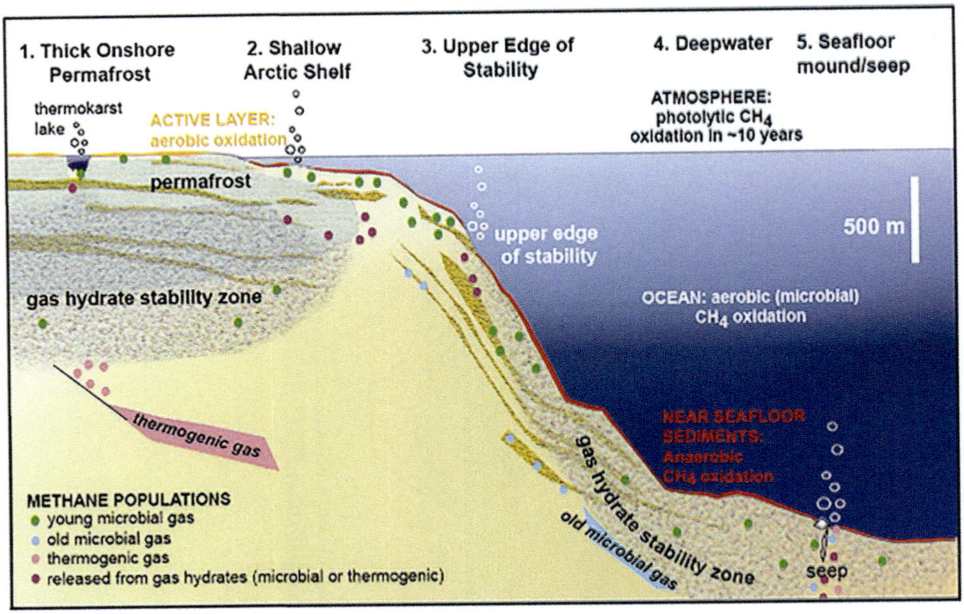

Figure 3.2 Schematic showing the formation of methane at continental margins at the depths of a few hundred meters below sea level. *Source:* [14] / U.S. Department of the Interior / Public Domain.

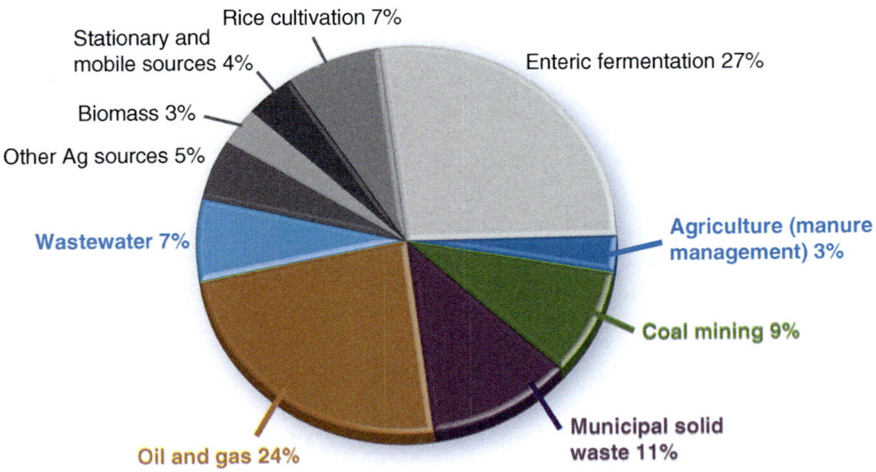

Figure 3.3 Estimated percentages of methane emissions from different anthropogenic activities by 2020. "Ag" stands for agriculture. *Source:* [18] / Global methane initiative / Public Domain.

The process of utilizing methanogens to convert original materials into CH_4 is termed bio-methanation. This process occurs in various settings, such as landfills, sewage treatment plants, wetlands, rice fields, oceans, human and livestock systems, termite habitats, and hydrothermal vents, as illustrated in Figure 3.4 [19]. Based on Figure 3.3, methane formed through enteric fermentation stands as the largest source of anthropogenic greenhouse CH_4 emissions. Statistical data suggests that a mature beef cow releases over 100 g of CH_4 per day [19–22]. Methanogenesis plays a crucial role in the global biogeochemical cycle, serving as an integral part of the energy and carbon cycle. Understanding the profound working mechanisms of methanogens enables the utilization of their potential to transform various organic wastes into CH_4.

Methanogenesis, an anaerobic respiration process culminating in CH_4 formation, is exclusively performed by methanogens, a type of archaea. It transforms simple substrates into CH_4, yet its biochemical mechanism is extraordinarily complex. Methanogenesis encompasses various pathways in methanogens, demonstrating two distinct features in biological methanogenesis: all methanogens belong to the archaea domain, and variations in pathways exist among them based on substrates. Three primary categories of methanogens include hydrogenotrophic, acetolactic, and methylotrophic. Hydrogenotrophic methanogens, the main category, transform CO_2, acetate, and H_2 to CH_4. They are found in diverse environments such as animal gastrointestinal tracts, termite systems, and deep-sea marine sediments, contributing to the formation of over one-quarter of methane produced

Figure 3.4 Sources of biological methane emissions to the atmosphere. Methanogens are found in a wide range of habitats, where they actively contribute 70% of the 500–600 Tg of CH_4 introduced to the atmosphere each year. Percentages are the contributions of each habitat to the biological portion of the global methane emission budget. The ecology of each habitat largely determines the dominant substrates and probably selects for the types of methanogens found therein. Hydrogenotrophic methanogens are found in H_2-rich habitats, acetolactic methanogens in acetate (Ac)-rich habitats, and methylotrophic methanogens in habitats rich in methyl compounds. *Source:* Reproduced with permission from [19] / Elsevier.

through biochemical processes. Methylotrophic methanogens transform methanol and dimethyl sulfate, predominantly in sulfate-rich sediments and habitats, contributing a minor portion to global CH_4 formed from biochemical processes. Acetolactic methanogens, the third type, convert acetate to CH_4 and CO_2, playing a dominant role in CH_4 production in anaerobic digesters, wetlands, and rice fields. Statistical data reported in the literature details the proportion of CH_4 produced among the total global methane production through biochemical processes, focusing on primary substrates [19, 23]. Figure 3.4 presents the portions of CH_4 formed from different habitats producing CH_4 through the above biochemical processes. The specific methanogens responsible for the biochemical transformation of substrates into CH_4 are delineated in the caption of Figure 3.4.

A fundamental understanding of the three types of methanogenesis at a molecular level is significant for better utilization of this unique biochemical production of CH_4 to transform various organic wastes to CH_4 since CH_4 is nearly the most important raw material for chemical industries. More importantly, elucidation of the biochemical mechanisms of methanogenesis is crucial for mitigating the emission of CH_4 from these sources to the atmosphere.

Figure 3.5 schematically presents methanogenic pathways of hydrogenotrophic, methylotrophic, and acetolactic methanogenesis [23]. Although the three pathways are different, there are some common features. One common feature is the same last step, which is the reduction of CH_3-S-CoM by a common enzyme CoB-SH to form the heterodisulfide (CoM-S-S-CoB). This step is shown at the bottom of Figure 3.5 [24]. Another common feature of the three pathways is to recycle the CoM-S-S-CoB back to CoB-SH and CoM-SH by a heterodisulfide reductase. This step is shown at the left-bottom corner of Figure 3.5. The heterodisulfide reductases are different in these pathways. In the pathway of most hydrogenotrophic methanogens, it is a cyclochrome-independent or Flavin-based electron bifurcation system (FBeB). However, it is a cyclochrome-dependent electron transfer system (CDeT) in methylotrophic and acetolactic methanogens. Other than these common features, other steps are different. The specific differences depend on substrates such as CO_2, acetate, or methyl-containing compounds.

As shown with blue lines and fonts in Figure 3.5, a methyl-containing molecule typically takes the methylotrophic pathway catalyzed by methylotrophic methanogens. In the methylotrophic pathway, the CH_3 of a methyl-containing molecule must be first transferred to a substrate-specific methyltransferases and then CH_3-S-CoM, which is the common intermediate of all the three pathways. If H_2 is available, CH_4 is produced since H_2 can act as a reductant, offering electrons to the formed CH_3-S-CoM to form CH_4. If neither H_2 nor external electron donor is available in the system, CH_3-S-CoM is oxidized by following steps marked with upward blue arrows (see the middle of Figure 3.5), which is the reverse process of CO_2-reducing pathway (marked with downward green arrows in the middle of Figure 3.5). The oxidation of a CH_3 of some CH_3-S-CoM to CO_2 along the upward blue arrows can provide several electrons to reduce the other CH_3-S-CoM to CH_4. Thus, the conversion of such a methyl-containing molecule to CH_4 is intricate since it involves oxidizing some intermediate species to generate electrons to reduce some intermediates to form CH_4.

Compared to the methylotrophic pathway catalyzed by methylotrophic methanogens, acetate takes the acetolactic pathway (marked with yellow arrows and yellow fonts on the right side in Figure 3.5). As shown in Figure 3.5, ATP activates acetate to form an enzyme, an acetyl CoA. The acetyl-CoA is decomposed into methyl and carbonyl. Then, the CH_3

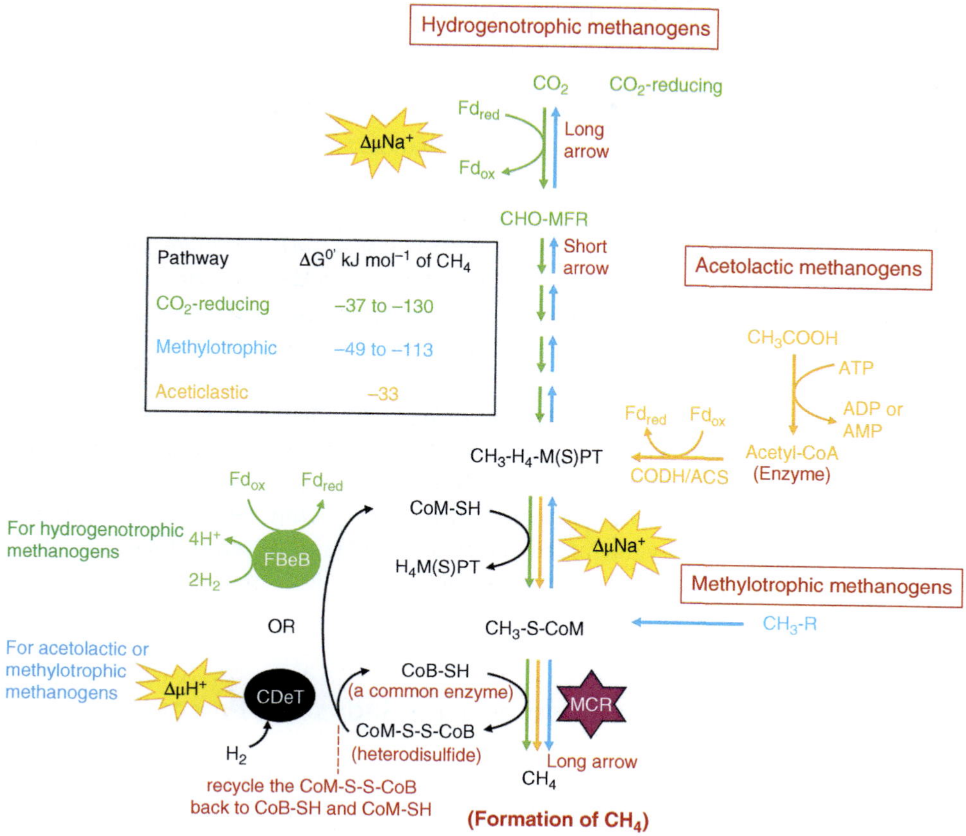

Figure 3.5 The three enzymatic pathways of methanogenesis. All methanogenic pathways contain three common steps: the transfer of the methyl group to coenzyme M (CoM-SH), the reduction of methyl-coenzyme M with coenzyme B (CoB-SH), and the recycling of the heterodisulfide CoM-S-S-CoB. Green, CO_2-reducing pathway; yellow, acetolactic pathway; blue, methylotrophic pathway; Abbreviations: MFR, methanofuran; H_4 MPT, tetrahydromethanopterin; H_4 SPT, tetrahydrosarcinapterin; $\Delta G^{o\prime}$, standard free energy change; CH_3-R, methyl-containing compounds such as methanol, methanethiol, dimethyl sulfide, monomethylamine, dimethylamine, trimethylamine, and tetramethylammonium; Fd_{red}, reduced form of ferredoxin; Fd_{ox}, oxidized form of ferredoxin; $\Delta\mu(Na^+)$, electrochemical sodium ion potential; $\Delta\mu(H^+)$, electrochemical proton potential; FBeB, flavin-based electron bifurcation; CDeT, cytochrome-dependent electron transfer; MCR, methyl-coenzyme M reductase; CODH/ACS, carbon monoxide dehydrogenase/acetylCoA synthase/decarbonlyase complex. *Source:* Reproduced with permission from [19] / Elsevier.

binds to tetrahydromethanopterin H_4MPT or tetrahydrosarcinapterin H_4SPT, forming CH_3-H_4MPT or CH_3-H_4SPT, designated as CH_3-$H_4M(S)PT$ in Figure 3.5. Subsequently, the CH_3 of CH_3-H_4MPT or CH_3-H_4SPT exchanges with H of CoM-SH, forming CH_3-S-CoM. CH_3S-CoM is the common intermediate as described earlier. Then, the CH_3 on CH_3-S-CoM is transformed to CH_4 upon it is reduced by a thiol, coenzyme B (CoB-SH), to form CoM-S-S-CoB. The carbonyl left from breaking the enzyme, acetyl-CoA, is oxidized to CO_2 along the pathway marked with upward blue arrows to CO_2 Figure 3.5. This oxidization generates electrons to reduce the intermediate CH_3-S-CoM to CH_4.

Other than the methylotrophic (blue arrows) and aceticlastic (yellow arrows) pathways, CO_2 reduction to CH_4 catalyzed by hydrogenotrophic methanogens is called the hydrogenotrophic pathway, marked with green arrows in Figure 3.5. As marked in the top panel of Figure 3.5, the CO_2 is bound to a CH_4 carrier called methanofuran (MFR). The CO_2 bound to MFR is reduced to formylmethanofuran (CHO-MFR) by Fd_{red}. The formed formyl is bound to H_4MPT or H_4SPT, which are also intermediates in the above aceticlastic pathway, forming CH_3-H_4MPT or CH_3-H_4SPT. The formed CH_3-H_4MPT or CH_3-H_4SPT can dehydrate to form CH_2=H_4MPT or CH_3-H_4MPT through the electron donation of deazaflavin coenzyme $F_{420}H_2$. Once the CH_3-H_4MPT is formed, it is transformed to CH_4 through the two steps shared by the acetolactic pathway (marked with long, downward, yellow arrows) and methylotrophic pathway (marked with long, downward, blue arrows).

Research in the literature has conducted thermodynamic analyses of the three transformations illustrated in the black box at the top-left corner of Figure 3.5. It appears that the conversion of CO_2 into CH_4 through H_2 is notably thermodynamically favorable. Moreover, methylotrophic methanogenesis also demonstrates thermodynamic favorability, exhibiting a decrease in Gibbs free energy ranging from −49 to −113 kJ mol^{-1}. Even acetolactic methanogenesis exhibits thermodynamic favorability, indicating a decrease in Gibbs free energy by 33 kJ mol^{-1}.

3.4 Methane Formed as a Byproduct in Industrial Processes

Industries typically never prioritize the production of CH_4 from other chemicals since CH_4 is the most economical raw materials globally in nature. However, CH_4 often emerges as a byproduct of various petrochemical and chemical processes in industries. For instance, in Fischer–Tropsch synthesis (FTS), although the primary products are long-chain hydrocarbons such as gasoline, diesels, and alcohol, CH_4 is generated as a secondary product.

The Sabatier reaction discovered by Paul Sabatier and Jean-Baptiste Senderens in 1897, utilized for removing CO or CO_2 in H_2 under certain circumstances, is a process yielding CH_4 as a primary product. This process is exothermic even at room temperature, as per its thermodynamics: $CO + 3H_2 \rightarrow CH_4 + H_2O$, $\Delta H_{298} = -206\ \mathrm{kJ\,mol^{-1}}$ and $CO_2 + 4H_2 \rightarrow CH_4 + 2H_2O$, $\Delta H_{298} = -165\ \mathrm{kJ\,mol^{-1}}$. This process stands as the reverse of steam reforming. Typically, industrial catalysts of Sabatier reaction are Ni supported on magnesia, calcium aluminates, or alumina, chosen for their reasonable activity and moderate pricing [24]. Only step sites on surface of supported Ni nanoparticles function as active sites for the direct dissociation of CO to C and O when H_2 is absent [25]. The dissociation of CO to a carbon atom and OH species through COH intermediate in the presence of H_2 is the rate-determining step [25]. In literature, Ru supported on Co_3O_4 was extensively studied for the hydrogenation of CO_2 to form CH_4 [26].

Co_3O_4, stable at relatively low temperatures (<300 °C), was used as a support. Loading 5% Ru on Co_3O_4 substantially enhances catalytic activity and selectivity for CO_2 methanation on Co_3O_4 within the range of 200–400 °C. Surface chemistries of Co_3O_4, both without and with supported Ru, were studied using ambient pressure X-ray photoelectron spectroscopy (AP–XPS) [26]. These AP-XPS studies revealed that the active phase of the Co_3O_4 surface is metallic cobalt, formed through the reduction of Co_3O_4. In the case of Co_3O_4 with supported Ru, the active phase constitutes a Co-Ru thin alloy film generated during the Sabatier reaction. The boosted catalytic activity in CO_2 methanation and the selectivity for CH_4 result from the alloyed Ru with Co.

As discussed in Chapter 5, various catalytic processes were developed to transform CH_4 to syngas, significant intermediate compounds of chemical industries (Figure 1.1). In parallel to these processes utilizing CH_4 as a reactant, coal gasification is another process for producing syngas. The coal gasification often produces CH_4 as one of the byproducts. For instance, CH_4 emerges as one of the byproducts in the gasification of low-grade lignite to produce syngas, particularly in countries endowed with abundant local resources of low-grade lignite.

References

1 Methane and Climate Change. IEA (2021), Methane Tracker 2021, IEA, Paris https://www.iea.org/reports/methane-tracker-2021, Licence: CC BY 4.0 and https://www.iea.org/reports/methane-tracker-2021/methane-and-climate-change

2 McCollom, T. M., Bach, W. 2009. "Thermodynamic constraints on hydrogen generation during serpentinization of ultramafic rocks." *Geochim. Cosmochim. Acta* 73, 856–875.

3 Neubeck, A., Duc, N. T., Bastviken, D., Crill, P., Holm, N. G. 2011. "Formation of H2 and CH4 by weathering of olivine at temperatures between 30 and 70 °C." *Geochem. Trans.* 12, 6–15.

4 Mayhew, L. E., Ellison, E. T., McCollom, T. M., Trainor, T. P., Templeton, A. S. 2013. "Hydrogen generation from low-temperature water-rock reactions." *Nat. Geosci.* 6, 478–484.

5 Vovk, I. F. 1987. "Radiolytic salt enrichment and brines in the crystalline basement of the East European Platform." *Geol. Assoc. Canada* 33, 197–210.

6 Lin, L., Hall, J., Lippmann-Pipke, J., Ward, J. A., Sherwood Lollar, B., DeFlaun, M., Rothmel, R., Moser, D., Gihring, T. M., Mislowack, B., Onstott, T. C. 2005. "Radiolytic H2 in continental crust: Nuclear power for deep subsurface microbial communities." *Geochem. Geophys. Geosyst.* 6, Q07003.

7 Lin, L.-H., Slater, G. F., Sherwood Lollar, B., Lacrampe-Couloume, G., Onstott, T. C. 2005. "The yield and isotopic composition of radiolytic H2, a potential energy source for the deep subsurface biosphere." *Geochim. Cosmochim. Acta* 69, 893–903.

8 Nandi, R., Sengupta, S. 1998. "Microbial production of hydrogen: An overview." *Crit. Rev. Microbiol.* 24, 61–84.

9 Wang, J., Wan, W. 2009. "Factors influencing fermentative hydrogen production: A review." *Int. J. Hydrogen Energy.* 34, 799–811.

10 Etiope, G., Whiticar, M. J. 2019. "Abiotic methane in continental ultramafic rock systems: Towards a genetic model." *Appl. Geochem.* 102, 139–152.

11 Kimura, H., Nashimoto, H., Shimizu, M., Hattori, S., Yamada, K., Koba, K., Yoshida, N., Kato, K. 2010. "Microbial methane production in deep aquifer associated with the accretionary prism in Southwest Japan." *ISME J.* 4, 531–541.

12 Ekendahl, S., O'Neill, A. H., Thomsson, E., Pedersen, K. 2003. "Characterization of yeasts isolated from deep igneous rock aquifers of the Fennoscandian shield." *Microb. Ecol.* 46, 416–428.

13 Maslin, M., Owen, M., Betts, R., Day, S., Jones, T. D., Ridgwell, A. 2010. "Gas hydrates: Past and future geohazard?." *Philos. Trans. R. Soc. A* 368, 2369–2393.

14 Ruppel, C. D. (2011) "Methane hydrates and contemporary climate change." *Nat. Educ. Knowledge* 3, 29–31

15 Boswell, R., Collett, T. S. 2011. "Current perspectives on gas hydrate resources." *Energy Environ. Sci.* 4, 1206–1215.

16 Dickens, G. R. 2011. "Down the rabbit hole: Toward appropriate discussion of methane release from gas hydrate systems during the Paleocene-Eocene thermal maximum and other past hyperthermal events." *Clim. Past* 7, 831–846.

17 U.S. Environmental Protection Agency, Climate Change Division. Overview of Greenhouse Gases. https://www.epa.gov/ghgemissions/overview-greenhouse-gases#methane

18 Global Methane Initiative. https://globalmethane.org/documents/gmi-mitigation-factsheet.pdf

19 Lyu, Z., Shao, N., Akinyemi, T., Whitman, W. B. 2018. "Methanogenesis." *Curr. Biol.* 28, R719–R736.

20 Eckard, R. J., Grainger, C., de Klein, C. A. M. 2010. "Options for the abatement of methane and nitrous oxide from ruminant production: A review." *Livestock Sci.* 130, 47–56.

21 Cottle, D. J., Eckard, R. J. 2018. "Global beef cattle methane emissions: Yield prediction by cluster and meta-analyses." *Anim. Prod. Sci.* 58, 2167–2177.

22 Grainger, C., Clarke, T., McGinn, S. M., Auldist, M. J., Beauchemin, K. A., Hannah, M. C., Waghorn, G. C., Clark, H., Eckard, R. J. 2007. "Methane emissions from dairy cows measured using the sulfur hexafluoride (SF6) tracer and chamber techniques." *J. Dairy Sci.* 90, 2755–2766.

23 Lyu, Z., Whitman, W. B. 2019. "Transplanting the pathway engineering toolbox to methanogens." *Curr. Opin. Biotechnol.* 59, 46–54.

24 Vannice, M. A. 1976. "The catalytic synthesis of hydrocarbons from mixtures over the group VIII metals: IV. The kinetic behavior of CO hydrogenation over Ni catalysts." *J. Catal.* 44, 152–162.

25 Andersson, M. P., Abild-Pedersen, F., Remediakis, I. N., Bligaard, T., Jones, G., Engbæk, J., Lytken, O., Horch, S., Nielsen, J. H., Sehested, J., Rostrup-Nielsen, J. R., Nørskov, J. K., Chorkendorff, I. 2008. "Structure sensitivity of the methanation reaction: H2-induced CO dissociation on nickel surfaces." *J. Catal.* 255, 6–19.

26 Zhu, Y., Zhang, S., Ye, Y., Zhang, X., Wang, L., Cheng, F., Tao, F. 2012. "Catalytic conversion of carbon dioxide to methane on cobalt-ruthenium bimetallic nanocatalysts and correlation between surface chemistry of catalysts under reaction conditions and catalytic performances", *ACS Catal.* 2, 2403–2408.

4

Extraction of Methane for Chemical Production

CH_4 is a raw material for the chemical and energy industries in our society [1]. While various anthropogenic activities are involved in the production of CH_4, most of the CH_4 used in the chemical and energy sectors is extracted from shale rocks. The CH_4 buried in shale rocks is formed naturally, as briefed in Chapter 3. The price of CH_4 (methane) in the United States has significantly declined since 2005, coinciding with the widespread adoption of advanced hydraulic fracturing and horizontal drilling techniques. These methods have enabled efficient extraction of shale gas from rock formations, including shale, sandstone, and carbonate rock. Although most gas wells are on land, over 2% of the CH_4 in the US market is produced from onshore gas wells [2]. In 2021, the United States extracted about 35 trillion cubic feet (Tcf), with 70% originating from 5 of the 34 states producing CH_4. These states are Texas, Pennsylvania, Louisiana, West Virginia, and Oklahoma, contributing 24.6%, 21.8%, 9.9%, 7.4%, and 6.7% of the CH_4 in the US market, respectively [2]. Figure 4.1 illustrates the quantities of extracted CH_4 (blue curve), consumed CH_4 (yellow curve), and exported CH_4 from the Unites States (green curve) over the last seven decades [2]. As shown in Figure 4.1, from 1990 to 2005, the CH_4 consumed in the Unites States was supplied domestically and internationally, with over a quarter of CH_4 consumed by the United States being imported before 2010. The large-scale application of hydraulic fracturing and horizontal drilling techniques in the last two decades filled the gap between consumption and production, as shown in Figure 4.1. In 2017, the United States started to export CH_4 to other countries. In recent years, the amount of exported CH_4 has sharply increased, although the climbing trend was slightly slowed down during the Covid pandemic period and was expected to jump up after the pandemic.

Hydraulic fracturing and horizontal drilling techniques are employed for the extraction of oil and gas from source formations. Figure 4.2 schematically illustrates the concept of hydraulic fracturing and horizontal drilling techniques [3–5] involving complex physical and chemical processes. Hydraulic fracturing has been employed for decades to enhance the extraction of oil and gas from conventional reservoir formations. This technique has even been used to extract natural gas from limestone and sandstone since the 1940s and from shale rocks as early as the 1970s [6]. Hydraulic fracturing has been extensively utilized for gas extraction from shale formations since the beginning of this century. Fracturing, a general term for

Low-temperature Activation and Catalytic Transformation of Methane to Non-CO_2 Products,
First Edition. Franklin Tao.

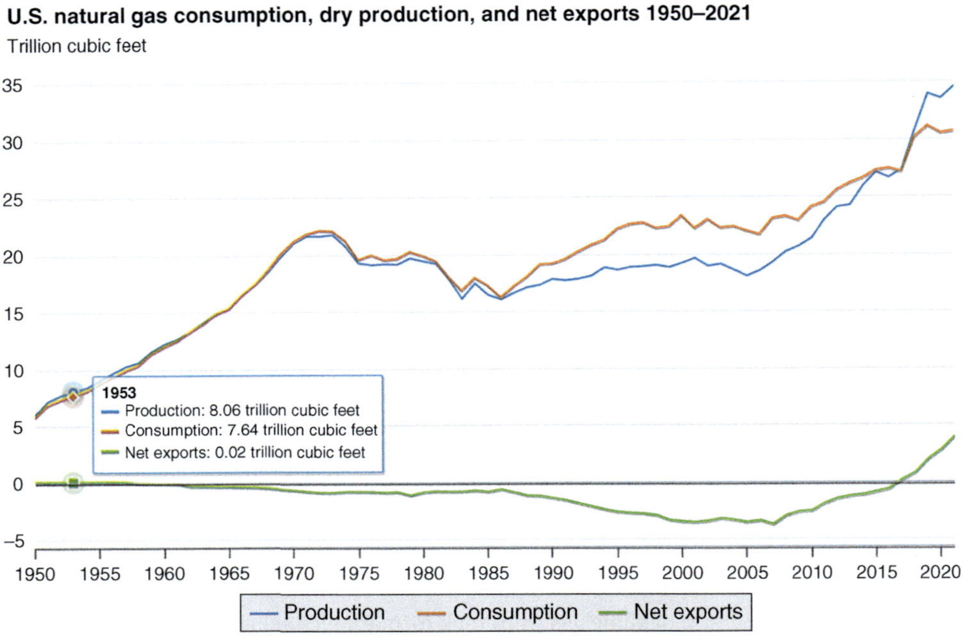

Figure 4.1 The United States' natural gas consumption, dry production, and net exports in 1950–2022. *Source:* [2] / U.S. Energy Information Administration / Public Domain.

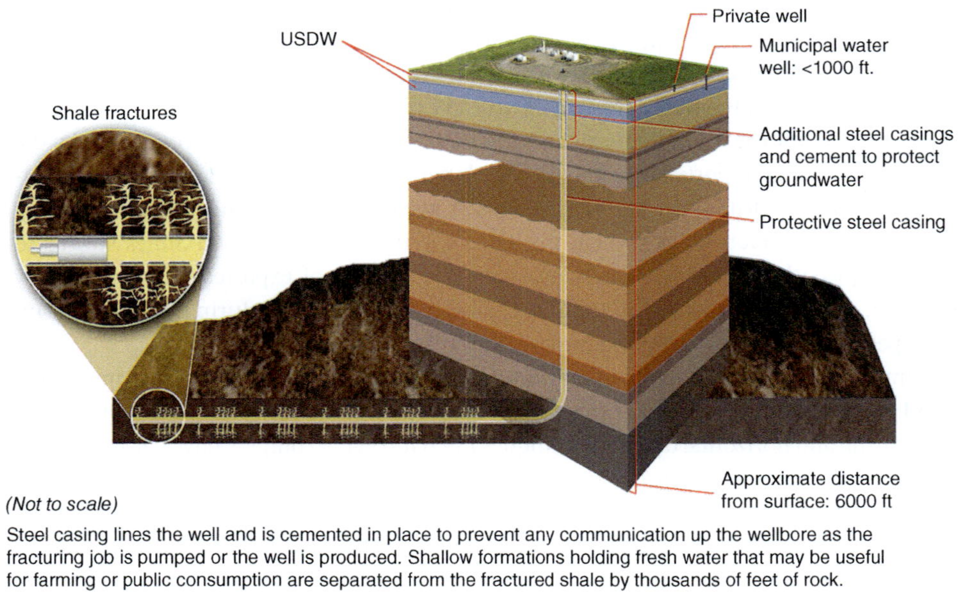

Figure 4.2 An example of a typical hydraulic fracturing configuration. *Source:* United States Department of Energy / Public Domain, National Energy Technology Laboratories, Shale Gas: Applying Technology to Solve America's Energy Challenges, March 2011. https://www.hsdl.org/?abstract&did=8711.

cracking formations to recover valuable resources such as oil or gas, involves a powerful technique to extract valuable substances stored in source formations with low porosity, such as shale rock. Horizontal drilling, a technique that creates horizontal exposure for stored substances instead of vertically, significantly reduces ecosystem disruption on the Earth but substantially increases the extraction of substances such as CH_4. The following is a brief description of these techniques, with more information available in literature references [7, 8]. The maturation of hydraulic fracturing and horizontal drilling techniques enables the use of fluid pressure to break down source formations. These techniques are particularly valuable for gas extraction from shale rock due to its high permeability despite extremely low porosity. Detailed information on hydraulic fracturing and horizontal drilling techniques can be found in literature references as they are beyond the scope of this book [7–9].

It is widely recognized that conventional oil and gas resources are typically extracted from reservoir formations. Trapping oil and gas diffuse from underlying organic-rich source formations. The oil and gas in reservoir formations have accumulated through upward diffusion from source formations over hundreds or thousands of years. One significant difference between reservoir formations and underlying source formations, aside from their geologic locations, is their porosity. Reservoir formations exhibit relatively high porosities, allowing oil and gas to diffuse easily to wells. Wells for reservoir formations are typically vertical. In contrast, source formations have extremely low porosities. Before the advent of advanced hydraulic fracturing and horizontal drilling techniques, extracting oil and gas from source formations was neither possible nor economical, despite awareness of vast gas reserves in these source formations since the 1820s [9, 10].

A shale gas well differs significantly from a conventional well. The extraction rate of a conventional well remains stable in terms of the amount of oil or gas extracted daily. In contrast, a shale gas well exhibits a much higher extraction rate in the first several months, followed by a subsequent decrease in productivity over a period of years. To maintain the original productivity, continuous drilling is required to create more pores in shale rocks and even open new wells nearby. Another distinction is that shale gas wells employ directional drilling to extend the wellbore as deep as thousands of feet to reach the subsurface formation [6, 11]. Directional drilling allows for the drilling of several or more wells without occupying multiple well pads. In addition, the use of the horizontal drilling technique in shale gas wells can prevent disruptions to surface ecosystems. This technique enables the exposure and extraction of gas from a large sector of the shale rock region without the need for multiple well pads and without impacting any parts of the Earth's surface [11–13]. Before shale gas can be transported to end-users, it must undergo a series of processes at the wellhead and processing facilities [14].

Successful hydraulic fracturing and horizontal drilling techniques have changed the map of oil and gas production since the never-touched source formations become the resources of oil and gas to be extracted. Certainly, it increases the estimated availability of oil and gas in many countries, particularly the United States. However, it has brought negative impacts on the environment [15]. One concern is the consumption of water since each gas well-extracting shale gas consumes a significant amount of water. There have been more concerned about the environmental pollution of the used water [16–18]. Water contains different chemicals that function in fracking, dissolution, or even chemical reactions with shale rocks. The management of the water pumped to the subsurface has been an important topic from environmental and geological points of view [19]. Direct emission of CH_4 to the ambient during such extraction unfortunately contributes to the greenhouse effect, as briefed in Chapter 5 [20]. In addition, the footprint of the shale gas well can largely disturb the landscape [20, 21]. These emerging environmental issues associated with methane extraction must be addressed, even though all energy technologies have their own limitations and drawbacks.

References

1 U.S. Department of Energy (US DOE). 2009. Modern shale gas development in the United States: A primer. Rep. DoE-FG26-04NT15455, Washington, DC. 116 pp.

2 U.S. Energy Information Administration. "Natural gas explained where our natural gas comes from.", 2022. https://www.eia.gov/energyexplained/natural-gas/where-our-natural-gas-comes-from.php

3 Aminzadeh, F. 2018. "Hydraulic fracturing, an overview." *J. Sustainable Energy Eng.* 6, 204–228.

4 Barati, R., Liang, J. T. 2014. "A review of fracturing fluid systems used for hydraulic fracturing of oil and gas wells." *J. Appl. Polym. Sci.* 131, 40735.

5 National Energy Technology Laboratories, Shale Gas: Applying Technology to Solve America's Energy Challenges, March 2011. https://www.hsdl.org/?abstract&did=8711

6 Soliman, M. Y., Dusterhoft, R. 2016. "*Fracturing Horizontal Wells.*" McGraw Hill.

7 Zendehboudi, S., Bahadori, A. 2016."*Shale Oil and Gas Handbook: Theory, Technologies, and Challenges.*" Gulf Professional Publishing. 1st ed.

8 Ma, Y. Z., Holditch, S. 2015. "*Unconventional Oil and Gas Resources Handbook: Evaluation and Development.*" Gulf Professional Publishing. 1st ed.

9 Montgomery, C. T., Smith, M. B. 2010. "Hydraulic fracturing: History of an enduring technology", *J. Petrol. Tech.* 62, 26–32.

10 Shellenberger, M., Nordhaus, T., Trembath, A., Jenkins, J. 2012. "*Where the Shale Gas Revolution Came From.*" The Breakthrough Institute. https://thebreakthrough.org/issues/energy/where-the-shale-gas-revolution-came-from

11 Short, J. A. 1993. "*Introduction to Directional and Horizontal Drilling.*" Pennwell Corp Publishing.

12 Pettus, D. 1992. "*Horizontal, High-angle, and Extended-reach Drilling.*" Southwest Geoservices.

13 Drohan, P. J., Brittingham, M., Bishop, J., Yoder, K. 2012. "Early trends in landcover change and forest fragmentation due to shale-gas development in Pennsylvania: A potential outcome for the northcentral Appalachians." *Environ. Manage.* 49, 1061–1075.

14 Goellner, J. F., Hamilton, B. A. 2012. "*Expanding the Shale Gas Infrastructure.*" American Institute of Chemical Engineers. 49–59. http://www.aiche.org/sites/default/files/cep/20120849.pdf.

15 Stephenson, E., Shaw, K. 2013. "A dilemma of abundance: Governance challenges of reconciling shale gas development and climate change mitigation." *Sustainability* 5, 2210–2232.

16 Osborn, S. G., Vengosh, A., Warner, N. R., Jackson, R. B. 2011. "Methane contamination of drinking water accompanying gas-well drilling and hydraulic fracturing." *PNAS.* 108, 8172–8176.

17 Jackson, R. B., Vengosh, A., Darrah, T. H., Karr, J. D. 2013. "Increased stray gas abundance in a subset of drinking water wells near Marcellus shale gas extraction." *PNAS.* 110, 11250–11255.

18 Gallegos, T. J., Varela, B. A., Haines, S. S., Engle, M. A. 2015. "Hydraulic fracturing water use variability in the United States and potential environmental implications." *Water Resour. Res.* 51, 5839–5845.

19 Colborn, T., Kwiatkowski, C., Schultz, K., Bachran, M. 2011. "Natural gas operations from a public health perspective." *Human Ecol. Risk Assess.* 12,1039–1056.

20 Schmidt, C. W. 2011. "Blind rush? Shale gas boom proceeds Amid Human Health Questions." *Environ. Health Perspect.* 119, a348–a353.

21 Slonecker, E. T., Milheim, L. E. 2015. "Landscape disturbance from unconventional and conventional oil and gas development." *Environments* 2, 200–220.

5

Methane Emission and Its Impact on Environment

The impact of greenhouse gas emissions on environment is a crucial global issue. Extensive literature, including numerous books and reviews, has been dedicated to exploring this topic [1–3]. While this chapter briefs the importance of this issue, it primarily serves as an introduction and does not delve deeply into the complexities of this environmental issue. Readers are encouraged to explore literature for more information.

5.1 Methane Emissions

According to the National Aeronautics and Space Administration (NASA) data, the concentration of CH_4 in the atmosphere has increased by over 150% in the last 150 years [4]. The impact of CH_4 on global warming is 28–34 times that of CO_2. Thus, the 2016 Paris Agreement set the goal of controlling climate change, aiming to limit "the increase in the global average temperature to well below 2 °C above preindustrial levels." This commitment marks the first international policy addressing CH_4 emissions. It does not imply eliminating the use of methane in industries and energy supply chains. The key is ensuring that the utilization of CH_4 does not result in the release of CH_4 or CO_2.

Unfortunately, sources of CH_4 emissions are widespread, ranging from coal mines, wet fields, petroleum industries, power plants, nonelectric vehicles, and biomass burning to the digestion of ruminant livestock. Understanding how anthropogenic activities contribute to CH_4 release is crucial for minimizing methane emissions to the atmosphere. According to the Global Methane Initiative (www.globalmethane.org), CH_4 emissions account for 20% of global greenhouse gas emissions [5].

Methane emissions occur through natural and anthropogenic processes. Natural methane emissions primarily come from methane hydrates stored under the seabed (discussed in Chapter 4), nonagricultural wetlands, and wildfires [6]. Enormous amounts of CH_4, estimated at 1.3×10^4 Gt carbon, are stored in the form of methane hydrates [7, 8]. Considering the predicted climate warming rate of 0.2 °C per decade [9], catastrophic methane emissions from methane hydrates are not anticipated over the next few hundred years [10]. This is because most methane hydrates are stored at depths of 300–500 m.

Low-temperature Activation and Catalytic Transformation of Methane to Non-CO_2 Products,
First Edition. Franklin Tao.

Currently, the contribution of CH_4 emissions from methane hydrates is relatively small, although literature predicts that methane released from the circum-Arctic Ocean continental shelves may occur [11–13].

Methane release from nonagricultural wetlands is generally regarded as a natural process. Approximately 40% of biogenic methane enters the atmosphere, while the remainder is consumed by aerobic and anaerobic methane-oxidizing microorganisms. Emissions from wetlands and wildfires are therefore categorized as natural methane sources. However, these sources exhibit substantial temporal and spatial variability, making their net contribution to climate change difficult to quantify. Consequently, current estimates focus primarily on anthropogenic methane emissions.

As presented in Figure 5.1a, gas and oil systems contribute approximately 20% to global anthropogenic methane emissions. CH_4 is a primary component of natural gas systems and a byproduct of petroleum industries. In the gas system, methane is emitted during the extraction, processing, storage, transportation, and distribution of natural gas. Numerous instances illustrate methane emissions within oil and gas systems. Power plants, even when equipped with high-temperature catalysts for catalyzing the complete oxidation of unburned methane after combustion process, still release a certain amount of CH_4 that cannot be burned. Another source is the CH_4 generated from the incomplete combustion of gasoline in automobile engines. It is reported that the concentration of CH_4 in the exhaust of gasoline engines is approximately 1000 ppm [15]. However, vehicles using natural gas or liquefied petroleum gas engines release more CH_4 compared to those with gasoline engines. The oil refinery system, referring to the manufacturing system of gasoline and other liquid fuels, is another source of CH_4 emission. CH_4 is a byproduct of the complex processes of oil refining.

Methane released during or after mining operations is termed coal mine methane, as the exhaust gas drained from the mine seam contains CH_4. The drained gas from a coal mine is carried by ventilation air. Typically, the CH_4 concentration in the ventilation air flowing through the coal mine ranges from 0.1% to 1% since the drained gas is diluted by the flowing air of the ventilation system. The collection or removal of methane in drained gas through ventilation air is crucial for the safety and working conditions in the mine, as CH_4 is a major trigger for coal mine explosions [16, 17]. Controlling CH_4 emissions is crucial for mine safety, and the surrounding environment, as well as for coal productivity. Establishing a ventilation system to remove or collect CH_4 before either releasing drained gas to the ambient or flaring is essential form mitigating CH_4 emission [18].

In general, coal mine methane has not been extensively utilized for chemical production or as an energy source although CH_4 can, in principle, be flown to a natural gas pipeline, for ultimate uses in power generation, electricity generation, fueling vehicles with natural gas engines, or for chemical transformation into value-added chemicals.

5.2 Fundamentals on Methane-relevant Environmental Issue

Earth is the only planet capable of sustaining human life, with a radius of 3958.8 miles. The space above Earth's surface is divided into five regions: the troposphere, stratosphere, mesosphere, thermosphere, and exosphere. The troposphere, where human habitation occurs,

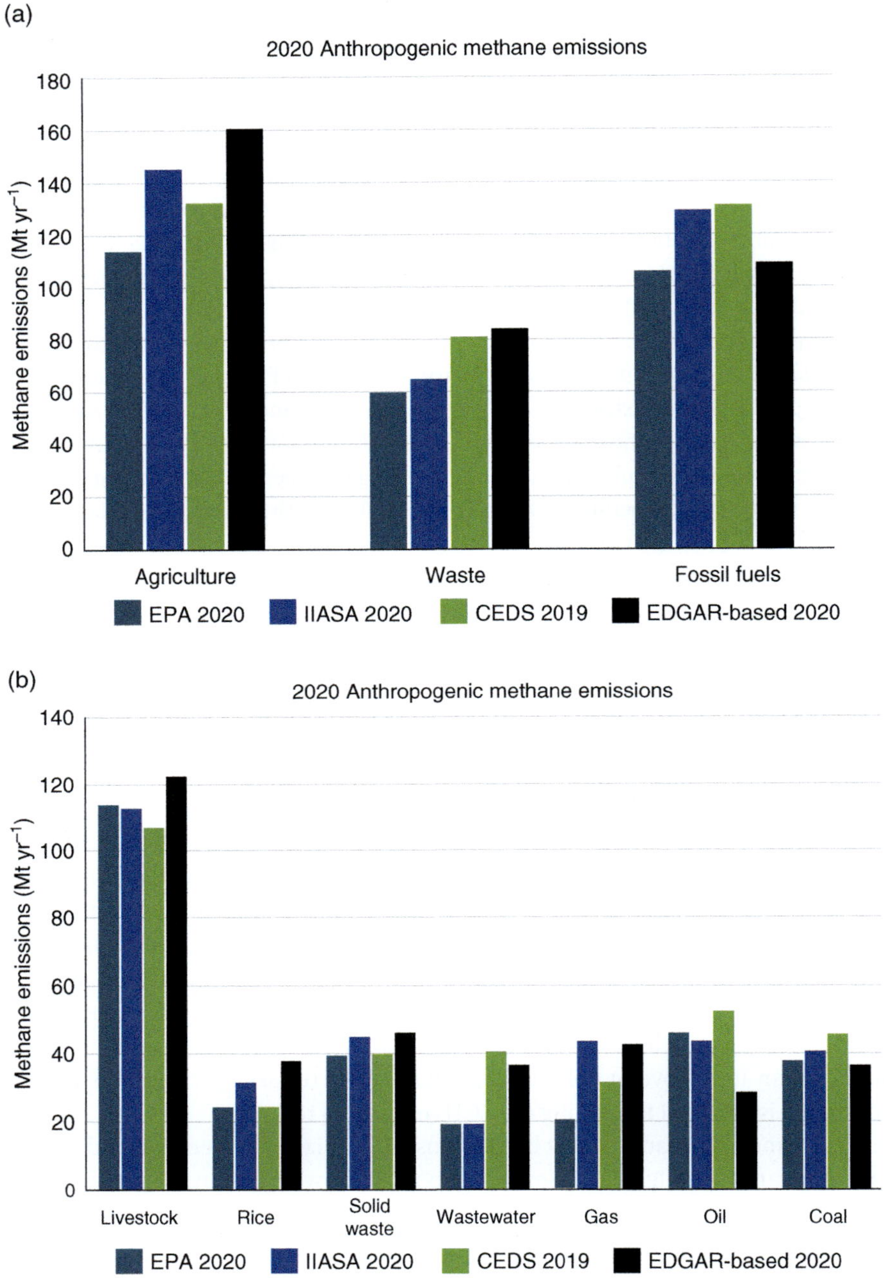

Figure 5.1 Estimated amount of CH_4 released from anthropogenic emissions by sector (a) and subsector (b) from the indicated inventories. *Source:* [14] / U.S. Department of the Interior / Public Domain.

extends from the Earth's surface up to 0 to 14.5 km. Beyond this, the stratosphere spans from 14.5 to 50 km, followed by the mesosphere from 50 to 85 km. The thermosphere encompasses a further region, while the exosphere extends even further. Although there is nearly a high vacuum in the mesosphere region, the Earth continuously releases a significant amount of gases into the atmosphere, which remains in the troposphere due to gravitational forces and does not escape into deeper space.

The Earth's temperature is intricately influenced by the interplay of sunlight absorption by the Earth's surface and atmosphere, coupled with the release of heat from the Earth's surface into space through radiation. Sunlight, the radiant energy emitted by the Sun, plays a pivotal role in this dynamic system. A fraction of the Sun radiation reaches the Earth's atmosphere, significantly impacting the planet's temperature. Fluctuations in solar radiation contribute to the natural variation in the Earth's temperature. According to NASA data, approximately 30% of incoming sunlight is reflected back into space by clouds and aerosol particles in the atmosphere and reflective surfaces such as ice on the Earth's surface. The remaining 70% is absorbed by the Earth surface. While the Earth's land surfaces and oceans absorb the majority, the atmosphere above the Earth's surface absorbs the rest. Simultaneously, the Earth's surface emits energy in the form of thermal infrared radiation, warming their surroundings. This thermal infrared radiation, a mechanism for transferring energy between objects, reaches the atmosphere from the Earth's surface. A significant portion of the energy released from the Earth's surface through this thermal infrared radiation is absorbed by water vapor and greenhouse gases like CO_2 and CH_4 in the atmosphere. The rest of the energy released transmitted through this thermal infrared radiation is ultimately released into space. This intricate process contributes to the complex and delicate balance regulating the Earth's temperature.

Upon absorbing radiation from Sun and being heated, water vapor and greenhouse gas molecules act as countless microscopic heaters at the micrometer or even nanometer scales, emitting energy in *all directions*. Consequently, a portion of the emitted energy is radiated back to Earth's surface, warming both the surface and the lower atmosphere. The absorption of this back-radiated energy by the lower atmosphere and Earth's surface helps maintain a comfortable temperature of 15 °C [19]. This absorption and back-radiation of radiation by greenhouse gases from the Earth's surface is beneficial for human beings, constituting a healthy greenhouse effect that is not only welcomed but also necessary for our survival. In this context, it is evident that the greenhouse effect is indispensable. In other words, it is important to mitigate greenhouse gas emissions rather than completely eliminate greenhouse gases from the atmosphere.

Concerns arise from the intensified greenhouse effect caused by the escalating accumulation of greenhouse gases over the past couple of centuries, a consequence of the Industrial Revolution and recent rapid development in some continents over the last several decades. According to NASA data (Figure 5.2), concentrations of CO_2 and CH_4 in the atmosphere have increased by nearly 38% and 148%, respectively, as of 2009, compared to their concentrations in 1750 [19]. When the concentrations of greenhouse gases exceed normal values, a larger portion of the energy released from Earth's surface to the atmosphere is redirected back to the Earth's surface through the radiation of these gases, resulting in a higher temperature on the Earth's surface than the normal value.

Figure 5.3 and Figure 5.4 present the global temperature anomaly over the last 140 years (1880–2020) compared to the average temperature in the period of 1951–1980. The temperature fluctuated in the range of −0.50 and +1.00 °C. The temperature from 1880 to 1940 is lower than the average temperature from 1951 to 1980, likely due to the recovery of manufacturing industries with an increasing release of greenhouse gases from 1951 to 1980. The obvious increase in temperature from 1980 to 2020 is attributed to the

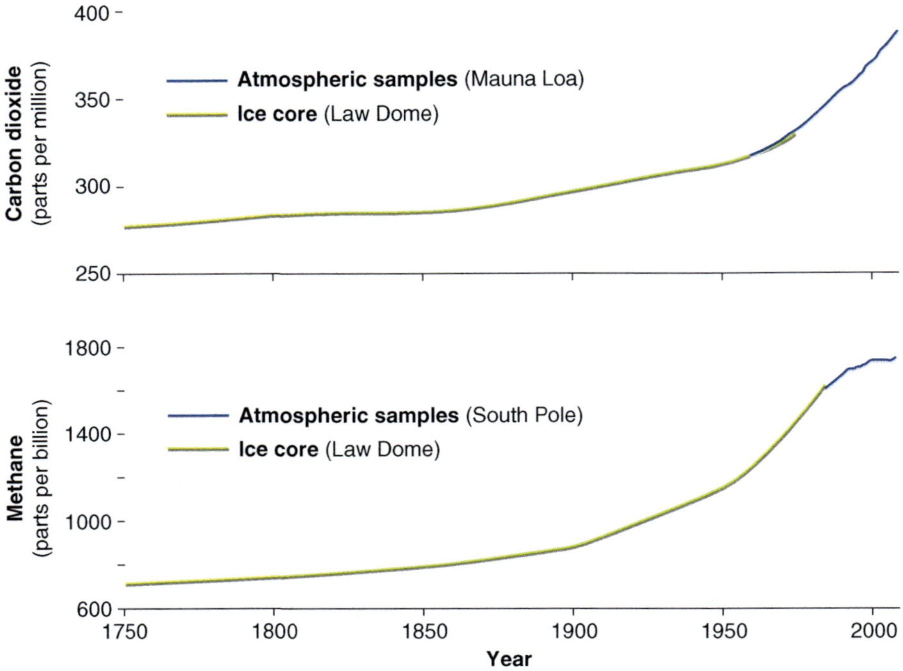

Figure 5.2 Concentrations of greenhouse gases CO_2 and CH_4 in the last two to three centuries. *Source:* [19] / NASA / Public Domain.

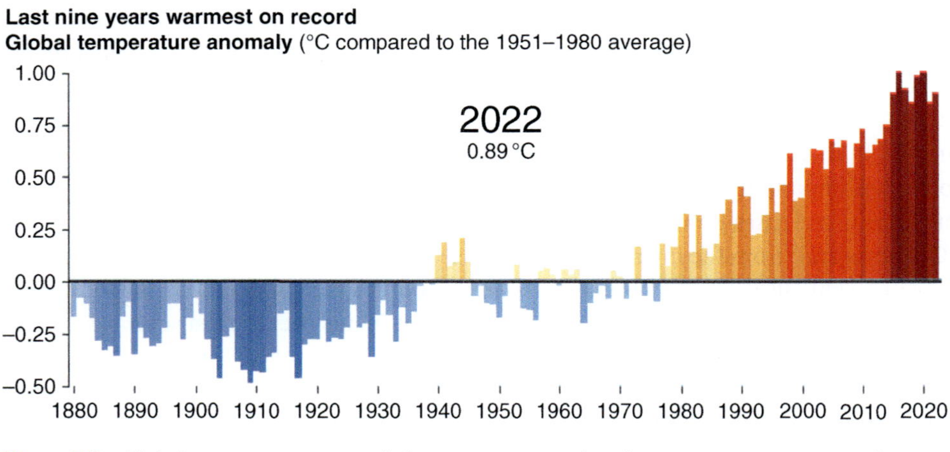

Figure 5.3 Global temperature anomaly in a way compared to the average temperature of 1951–1980 in the last 140 years. *Source:* Reproduced with permission from [20] / NASA / Public Domain.

increasing release of greenhouse gases in the last 4–5 decades (Figures 5.3 and 5.4). The leveling off temperature from 1940 to 1980 could be rationalized by countering effects, including the warming effect of greenhouse gases released from burgeoning industrial development after World War II and the cooling effect of aerosols released from

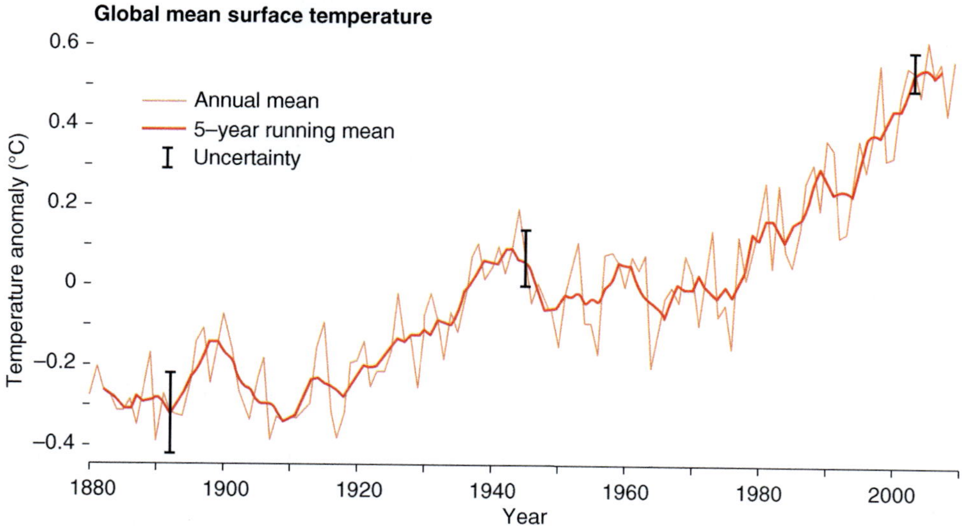

Figure 5.4 Plot of each year's global temperature anomaly in a way compared to the average temperature of 1951–1980 in the last 140 years. *Source:* [19] / NASA / Public Domain.

manufacturing facilities, coal power plants, and transportation vehicle exhaust pipelines, which reflect some sunlight back to space.

The impact of global temperature change is profound. Even a one-degree change can have severe consequences, such as melting a vast amount of glaciers and elevating sea levels.

CO_2 mitigation often necessitates alternative, non–CO_2-emitting processes, since CO_2 is an inherent byproduct of many manufacturing operations, particularly in the petrochemical industry. In contrast to CO_2, in general CH_4 is not a byproduct of the currently major manufacturing processes. Thus, mitigating CH_4 emissions does not require the development of entirely new production pathways to replace existing ones. In fact, CH_4 has progressively substituted petroleum in the production of major chemical feedstocks in industries. The advancement of CH_4-based conversion processes that avoid emissions of both CH_4 and CO_2 is therefore consistent with the broader goal of reducing greenhouse gas emissions. For example, the development of highly selective catalytic routes for CH_4 reforming and low-temperature catalytic oxidation of CH_4 offer pathways to transform CH_4 into valuable intermediates without generating CO_2 byproducts.

References

1 Stille, D. R. 2006. "*The Greenhouse Effect: Warming the Planet.*" Compass Point Books.

2 Ziska, L. H. 2022. "*Greenhouse Planet: How Rising CO_2 Changes Plants and Life as We Know It.*" Columbia University Press.

3 Smirnov, B. M. 2020. *"Global Atmospheric Phenomena Involving Water: Water Circulation, Atmospheric Electricity, and the Greenhouse Effect."* Springer Nature. 1st ed.

4 https://earthobservatory.nasa.gov/features/GlobalWarming/page1.php

5 www.globalmethane.org

6 Ruppel, C. D. 2011. Methane Hydrates and Contemporary Climate Change. *Nature Education Knowledge* 3, 29.

7 Boswell, R., Collett, T. S. 2011. “Current perspectives on gas hydrate resources.” *Energ. Environ. Sci.* 4, 1206–1215.

8 Dickens, G. R. 2011. “Down the rabbit hole: Toward appropriate discussion of methane release from gas hydrate systems during the Paleocene-Eocene thermal maximum and other past hyperthermal events.” *Clim. Past* 7, 831–846.

9 Intergovernmental Panel on Climate Change. 2007. “*Climate Change 2007: The Physical Basis.*” New York, Cambridge University Press.

10 Ruppel, C. D.; Kessler, J. D. 2016. “The interaction of climate change and methane hydrates.” *Rev. Geophysics* 55, 126–168.

11 Maslin, M., Owen, M., Betts, R., Day, S., Jones, T. D., Ridgwell, A. 2010. “Gas hydrates: Past and future geohazard.” *Philos. Trans. R. Soc. A* 368, 2369–2393.

12 Macdonald, G. 1990. “Role of methane clathrates in past and future climate.” *Clim. Change* 16, 247–281.

13 Lachenbruch, A. H. 1990. “*Permafrost, the Active Layer, and Changing Climate.*” United States Geological Survey, Open File Report No.: 94-694. https://pubs.usgs.gov/of/1994/0694/report.pdf

14 Ruppel, C. D. 2011. “Methane hydrates and contemporary climate change.” *Nat. Educ. Knowl.* 3, 29–31.

15 Gelin, P., Primet, M. 2002. “Complete oxidation of methane at low temperature over noble metal based catalysts: A review.” *Appl Catal B* 39, 1–37.

16 Coward, H. F., Jones, G. W. 1952. “Limits of flammability of gases and vapors.” US Bureau of Mines Information Circular No. 503. 155 pp.

17 U.S. EPA. 2003. “Assessment of the worldwide market potential for oxidizing coal mine ventilation air methane.” United States Environmental Protection Agency. EPA 430- R-03-002.

18 Karacan, C. Ö., Ruiz, F. A., Cotè, M., Phipps, S. 2011. “Coal mine methane: A review of capture and utilization practices with benefits to mining safety and to greenhouse gas reduction.” *Int. J. Coal Geol.* 86, 121–156.

19 Riebeek, H., Simmon, R. 2010. “*Global Warming.*” NASA Earth Observatory. https://earthobservatory.nasa.gov/features/GlobalWarming/page1.php

20 NASA Earth Observatory. https://earthobservatory.nasa.gov/world-of-change/global-temperatures#:~:text=Air%20temperatures%20on%20Earth%20have,for%20making%20our%20planet%20warmer.

6

Brief of High-Temperature Catalytic and Noncatalytic Transformation of Methane

While methane has been used as a raw material for decades, one of its industrial applications has remained predominantly confined to the generation of synthesis gas, a precursor to produce value-added chemicals via well-established downstream processes (Figure 1.1). Certainly, there has been considerable effort in exploring low-temperature transformations, given their potential benefits, including reduced energy consumption, cost-effective manufacturing, and the ability for local production without the need for expensive transportation. Before delving into those low-temperature transformation routes, this chapter provides an overview and discussion of these main high-temperature processes for transforming CH_4 to significant intermediate compounds, as reported in the literature. This chapter aims to offer a brief understanding of CH_4 catalytic processes at high temperature although the primary focus of this book lies on the aspect of low-temperature transformation of CH_4.

6.1 Steam Reforming of Methane

Steam reforming of CH_4 has been an industrial practice for several decades. Literature predominantly addresses the design and improvement of catalysts, with less focus on the industrial processes themselves. Given its main function of producing important intermediate chemicals, syngas for the Fisher–Tropsch synthesis or ammonia synthesis, the steam reforming of CH_4 is often discussed in conjunction with downstream processes such as water–gas shift, FTS, methanol synthesis, or ammonia synthesis. This section concentrates on the design and improvement of catalysts, emphasizing high activity and selectivity in the steam reforming of CH_4 to produce syngas. Steam reforming of CH_4 is a highly endothermic process at room temperature, represented by the equation $CH_4 + H_2O \rightarrow CO + 3H_2$, with an enthalpy change of 206 kJ mol^{-1}. While this reaction becomes favorable at high temperatures and low pressures, the elevated temperature introduces a common challenge known as coke formation, which will be discussed further below. Notably, most Group VIII metals exhibit activity for this reaction, with Ru and Rh identified as the most active elements within Group VIII. Co and Ir were recognized as active elements, albeit presumably existing as oxides under steam reforming conditions in the 1970s [1].

Low-temperature Activation and Catalytic Transformation of Methane to Non-CO_2 Products,
First Edition. Franklin Tao.

6.1.1 Brief of Activity of Transition Metals

Various studies have ranked Pt, Ir, Ru, and Ni as the most active elements for CH_4 steam reforming [2]. A recent comparison of activities among Group VIII elements, based on turnover frequencies at 500 °C and 1 bar with a CH_4 conversion of 10%, revealed that Ru, Rh, and Ni are the most active metals. Turnover frequencies (TOFs) were predicted in literature by considering adsorption energies of C and O bound to step sites on the (221) surface of pure metals, including Fe, Co, Ni, Cu, Ru, Rh, Pd, Ag, Ir, Pt, and Au [3]. Plotting TOFs against C* and O* adsorption energies [$\Delta E(O^*)$] and [$\Delta E(C^*)$], a two-dimensional volcano curve was generated (see Figure 6.1). This theoretical study supports the experimental findings from the 1970s to the 1990s, indicating that Ru, Rh, and Ni are indeed the most active Group VIII metals [3]. The choice of the (221) surface in these calculations is justified by the stronger adsorption of all intermediates of CH_4 steam reforming on (221) compared to low-Miller index surfaces such as (111) [4]. Density functional theory (DFT) calculations further revealed that the activation barrier for the reaction channel of CH_4 steam reforming on a stepped surface like Ni(221) is lower than that on Ni(111), as presented in Figure 6.2. On a metallic Ni surface, the reaction pathway involves continuous dehydrogenation of CH_4 to C, followed by the coupling between the C atom and the O atom dissociated from H_2O to form CO; the dissociated H atoms couple to form molecular H_2. To support the catalyst Ni nanoparticles (NPs), nonreducible metal oxides with a rigid lattice,

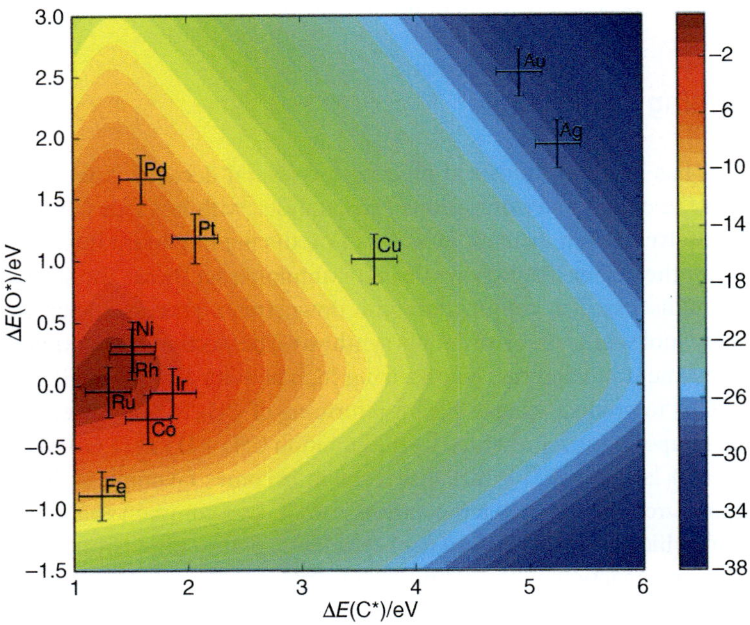

Figure 6.1 Two-dimensional volcano-curve of the turnover frequencies ($\log_{10}$) as functions of O* and C* adsorption energies. $T = 773$ K, $P = 1$ bar; 10% conversion. The error bars include an estimated 0.2 eV uncertainties in the adsorption energies. *Source:* Reproduced with permission from [3] / Elsevier.

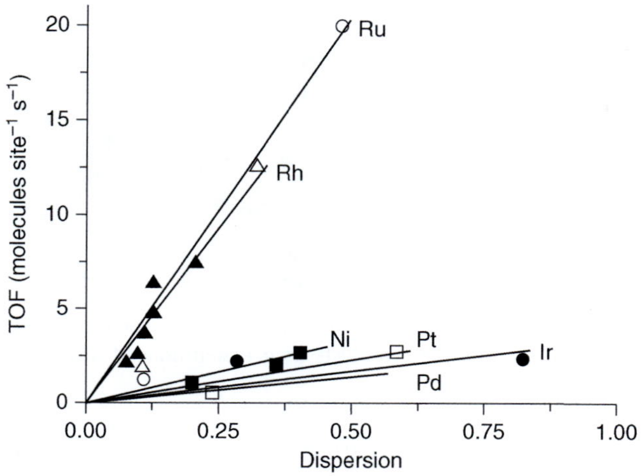

Figure 6.2 Reaction rate in terms of turnover frequency as a function of metal dispersion for CH_4 steam reforming under a condition (773 K, 0.19 bar CH_4, 0.74 bar H_2O, 0.07 bar H_2) on Ru, Rh, Ni, Pt, Ir and Pd. Ru (○), 5 wt% Rh (▲), 1 wt% Rh (Δ), Ni (■), Pt (□), Ir (●), and Pd (◆). *Source:* Reproduced with permission from [3] / Elsevier.

characterized by a high dissociation energy of the M–O bond, such as Al_2O_3, $MgAl_2O_4$, MgO, or CaO, serve as supports for Ni metal.

6.1.2 Deactivation of Metal Catalysts

Extensive effort has focused on understanding the deactivation and lifespan of steam reforming catalysts [5]. A multitude of factors, including coke formation, sulfur poisoning, metal segregation, alloy formation, and sintering, can contribute to the deactivation of catalysts employed in high-temperature methane steam reforming. Among these factors, literature predominantly addresses sintering, coke formation, and sulfur poisoning. The following provides a brief overview of these three critical deactivation factors.

Coke formation poses a near-universal challenge for supported metal NP catalysts employed in high-temperature CH_4-involved reactions. This phenomenon in supported metal catalysts is intricately linked to fundamental steps in the catalytic mechanism of CH_4 steam reforming. Generally, the continuous dissociation of C–H in CH_4 to carbon atoms consists of essential elementary steps in methane steam reforming. This dissociation leads to the coupling of C atoms or CH_x species on the surface of a metal NP, initiating the formation of carbon chains, a parallel unwanted reaction path on the metal catalyst. Based on experimental findings, three types of coke have been identified, namely, pyrolytic carbon, encapsulating carbon, and whisker carbon. Figure 6.3a illustrates pyrolytic carbon, formed through the thermal cracking of hydrocarbons at high temperatures [6]. Encapsulated carbon, shown in Figure 6.3b, is characterized by metal NPs enclosed in multiple layers of graphene. This type of carbon is formed when aromatics are present in the feed gas or when the steam-to-carbon ratio in the feed gas is low. Whisker carbon, another type, results from the thermal cracking of hydrocarbons at high temperatures, akin to pyrolytic carbon [7]. Whisker carbon, shown in Figure 6.3c, is formed by the initial creation of

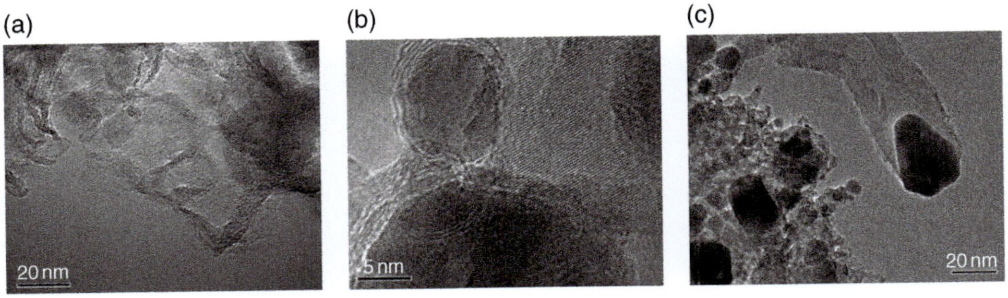

Figure 6.3 Electron microscopy images of pyrolytic carbon on a $MgAl_2O_4$ carrier (a), encapsulating carbon (b), and whisker carbon (c) on Ni/$MgAl_2O_4$ reforming catalysts. *Source:* Reproduced with permission from [6] / Elsevier.

atomic carbon on one side of a metal NP. Subsequently, carbon atoms migrate to the bulk of the metal NP, diffuse to the opposite end, and aggregate to form carbon layers [8–11].

Sulfur Poisoning in Catalysts

In terms of fundamental studies, more emphasis has been placed on the development of new catalysts with high activity and selectivity, along with a targeted understanding at the molecular level of the reaction mechanism, as opposed to addressing the technical challenges associated with sulfur poisoning. Sulfur forms stable compounds with most metals, leading to the incapability of catalyst atoms in binding with reactant molecules. Typically, the source of sulfur in CH_4 is H_2S [12]. The reaction of H_2S on Ni metal results in the formation of a robust NiS and H_2. The activity of the Ni catalyst experiences an exponential decrease with the increase in sulfur atom coverage (θ_S) by following the equation, $\boldsymbol{r}_i(\theta_S) = (1-\theta_S)^3 \times \boldsymbol{r}_0$ [7]. Here $\boldsymbol{r}_0$ is the activity of the Ni catalyst without any sulfur atoms. Clearly, the poisoning effect of sulfur on the Ni catalyst is severe. Sulfur exhibits a similar poisoning effect on other metal catalysts involved in CH_4 steam reforming.

Sintering

In the context of the reported metal catalysts for CH_4 steam reforming, where the active phases are metallic NPs supported on metal oxides, sintering emerges as a common feature of most metal catalysts at high temperatures. The primary mechanisms driving the sintering of metal NPs in steam reforming at relatively low temperatures include migration and coalescence [13]. As the temperature exceeds 600 °C, sintering typically occurs through Ostwald ripening [14]. From a broader perspective, the sintering of metal NPs represents a prevalent phenomenon in supported metal catalysts exposed to certain reducing gas environments. For a more in-depth understanding of this topic, additional information can be found in literature such as [14].

6.1.3 Catalyst of Singly Dispersed Sites

The above discussions make it evident that the continuous packing of metal atoms on the surface of a metal NP leads to coke formation since these carbon atoms formed on continuously packed metal atoms can readily couple to form precursors of coke. To address these drawbacks associated with Ni metal NPs, a single-atom catalyst, $Ce_{0.90}Ni_{0.05}Ru_{0.05}O_2$ (CNR), was developed (Figure 6.4) [15]. In this catalyst, both Ni and Ru atoms are individually

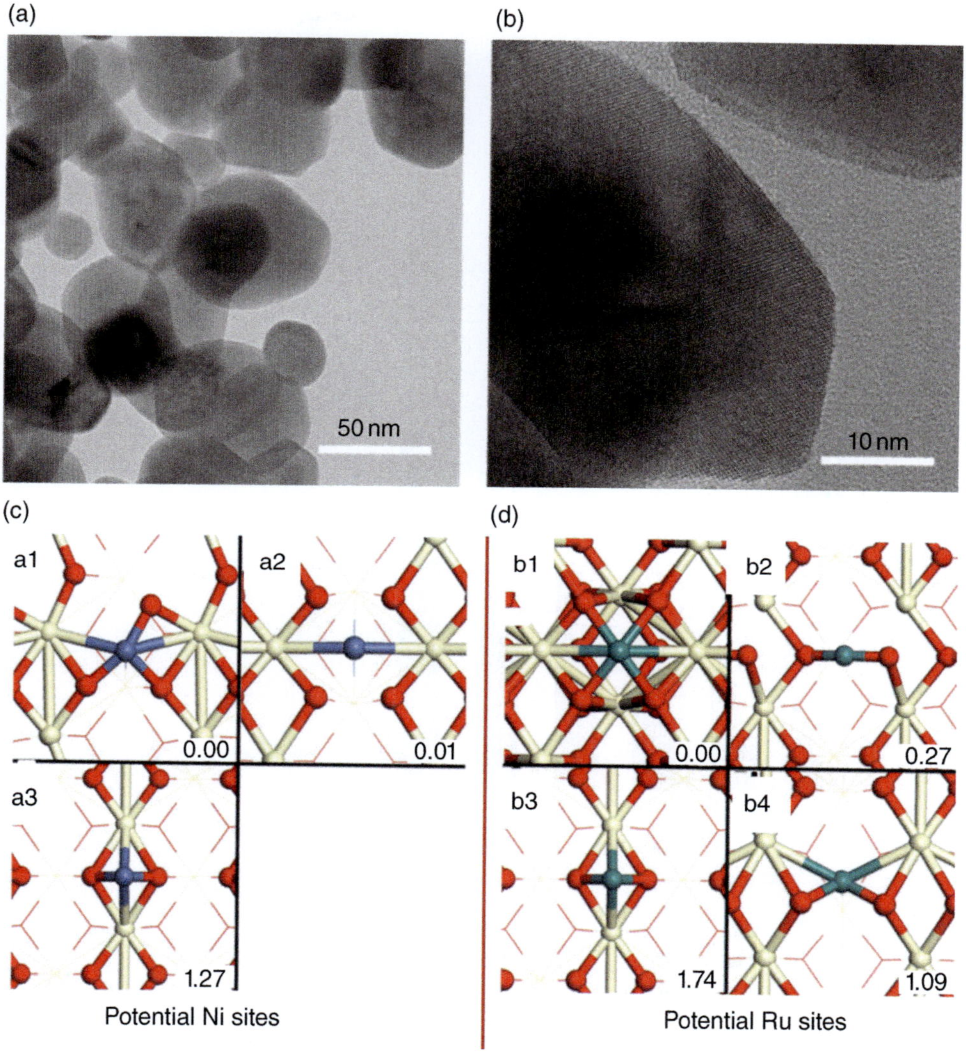

Figure 6.4 TEM images and optimized catalyst surface structure of $Ce_{0.90}Ni_{0.05}Ru_{0.05}O_2$. (a and b) TEM image of $Ce_{0.90}Ni_{0.05}Ru_{0.05}O_2$ catalyst after test at 500 °C in a fuel cell for 250 h. Neither metallic Ni nor Ru nanoclusters on CeO_2 nanoparticles were found. In addition, neither NiO nor RuO_x nanoclusters were found on CeO_2 nanoparticles. (c and d) DFT calculation proposed potential sites of Ni_1 and Ru_1 anchored on CeO_2 (110). Red, yellow, blue, and green balls stand for oxygen, cerium, nickel, and ruthenium, respectively. The most stable structures for Ni_1 and Ru_1 sites are (a1) and (b1), respectively. The values listed at the right bottom side of each figure are total energy of the corresponding structure; the unit of all the values are in eV. *Source:* Reproduced with permission from [15] / Springer Nature.

dispersed on CeO_2 during steam reforming of CH_4, a configuration confirmed through *in situ* extended X-ray absorption fine structure (EXAFS) studies [16]. The single-atom sites, Ni_1 and Ru_1, exhibit synergistic roles in the steam reforming of CH_4, as demonstrated by kinetic studies measuring turnover frequency and apparent activation barrier (Figure 6.5) [15]. DFT calculations further indicate that Ni_1 and Ru_1 are responsible for CH_4 activation and H_2O dissociation, respectively (Figure 6.6) [15].

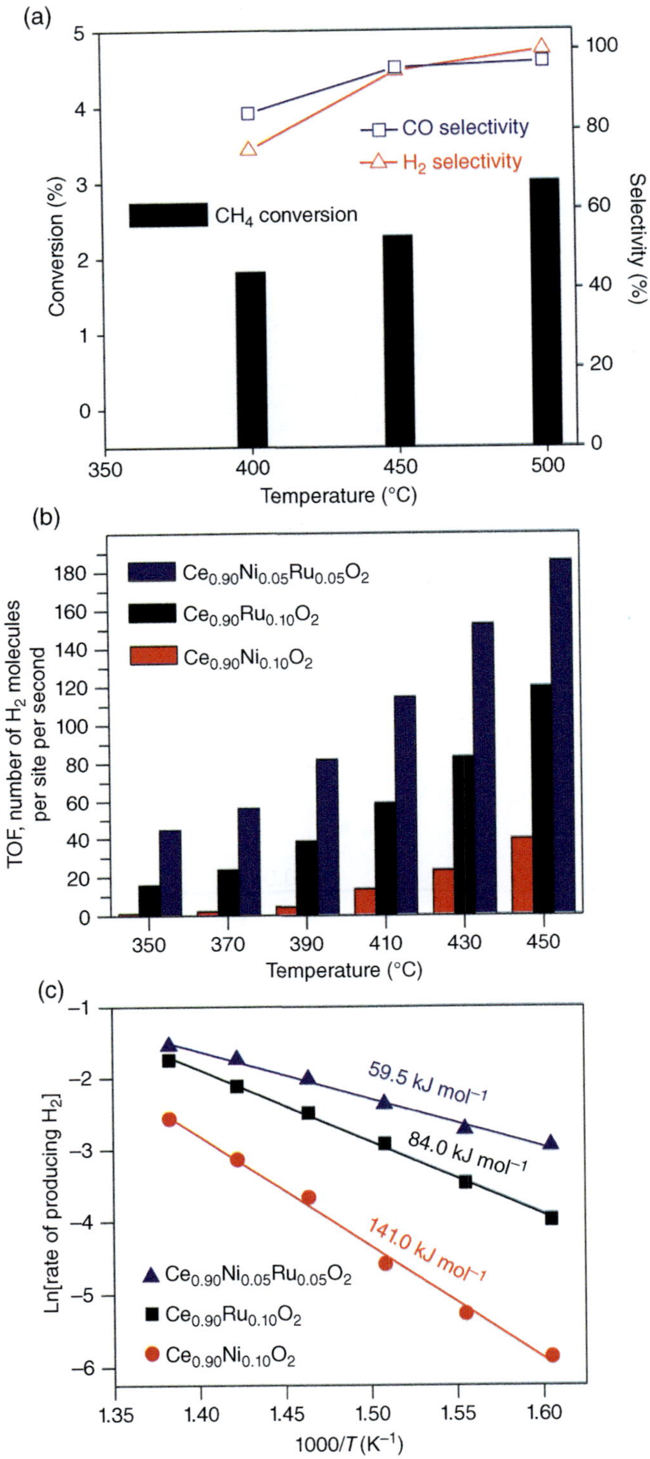

Figure 6.5 Catalytic performances of steam reforming on $Ce_{0.90}Ni_{0.05}Ru_{0.05}O_2$, $Ce_{0.90}Ni_{0.10}O_2$ and $Ce_{0.90}Ru_{0.10}O_2$ catalysts. (a) Conversion of methane and selectivity for hydrogen as a function of catalysis temperature for steam reforming on $Ce_{0.90}Ni_{0.05}Ru_{0.05}O_2$. (b) Number of produced H_2 molecules per second per Ni atom for $Ce_{0.90}Ni_{0.10}O_2$, per Ru atom for $Ce_{0.90}Ru_{0.10}O_2$, and per Ru and Ni atom for $Ce_{0.90}Ni_{0.05}Ru_{0.05}O_2$. (c) Arrhenius plots for producing H_2 on the three catalysts. Pure CH_4 (99%, Matheson) with a flow rate of 50 ml min^{-1} was introduced into a bubbling bottle containing deionized water kept at room temperature (3.5% H_2O). These linear lines were generated through the least-square fitting of data points for each catalyst. *Source:* Reproduced with permission from [15] / Springer Nature.

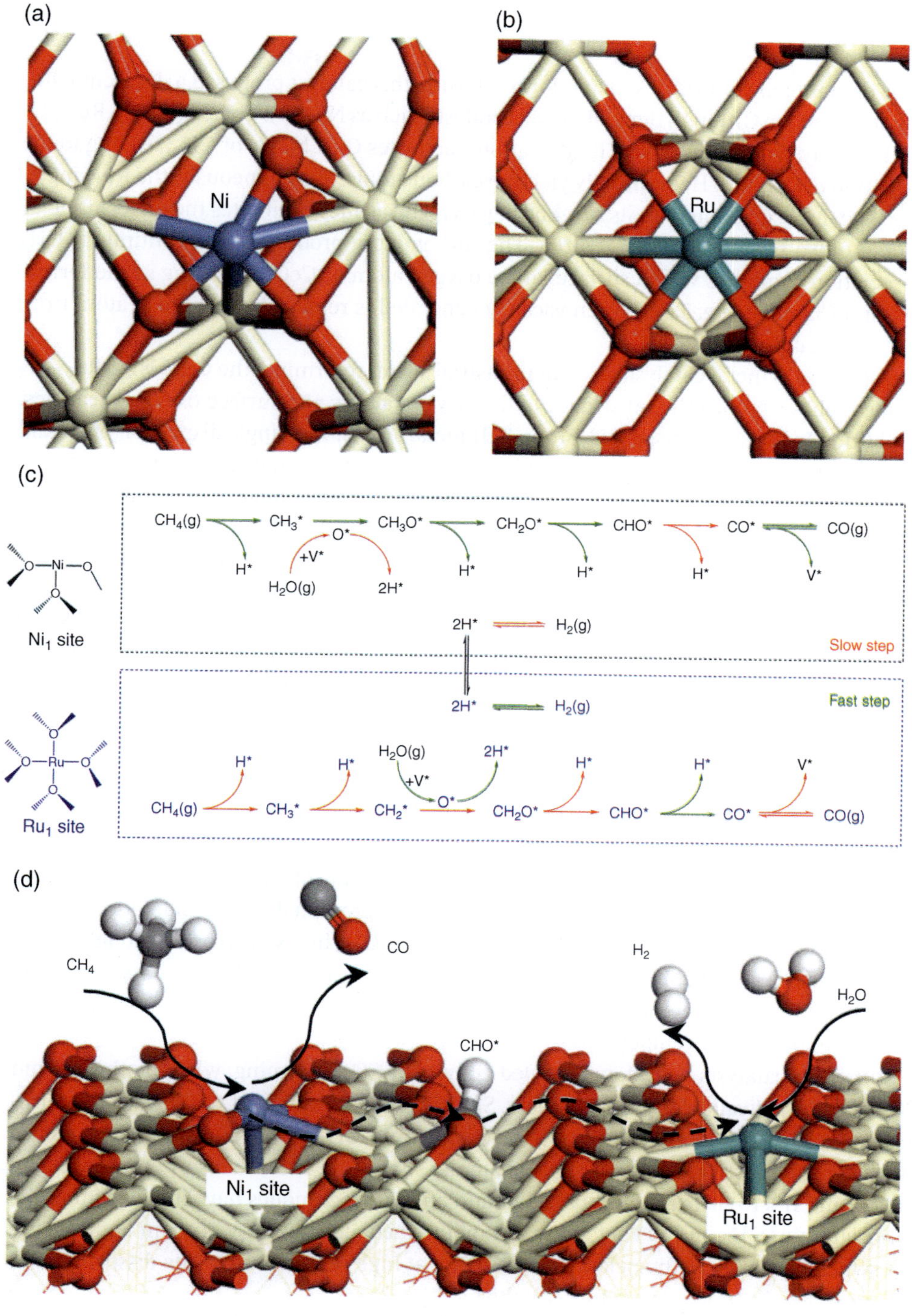

Figure 6.6 Understanding of reactions on the catalyst layer. (a–d) Optimized Ni_1 (a) and Ru_1 (b) sites on CeO_2 (110) of $Ce_{0.90}Ni_{0.05}Ru_{0.05}O_2$ during steam reforming of CH_4 and the proposed reaction pathway (c, d) from DFT calculations. In (a), Ni_1 atom bonds with three oxygen atoms directly. In (b), Ru_1 atom bonds with four oxygen atoms. Transformation of CH_4 to CO, and the activation of H_2O, were simulated; the results are arranged in the upper box (c for Ni_1 site) and lower box (c for Ru_1 site). The kinetically favorable and unfavorable steps are marked with green and red arrows, respectively. Cerium, carbon, oxygen, and hydrogen atoms are shown in yellow, grey, red, and white, respectively. *Source:* Reproduced with permission from [15] / Springer Nature.

Computational studies on steam reforming using this catalyst proposed a different pathway compared to Group VIII metal model catalysts such as Ni(111). The $Ce_{0.90}Ni_{0.05}Ru_{0.05}O_2$ (CNR) catalyst, according to DFT calculations, activates C–H bonds of CH_4, leading to the formation of CH_3 and H, eventually yielding carbon atoms. Simultaneously, the Ru_1 single-atom site, in association with its neighboring oxygen vacancy, facilitates the dissociation of H_2O into two H atoms and an O atom. The carbon atom produced from continuous CH_4 dissociation combines with a surface lattice oxygen atom of CeO_2, resulting in the formation of a CO molecule. The oxygen vacancy generated is replenished by the O atom from H_2O dissociation [15].

In contrast to metal catalysts used in industrial steam reforming, the $Ce_{0.90}Ni_{0.05}Ru_{0.05}O_2$ catalyst relies significantly on surface lattice oxygen atoms and surface oxygen vacancies, in addition to the singly dispersed Group VIII metal atoms. The single dispersion of Ni and Ru atoms is anticipated to effectively minimize the deactivation of the Ni metal catalyst. This catalyst has been used as a crucial component in a robust intermediate fuel cell [15]. The fuel cell converts the chemical energy of CH_4 into electricity by initially reforming CH_4 with steam at an intermediate temperature of 550 °C and subsequently undergoing electrooxidation of H_2 at the anode to generate electric current [15].

Other than $Ce_{0.90}Ni_{0.05}Ru_{0.05}O_2$, a number of other single-atom catalysts were reported to be active for CH_4 steam reforming. The structural characterization and reaction mechanism of these single-atom catalysts of CH_4 steam reforming can be found in literature.

6.2 Reforming of Methane by Consumption of CO_2

H_2O is readily available at a low cost in most regions. Consequently, steam reforming stands out as a preferable option compared to CO_2 reforming of CH_4. However, the growing concern over the global warming impact of CO_2 has prompted a reconsideration of CO_2 reforming of CH_4 as a viable method for transforming CH_4 into syngas, even though industries have not widely adopted this process thus far. The reaction enthalpy of dry reforming is 247 kJ mol^{-1} at room temperature. Due to its highly endothermic nature, the process is carried out at elevated temperatures.

Numerous catalysts have demonstrated activity in CO_2 reforming, where both CH_4 and CO_2 undergo activation on metal surfaces. Similar to the activation of CH_4 in steam reforming on Group VIII metal catalysts, the activation of CH_4 by CO_2 on transition metals involves a high activation barrier. For example, studies on the chemisorption of CH_4 on Ir(111) and Ir(110) using molecular beam and bulb gas techniques have indicated that the dissociation of CH_4 on Ir(111) follows a trapping-mediated pathway at room temperature but switches to a direct dissociation mechanism at high temperatures. The dissociation of CH_4 on a metal surface is structure-sensitive [17].

In general, the rate-determining step for dry reforming on transition metals is the dissociation of CH_4 to CH_3 [18]. The establishment of a Brønsted–Evans–Polanyi (BEP) correlation between kinetics and corresponding thermodynamic characteristics for $CH_4(ads) \rightarrow CH_3(ads) + H(ads)$ on transition metals, including Ni, Cu, Ru, Rh, Pd, Ag, Ir, Pt, and Au, has been reported in the literature [18]. This correlation indicates that the activation

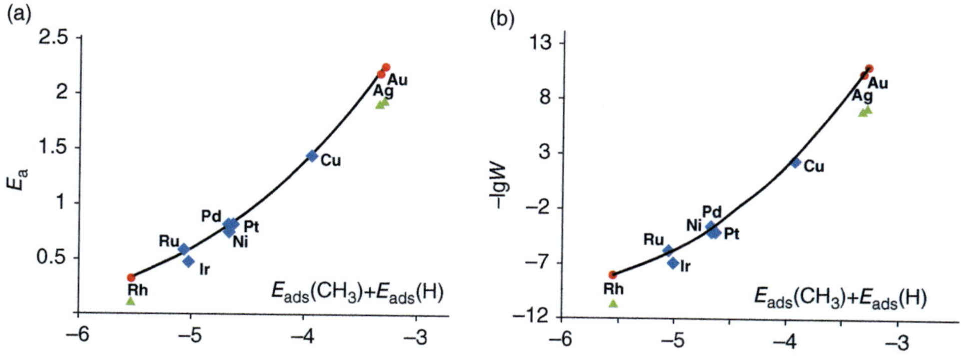

Figure 6.7 BEP relationships between the activation energy E_a (a) or the rate constant $-\log W$ at $T = 500$ K (b) and the sum of the H and CH_3 adsorption energies $E^t_{ads} = E_{ads}(H) + E_{ads}(CH_3)$, energies in eV, W in s^{-1}. *Source:* Reproduced with permission from [18] / American Chemical Society.

barriers (E_a) and rate constant ($-\lg W$) for CH_4(ads)→ CH_3(ads)+ H(ads) on transition metals depend on the surface reaction energy ΔI contributed from C–H dissociation energy (D), CH_3 adsorption energy [$E_{ads}(CH_3)$], and H adsorption energy [$E_{ads}(H)$]. Generally, D and $E_{ads}(CH_4)$ are similar for the transition metals. As illustrated in Figure 6.7 [18], the activation barrier E_a and rate constant of the rate-determining step of CH_4 activation, $-\lg W$, are mainly determined by the total adsorption energies of CH_3 and H, $E_{ads}(CH_3)$ + $E_{ads}(H)$, following a parabolic curve. This correlation, depicted in Figure 6.7, suggests that Rh, Ru, Ir, and Pd are among the most active metals for activating CH4 in dry reforming on transition metals.

6.2.1 Brief of Dry Reforming on Transition Metal Single-Crystal Model Catalyst Ni(111)

Numerous studies have investigated the dry reforming of methane on various transition metals, including Rh, Ir, Ru, Ni, Pd, and Pt [19–32]. According to the Brønsted–Evans–Polanyi correlation, Ni stands out for its high activity, featuring a low activation barrier for the rate-determining step of CH_4 activation. In this context, Ni(111) serves as an illustrative example to elucidate the reaction mechanism of dry reforming on the surface of a metal catalyst.

The inset of Figure 6.8 presents a simplified mechanism for methane dry reforming ($CH_4 + CO_2 \rightarrow 2CO + 2H_2$) catalyzed by Ni(111), as proposed in computational studies [33]. The mechanism suggests that, in the initial stages of the catalytic cycle, activations of CO_2 and CH_4 occur concurrently [33]. CH is formed through C–H activations of CH_4, yielding three H atoms – one forming an H_2 molecule and the other existing as a separate H atom. Simultaneously, O atoms are generated from the dissociation of CO_2, with the first CO molecule also formed in the process. The two intermediates (CH and O) couple on Ni(111) to form an intermediate CHO. Subsequent C–H bond dissociation of CHO produces the second CO molecule, and the H atoms from two CHO species combine to form the second H_2 molecule. Figure 6.8 provides a comprehensive overview of all elementary steps,

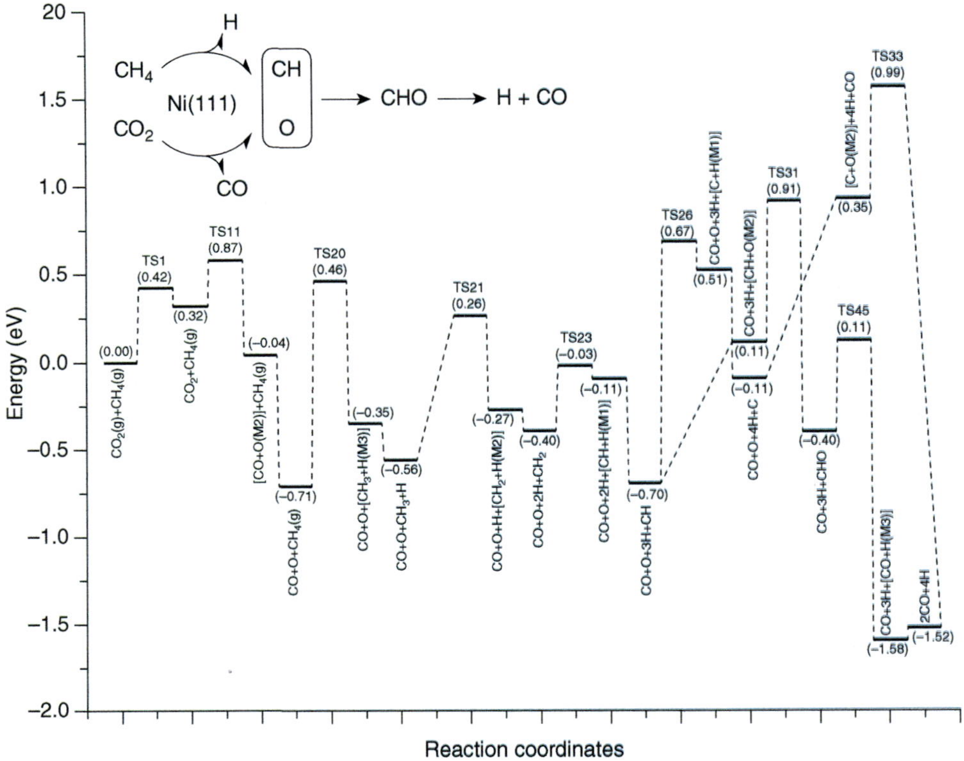

Figure 6.8 Potential energy of CO_2 reforming of CH_4 on Ni(111). [A+B(Mn)] denotes the co-adsorbed A and B in mode, CO_2(g) and CH_4(g) denote the free CO_2 and CH_4 molecules, respectively, and the other nomenclatures without any mark denote the separately adsorbed intermediates. The insert shows the activations of CH_4 to CH and 3H and CO_2 to CO and O are in parallel. *Source:* Reproduced with permission from [33] / Elsevier.

including intermediates and transition states, in the proposed mechanism for steam reforming of methane on Ni(111). CO_2 dissociates on Ni(111) to yield adsorbed CO and O, while CH_4 undergoes continuous dissociation to CH_3, CH_2, and ultimately CH. On the surface, the formed CH couples with the atomic oxygen to create an intermediate CHO. C–H dissociation of the intermediate CHO leads to the formation of CO and H, with CO desorbing into the gas phase as a product molecule. The H atoms combine to form an H_2 molecule, which then desorbs.

Activation of CO_2 on the metal surface of a single-crystal model catalyst can adopt different binding configurations, as illustrated in Figure 6.9 [33]. The dissociation of CO_2 to CO and O on Ni(111) occurs through electron transfer from the d orbital of a transition metal to an antibonding orbital of CO_2, weakening the bonding between C and O in CO_2 [33]. Stepped edge atoms, exhibiting a large dipole compared to metal atoms within a terrace, are typically highly undercoordinated. As a result, these stepped atoms can transfer more electron density to CO_2, rendering the dissociation structure-sensitive on a metal catalyst surface.

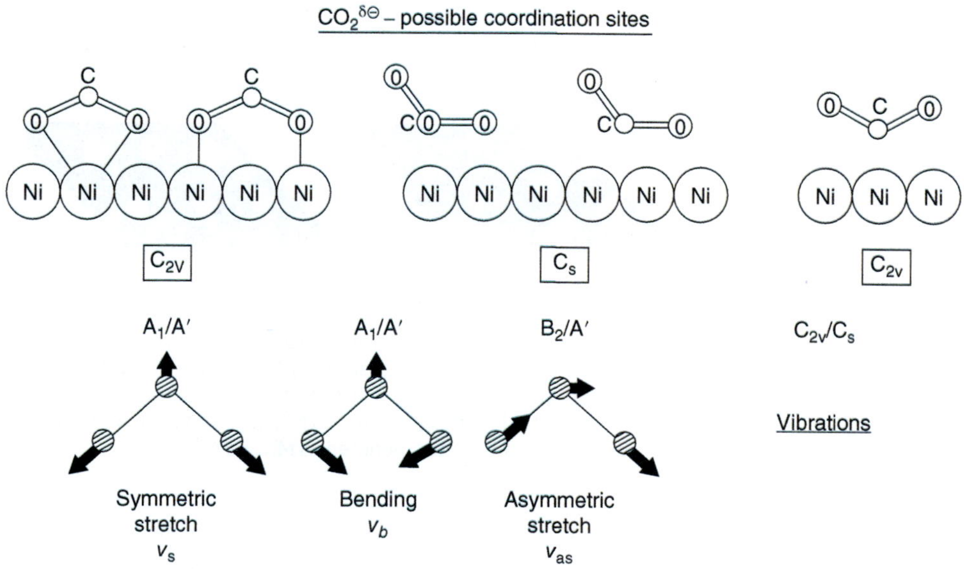

Figure 6.9 Schematic representation of the stretching and bending vibrations in bent $CO_2^{\bullet}$ for the three types of $CO_2^{\bullet}$ coordination. The irreducible representations of the vibrations are given for the sites of C_{2v} and C_s symmetries. *Source:* Reproduced with permission from [33] / Elsevier.

6.2.2 Catalyst of Singly Dispersed Sites

Apart from a transition metal supported on an insert oxide such as silica and alumina, a bifunctional catalyst consisting of two sets of metal NPs loaded on a reducible oxide has been reported. However, similar to the deactivation issues faced by Ni catalysts used in industrial steam reforming, problems like carbon deposition, sulfur poisoning, and metal NP sintering persist for these supported transition metal catalysts employed in dry reforming of CH_4. This is particularly problematic since both reactants, CH_4 and CO_2, can serve as carbon sources, providing carbon atoms through dissociation on a metal surface. Once again, on the surface of a metal NP, two CH_x species formed on two continuously packed metal atoms, such as two atoms of an element M^{α} on a metal surface (Figure 6.10a), readily couple to form a C–C chain and then a longer chain, accelerating coke formation. In contrast, compared to the continuous packing of metal sites on the surface of a metal NP (Figure 6.10a), the single dispersion of metal sites on a support of an oxide, schematically shown in Figure 6.10b, effectively prevents CH_x species on these metal sites from coupling with each other.

A recent literature report introduced a catalyst with dual single-atom sites (Figure 6.10c). On the surface of this single-atom catalyst, $Ce_{0.95}Ni_{0.025}Ru_{0.025}O_2$, both Ni_1 and Ru_1 atoms are individually dispersed on CeO_2, bonding to surface oxygen atoms of CeO_2. The Ni_1 single-atom site is responsible for activating CH_4 since the activation barrier for $CH_4(ad) \rightarrow CH_3(ad) + H(ad)$ on Ni_1 is lower than that for Ru_1. Computational studies suggest that CH_4 continuously dehydrogenates to form a carbon atom, which then couples with a surface lattice oxygen atom to form a CO molecule ($C^* + O \rightarrow CO$). However, as described in

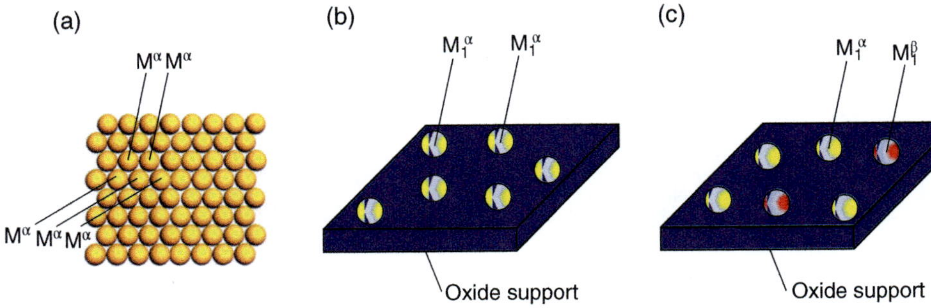

Figure 6.10 Schematic showing the difference in packing of metal atoms on surface of a metal nanoparticle and surface of a supported single-atom catalyst. (a) The surface of a metal nanoparticle where metal M^α sites are continuously packed ($\cdots M^\alpha M^\alpha M^\alpha M^\alpha \cdots$). (b) Surface of an oxide with supported, singly dispersed metal M^α sites (denoted as M_1^α). (c) Surface of an oxide with supported, two sets of singly dispersed metallic sites (denoted as M_1^α and M_1^β).

Section 6.2.2, on Ni(111), CH couples with one O atom to form an intermediate CHO. Then, the CHO dissociates to form CO and an H atom on the metal catalyst Ni(111), $CH + O^* \rightarrow CHO^*$, $CHO \rightarrow CO + H$ [34]. Evidently, the activation of CH_4 on the singly dispersed Ni_1 site anchored on CeO_2 is different from that on a Ni atom of a Ni metal catalyst.

Another distinctive feature of this single-atom catalyst, $Ce_{0.95}Ni_{0.025}Ru_{0.025}O_2$, is that activation of CO_2 is performed by Ru_1 and its adjacent oxygen vacancy (O_{vac}), $CO_2 + O_{vac} \rightarrow CO + O_{suface\ lattice}$ [16]. Ru_1 represents another set of sites on the surface. The O_{vac} on CeO_2 is created since the formation of CO from carbon atoms formed from dissociating CH_4 to C atom on Ni_1 consumes one surface lattice oxygen atom. Computational studies suggest that the activation barrier for $CO_2 + O_{vac} \rightarrow CO + O_{suface\ lattice}$ on $Ru_1 + O_{vac}$ is 0.01 eV, which is much lower than 1.23 eV on $Ni_1 + O_{vac}$. Thus, certainly, the role of the Ru_1 single-atom site of this catalyst is the activation of CO_2 to form CO and an O atom. This oxygen atom generated from CO_2 dissociation fills the O_{vac} created in the coupling of carbon atoms and a lattice oxygen atom of CeO_2 surface. Based on computational studies, the step of $CO_2 + O_{vac} \rightarrow CO + O_{suface\ lattice}$ on Ru_1 is kinetically favorable in contrast to Ni_1 site. Thermodynamically, the Gibbs free energy changes in dissociation of CO_2 to CO and O on $Ni_1 + O_{vac}$ and $Ru_1 + O_{vac}$ are 0.76 and −0.99 eV, respectively, suggesting that dissociation of CO_2 on $Ru_1 + O_{vac}$ is thermodynamically favorable compared to $Ni_1 + O_{vac}$. Thus, the activation of CO_2 on $Ru_1 + O_{vac}$ is both thermodynamically and kinetically favorable. Clearly, the Ni_1 and $Ru_1 + O_{vac}$ on the same CeO_2 NP surface play a synergistic role in activating CH_4 and CO_2, respectively, forming two CO molecules and two H_2 molecules [16].

6.3 Partial Oxidation of Methane

6.3.1 Brief of Catalytic Methane Partial Oxidation

Methane partial oxidation (MPO), while resulting in a mixture of CO and H_2 similar to steam and dry reforming, deviates from reforming significantly in several aspects. A notable distinction is the varied catalytic mechanisms employed by each catalyst active in MPO, contrasting with the similar reaction mechanisms observed for steam or dry reforming on different catalysts. Furthermore, MPO catalysts may adopt different pathways under varying catalytic conditions, adding complexity to the process. Unlike the sections on steam and dry reforming (Sections 6.1 and 6.2), Section 6.3 commences by introducing common features of MPO, addressing the challenges faced by MPO catalysts. The focus of Section 6.3 then shifts to exploring the specific catalytic performances, structural features, and MPO mechanisms unique to each type of MPO catalyst, including Co, Ni, Rh, Ru, and single-atom catalysts.

MPO traces its origins back to 1933, driven by the demand for hydrogen production from CH_4 during the early ammonia synthesis era of the 1920s–1930s [35]. In contrast to the high endothermic nature of room temperature CH_4 reforming, MPO exhibits a slightly exothermic character, presenting different challenges in contrast to reforming CH_4 with H_2O or CO_2. Exothermicity gives rise to significant temperature gradients and hotspots during catalysis, complicating the assessment of MPO catalytic performance and the exploration of its catalytic mechanism. The substantial temperature difference between the catalyst and the gas, often reaching several hundred degrees Celsius, poses a considerable challenge. To mitigate this, the utilization of a highly thermally conductive support material proves beneficial in dissipating heat to the surroundings, reducing temperature gradients, and minimizing hotspot formation. Overcoming the hindrances posed by hotspots and gas temperature gradients in fixed-bed reactors is achievable through the implementation of fluidized-bed reactors, as reported in the literature [36–40]. Other than these solutions, literature sources suggest that coupling exothermic MPO with endothermic CH_4 reforming can effectively address the issues associated with hotspot formation and gas temperature gradients [41–56].

Other than these issues, MPO also faces the challenge of catalyst deactivation driven from coke formation, sintering and even sulfur poisoning. In general, such deactivations are observed in high-temperature transformations of methane through any catalytic processes including steam reforming, dry reforming, and MPO catalyzed by metal catalysts.

6.3.2 Supported Ni Catalyst

Catalytic performances of numerous MPO catalysts employing diverse supports have been extensively studied. Various support materials, including Al_2O_3 [36–39, 53, 54, 57–76] mixed oxides Al_2O_3–ZrO_2 [77], Al_2O_3–SiO_2–SiC [41, 78, 79], Al_2O_3–SiO_2 [80], Al_2O_3–MgO [81–85], CaO, SiO, BaO [78, 82, 86–92], SiO_2–ZrO_2 [58, 93], and La_2O_3 [76, 86, 94–100] among others, have been investigated in literature. These studies aim to uncover ways to enhance catalytic performance and prevent deactivation. While these investigations suggest the potential for tuning catalytic performance through variations in supports and preparation methodologies, it is noteworthy that no specific correlation between the catalyst surface during MPO and its corresponding catalytic performance has been reported. Moreover, the introduction of secondary metals, such as Pt [60, 101–104], Pd [102, 105–107], Cr [108], Ru [102, 109], Fe [110], Rh [79, 111, 112], and Ir [60, 113] onto the support have been explored as a means to enhance the catalytic performance of MPO.

Ni-supported catalysts on Al_2O_3 surfaces demonstrate remarkable activity in MPO, achieving yields comparable to equilibrium values [57, 114, 115]. However, literature attributes the deactivation of Ni/Al_2O_3 to the formation of $NiAl_2O_4$, evidenced by the observed diffraction patterns in XRD. The emergence of $NiAl_2O_4$ may alter the specific chemical and coordination environments of Ni atoms on the surface layer. Confirmation of the deactivation cause requires in situ/operando characterization of the surface phase during catalysis.

Mixed oxide catalysts, NiO–MO (M = Mg, Ca, Sr, Ba) have been extensively studied [54, 69, 74, 91, 116–120]. While XRD confirms the formation of a Ni–Mg–O solid solution after catalysis, XPS studies suggest that the surface composition of NiO–MO remains consistent with the bulk composition before catalysis. However, these studies lack critical information on the surface state during or even after catalysis. High-temperature catalytic conditions in a gas phase comprising O_2, CO, H_2, and CH_4 likely induce substantial restructuring of the NiO–MgO surface. The literature proposes that the formation of a solid solution of Ni-Mg-O contributes to the high resist of this catalyst to carbon formation [88, 121, 122].

In addition to the formation of mixed oxides, encapsulation of the active phase in porous silica proves effective in resisting deactivation caused by carbon formation. Literature reports the coating of Ni NPs with silica layers (5–10 nm thickness) through a water-in-oil microemulsion, followed by high-temperature calcination to transform the silica into a mesoporous shell. The porous silica shell prevents Ni NP sintering, while the mesopores and interconnected channels enable reactants to access the Ni catalyst surface. Ni NPs enclosed in the mesoporous shell exhibit high activity for MPO and demonstrate superior thermal stability compared to Ni NPs supported externally on silica particles, which are prone to rapid deactivation.

Literature underscores the significant impact of different catalyst preparation methods on the catalytic performance of Ni-based catalysts. For instance, Ni supported on α-Al_2O_3 was prepared using two distinct methods [123]. One method employed $Ni(NO_3)_3$ as a precursor, while the other utilized Ni-acetylacetonate. Interestingly, carbon formation during MPO occurred on the Ni/α-Al_2O_3 prepared with $Ni(NO_3)_3$ but not on the counterpart prepared with Ni-acetylacetonate. The different mechanisms for anchoring Ni atoms on the α-Al_2O_3 surface led to distinct chemical and coordination environments for Ni atoms on these seemingly similar catalysts, both named Ni/α-Al_2O_3. Another example involves Ni/Al_2O_3 prepared with either sol–gel impregnation or conventional impregnation methods [124–126]. Ni in the impregnation-prepared Ni/Al_2O_3 exists in the form of NPs, evident from the average coordination number of Ni to Ni atoms at nearly 12. Conversely, the sol–gel-prepared Ni/Al_2O_3 showcases smaller Ni NPs, evidenced by an average CN(Ni-Ni) at 9 instead of nearly 12. Despite sharing the same name, these catalysts exhibit notable differences in their NP sizes, influencing their resistance to carbon formation and sintering. The Ni/Al_2O_3 with smaller nanoclusters demonstrates heightened resistance to these deactivation processes. In addition, the loading of an active metal, such as nickel, proves crucial for catalytic performance [127]. When considering a Ni catalyst prepared with a hydrotalcite-type precursor, the literature reports loading-dependent activation temperatures and thermal stability. Lower Ni loading necessitates a higher activation temperature. A higher activation temperature facilitates the segregation of more Ni atoms from the subsurface or bulk to the support's surface. Interestingly, a catalyst with lower Ni loading exhibits enhanced stability, likely due to the lower loading density of Ni nanoclusters resulting in reduced surface density of Ni NPs on the support, consequently minimizing the opportunity for sintering of these small Ni NPs.

The vast number of studies suggest that both a noble metal and a transition metal oxide can promote the catalytic performance of a supported Ni catalyst. An illustrative example involves the promotion of La_2O_3 effect to the Ni-based catalyst NiO/Al_2O_3 [128]. The $La_2O_3{\cdot}NiO/Al_2O_3$ was prepared through a double impregnation. First, $La(NO_3)_2$ and then $Ni(NO_3)_2$ were impregnated onto Al_2O_3. It was dried at 110 °C and then calcined at 600–800 °C in air. The $La_2O_3{\cdot}NiO/Al_2O_3$ exhibits an obvious low-ignition temperature than NiO/Al_2O_3. XRD analysis revealed the formation of $LaNiO_3$ on $La_2O_3{\cdot}NiO/Al_2O_3$ after calcination at 800 °C in air. The catalytic performance enhancement attributed to La_2O_3 was linked to the introduction of the new oxide phase, $LaNiO_3$ formed from the La_2O_3 and NiO during calculation. However, XRD patterns before or after MPO fail to provide insights into the catalyst surfaces during catalysis. To unravel the promotion mechanism, further investigations employing in situ/operando surface characterization techniques are imperative.

Alloying another metal with a host metal presents a promising avenue for improving catalytic performance and enhancing resistance to deactivation [112, 129]. For instance, Ni–Rh alloy NPs supported on Al_2O_3 were prepared through the co-precipitation of nickel and rhodium cations in an aqueous solution, followed by filtration, washing, calcination at 900 °C in air, and reduction at 750 °C in H_2. The Ni–Rh catalyst exhibits higher MPO activity and superior resistance to deactivation from carbon formation compared to Ni/Al_2O_3. It exemplifies the potential benefits of alloying strategies in catalyst design.

Understanding the promotion mechanism of a second metal (M) in Ni-based catalysts is more intricate compared to the promotion by oxide, primarily due to the potential coexistence of different states of Ni and M, including metallic states as separated Ni and M metal NPs or Ni-M bimetallic NPs, as well as a cationic state in a mixed metal oxide, Ni–M–O. An illustrative case involves the promotion of Cr to Ni/Al_2O_3 for MPO [108, 130]. The inclusion of Cr enhances the resistance of Ni/Al_2O_3 to carbon formation through the formation of Ni–Cr alloy NPs during catalyst preparation.

The ability of a second metal (M) to play a promotional role in Ni-based catalysts may be dependent on the oxide support of a Ni catalyst in certain cases. Consider the promotion effect of Ir to Ni/La_2O_3. A catalyst with 0.25 wt% Ir and 0.5 wt% Ni on La_2O_3 demonstrates the ability to convert 25% CH_4 with a selectivity of >80% for syngas at 600 °C, while 0.25 wt% Ir/La_2O_3 and 0.5 wt% Ni/La_2O_3 show no activity for MPO under the same conditions. Ir significantly promotes the catalytic activity of Ni/La_2O_3 in this case. However, such a promotion effect of Ir to Ni in Ni/La_2O_3 is not observed when Ni is loaded on Y_2O_3, ZrO_2, Al_2O_3, MgO, and SiO_2. The reasons for this difference remain unexplored in the literature.

As highlighted earlier, there is no uniform catalytic mechanism for MPO across different catalysts, and even on the same catalyst, the mechanism could be different under different catalytic conditions. Isotope exchange experiments, involving the use of a mixture of CH_4 and CD_4 to replace CH_4 during MPO on NiO/SiO_2, have identified the formation of CH_3D, CH_2D_2, and CHD_3 intermediates, indicating that CH_3, CH_2, and CH intermediates must have been formed on the catalyst prepared through pyrolysis [131, 132]. This suggests that

MPO on NiO/SiO_2 occurs through the continuous dissociation of CH_4 to CH_3, CH_2, and eventually CH before coupling with one O atom. Despite the significant number of reported MPO catalysts, few correlations between an active surface phase and its corresponding catalytic performance have been established. Some studies have proposed active sites for MPO, often utilizing XRD and TEM to analyze changes in the bulk phase of catalyst particles after MPO catalysis. However, for a fundamental understanding of changes in catalytic performance, it is critical to track the surface chemistry of the catalyst during catalysis.

6.3.3 Supported Co Catalyst

Numerous Co-based catalysts have been extensively reported in the literature [55, 73–76, 96, 133–152]. These catalysts employ various supports, including MgO [74, 149, 151, 153–156], Al_2O_3 [55, 71, 73, 74, 143, 145, 146, 149, 157–166], SiO_2 [73, 74, 151, 167–169], CaO [151, 154, 170, 171], TiO_2 [74, 149, 172, 173], SrO [154, 174], BaO [174], La_2O3 [74, 96, 149, 175–178], or CeO_2 [133, 149, 159, 179]. Additionally, guest metals such as Ni, Pd, Re, Rh, or Ru have been introduced to Co to enhance their catalytic performance in MPO. Similar to Ni-based catalysts, supported Co experiences deactivation, although it occurs at a higher temperature due to the oxidation of metallic Co, as indicated by H_2-TPR analysis [138, 180]. The utilization of MgO as a support leads to the formation of a Co–Mo–O mixed oxide upon calcination at 800 °C, as confirmed by XRD [134, 181]. Remarkably, this mixed oxide exhibits high activity and selectivity for syngas, unlike individual MgO or CoO, which cannot catalyze MPO for syngas production [151, 182]. In comparison to Co/SiO_2, Co–Mg–O mixed oxide demonstrates superior thermal stability, attributed to the reduced reducibility of Co incorporated into the MgO lattice and the high dispersion or even single-atom dispersion of Co atoms in MgO lattice, as opposed to Co/SiO_2. Despite the higher activity, selectivity, and relatively greater stability of Co–Mo–O mixed oxide, Co/CaO, Co/SrO, and Co/BaO display rapid deactivation within a few to several tens of hours. To mitigate the oxidation of cobalt metal to oxide, the introduction of a second metal, such as Pt, proves effective. Through the formation of Co–Pt bimetallic NPs, the Co–Pt catalyst exhibits enhanced conversion and selectivity for MPO, leading to syngas formation compared to the pure Co catalyst. Notably, Co demonstrates higher reducibility upon alloying with Pt in the Co–Pt catalyst [146, 183, 184].

Literature extensively documents numerous investigations into the influence of a second metal (0.1 wt% of Ni, Rh, Pd, or Ru) or an oxide (<4 wt% Fe_3O_4/Cr_2O_3, La_2O_3, SnO_2, or K_2O) on the catalytic performance and stability of Co-based catalysts, aiming to develop a catalyst with high activity, selectivity, and thermal stability [71, 185]. It has been observed that adding 0.1 wt% Pt or 0.1 wt% Rh to Co/Al_2O_3 allows the activity exhibited at a high temperature to be maintained at a lower temperature [71, 185]. However, the introduction of Ni to Co/Al_2O_3 has been found to promote carbon deposition and catalyst deactivation. The incorporation of Fe_3O_4 into Co/Al_2O_3 activates the complete oxidation of CH_4 to form CO_2 and H_2O across the entire temperature range.

While these modifications have been reported [71, 185], there is a crucial need to characterize the surfaces of Co/Al_2O_3 and these catalysts with promoters (metal oxide) during catalysis. Subsequently, establishing a correlation between the surface chemistry of the promoter and the promotion in catalytic activity and selectivity measured under a kinetics-controlled regime, is essential. Through such efforts, a comprehensive understanding of the profound promotion or suppression effects of a second metal or oxide is expected to be achieved in the future.

6.3.4 Supported Pt Catalyst

Numerous Pt-based catalysts have been investigated, and this section only provides a concise brief of a few representative examples. Pt/MgO is recognized as an active catalyst for MPO. A kinetic model for MPO on Pt/MgO has been developed, indicating an indirect mechanism involving first a primary complete oxidation of CH_4 to CO_2 and H_2O, followed by steam reforming of CH_4 with H_2O and dry reforming of CH_4 with CO_2 to produce CO and H_2 [186].

Pt/Al_2O_3 is another extensively studied MPO catalyst. Its catalytic performance can be enhanced by a modification with CeO_2. The $Pt/CeO_2/Al_2O_3$ catalyst, prepared through first impregnation of $Ce(NO_3)_3$ onto Al_2O_3, drying at 110 °C, calcination at 600 °C, and subsequent impregnation of $Pt(NH_3)_4(NO_3)_2$ onto the prepared CeO_2/Al_2O_3 with further drying at 150 °C, exhibits superior catalytic activity [187]. Compared to Pt supported on Al_2O_3 [187, 188], Pt on CeO_2/Al_2O_3 demonstrates higher dispersion, attributed to the largely increased surface defects on CeO_2 compared to Al_2O_3. These surface defects act as anchoring points for the Pt precursor. Interestingly, CeO_2 or CeO_2/Al_2O_3 alone shows limited reactivity for MPO, suggesting the key role of Pt in these catalysts [189]. However, upon Pt loading on CeO_2/Al_2O_3, a highly active MPO catalyst is formed as Pt facilitates CH_4 dissociation, and CeO_2 contributes surface lattice oxygen. The rate-determining step on CeO_2/Al_2O_3 is expected to be the dissociation of C–H in CH_4 [190], while literature suggests that, for $Pt/CeO_2/Al_2O_3$, the rate-determining step is the coupling of C and O to form CO. The addition of CeO_2 to Pt/Al_2O_3 modifies the reaction pathway, and the lattice oxygen of CeO_2 plays a crucial role in coupling with surface carbon atoms to form CO molecules to resist carbon formation, a unique function of reducible oxides compared to nonreducible oxide supports such as Al_2O_3 and SiO_2.

6.3.5 Supported Rh Catalyst

Rhodium-based catalysts supported on various oxides have been extensively studied, driven in part by the lack of a conclusive understanding of the nature of the active phase and specific reaction mechanisms. Studies on 1 wt% Rh/γ-Al_2O_3, utilizing a temperature programmed reaction (TAP) reactor, suggest that MPO on this catalyst occurs through a redox path, where surface lattice oxygen on RhO_x couples with carbon atoms to form CO [191]. This observation is supported by findings indicating a lack of CH_4 dissociation on either Rh surfaces saturated with oxygen atoms or reduced Rh metal [192]. Another study proposed that the active Rh atoms could be a partially reduced cation, Rh^{3+} [192]. DRIFTS studies on MPO with Rh/γ-Al_2O_3 and Rh/γ-Al_2O_3 only detected CO_2, not CO, from a fully oxidized Rh surface under MPO conditions [192, 193]. Isotope-labeled experiments on Rh/γ-Al_2O_3 indicated that at a low Rh loading of 0.05 wt%, MPO follows a two-step process involving initial complete oxidation to form CO_2 and H_2O, followed by reforming CH_4 with H_2O and CO_2 to produce syngas [194, 195]. This aligns with the proposed mechanism for

MPO on Rh/SiO_2 [196]. Notably, the mechanism suggested for $Rh/\gamma\text{-}Al_2O_3$ [191–193] and Rh/SiO_2 [196] differs from the pyrolysis mechanism reported for Rh/TiO_2. Thus, there is no uniform mechanism for MPO, even for the same metal catalyst supported on different oxides. Furthermore, it was discovered that the MPO mechanism on 1.0 wt% $Rh/\gamma\text{-}Al_2O_3$ shifts at temperatures above 650 °C, where a pyrolysis mechanism becomes dominant. These differences in the reaction mechanism highlight the need for advanced in situ/operando characterizations coupled with high-level computational studies to propose a reasonable MPO reaction mechanism.

The pulse transient technique was employed in studies investigating the mechanism of MPO on Rh sponge [197]. The literature [197] proposed a Mars-van Krevelen reaction mechanism. Studies using a TAP reactor revealed CO and H_2 as the main products. In pulse experiments with CH_4 and $^{18}O_2$, only $C^{16}O$ was observed, not $C^{18}O$, suggesting that the oxygen atom in the product CO originates from the catalyst RhO_x in the MPO pathway. In a side reaction involving the complete oxidation of CH_4, both $C^{16}O^{18}O$ and $C^{16}O^{16}O$ were detected, indicating that both Rh_1O_x and $^{18}O_2$ serve as sources of oxygen atoms for the byproduct CO_2. Based on these isotope-labeled studies, it is suggested that CH_4 dissociates on Rh sponge through a pyrolysis mechanism different from the indirect mechanism proposed for Pt/MgO [186]. The formed carbon atom or CH_x takes oxygen atom(s) from RhO_x to form CH_xO intermediate, and the CH_xO dissociates to produce CO [197].

6.3.6 Supported Ru Catalyst

Supported Ru catalysts exhibit activity in MPO, with various metal oxides, including Al_2O_3, TiO_2, and La_2O_3, serving as supports for Ru catalysts. Ru/Al_2O_3 catalyzes complete oxidation at temperatures below 450 °C, producing only CO_2 and H_2O, with ruthenium existing as RuO_2 [198]. However, at temperatures exceeding 450 °C, syngas is observed. A size-dependent catalytic mechanism was proposed based on MPO studies on small-sized Ru NPs supported on Al_2O_3 and larger Ru NPs prepared via colloidal chemistry. For large Ru NPs (15 nm) supported on Al_2O_3, syngas is the predominant product compared to the complete oxidation of CH_4. This size-dependent mechanism may stem from the larger Ru NP maintaining its metallic state more readily than the smaller Ru nanocluster, suggesting the favorability of metallic Ru of large Ru NPs for a direct pyrolysis mechanism. This deduction is supported by FTIR and XAS characterizations of Ru/Al_2O_3 [199]. These characterizations, along with the evaluation of catalytic performances at different temperatures, suggest that MPO is performed through a direct pyrolysis mechanism on a reduced Ru metal surface. However, when the surface is partially oxidized, the dominant pathway involves the first complete oxidation followed by reforming of CH_4 using H_2O and CO_2, performing an indirect mechanism similar to Pt/MgO [186]. These studies suggest that MPO on Ru catalysts adopt different mechanisms along with the variation of the oxidation state of ruthenium catalysts.

An intriguing observation is that variations in doping elements (W^{6+} versus Ca^{2+}) in the support material of a Ru catalyst, such as TiO_2 can be employed to adjust the catalytic performance of the supported Ru catalyst [200, 201]. Ru/TiO_2 demonstrates activity in MPO, with CO chemisorbed on the Ru/TiO_2 catalyst, even at temperatures as high as 800 °C, as confirmed by situ FTIR. However, no CO could be chemisorbed on TiO_2 modified with

W^{6+} [201], indicating with a consequence from the modification of charge transfer. The introduction of W^{6+} to TiO_2 lattice enhances the charge transfer ability of TiO_2 to Ru, thereby promoting the oxidation of Ru. Conversely, the introduction of Ca^{2+} with a lower valence than Ti^{4+} enhances the resistance of Ru to oxidation. These alterations in the electron transfer ability of the TiO_2 supporting the Ru metal NPs effectively vary the adsorption energy of molecules on the catalytic metal, Rh, such as CO, or atoms, such as carbon atoms. The introduction of Ca^{2+} to TiO_2 reduces its capability to transfer electrons to Ru metal, thus weakening the binding of CO or carbon atoms to Ru metal NPs, by which the carbon deposition on TiO_2 with doped Ca^{2+} was found to be limited.

6.3.7 Supported Single-Atom Rh Catalyst

Recently, a Rh single-atom catalyst demonstrated high activity and selectivity for syngas production at 650 °C through MPO (Figure 6.11) [202]. This catalyst was synthesized by anchoring Rh atom precursors on the surface of TiO_2. Subsequent annealing in air at 650 °C

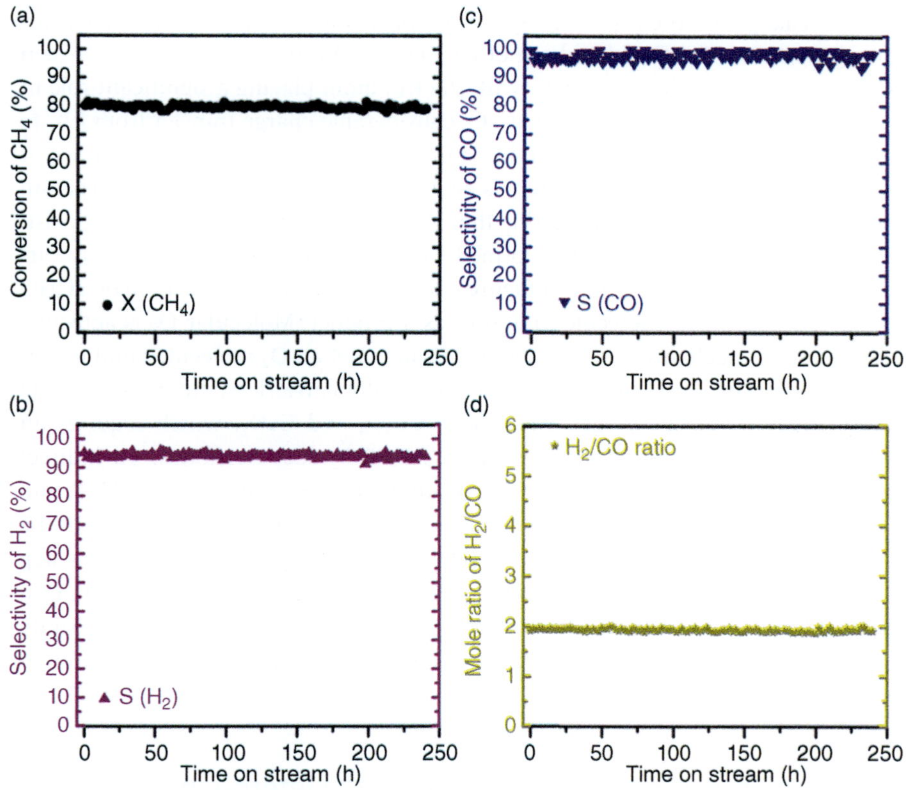

Figure 6.11 Catalytic durability of the Rh_1/TiO_2 catalyst (0.037 wt% Rh/TiO_2) for MPO in a period of 10 days. (a) Conversion of CH_4 at 650 °C; (b) selectivity for producing H_2 at 650 °C; (c) selectivity for producing CO at 650 °C; (d) H_2/CO molar ratio at 650 °C. Catalysis conditions: catalyst is 50 mg of 0.037 wt% Rh/TiO_2; the gas fed to the reactor is a mixture of 25 mL min^{-1} of 10% CH_4 and 25 mL min^{-1} of 5% O_2; catalysis temperature is 650 °C. *Source:* Reproduced with permission from [202] / American Chemical Society.

led to the replacement of Ti atoms on the TiO_2 surface by Rh atoms. *In situ* EXAFS characterization of the catalyst in a mixture of CH_4 and O_2 (2:1) after catalysis at 650 °C suggested that Rh atoms were singly dispersed on the TiO_2 surface, as evidenced by the absence of peaks corresponding to Rh–O–Rh bonds of RhO_x and Rh–Rh bonds of metallic Rh NPs in the r-space spectrum of the Rh K-edge. In addition, the single dispersion of Rh atoms was confirmed through diffuse reflectance infrared Fourier transform studies of the catalyst with adsorbed CO. Fitting the r-space spectrum of the Rh K-edge suggested that, on average, about five oxygen atoms coordinate with each Rh_1 atom. This coordination shell of oxygen atoms around Rh_1 is consistent with the oxygen atom shell around Ti atoms on the (101) surface of TiO_2. Consequently, a Rh_1 single-atom site was formed by substituting for Ti_1 on the (101) surface of TiO_2 support. The cationic nature of Rh_1 was confirmed by ambient pressure X-ray photoelectron spectroscopy (AP–XPS) studies of the catalyst at 650 °C in a mixture of CH_4 and O_2 (2:1).

The MPO pathway on the Rh_1/TiO_2 single-atom catalyst follows a mechanism that differs from that on Rh NPs supported on oxides [202]. The complete catalytic cycle involves 26 elemental reactions (Figure 6.12). Essentially, CH_4 continuously dissociates into CH through C–H breakages on the Rh_1O_5 catalytic site, consisting of Rh_1 atoms and five coordinating oxygen atoms. The C–H breakage of CH_4 occurs through the coupling between C–H of CH_4 and Rh_1–O of the Rh_1O_5 site, with the Rh_1 atom playing a significant role in CH_4 activation. Computational studies suggest a considerable charge transfer from Rh_1 to the C atom of CH_4, weakening the C–H bond. Transition state TS_1 leads to the cleavage of the first C–H bond of CH_4. In contrast, there is negligible charge transfer from the Ti atom of bare TiO_2 to the C atom of CH_4, indicating that bare TiO_2 is not active for MPO, consistent with experimental observations. The formed CH species inserts between Rh_1 and its adjacent O atom (structure e in panel 2 in Figure 6.12), resulting in the formation of a CHO intermediate and then a CO molecule to desorb as a product. Molecular O_2 is activated through an insertion mechanism, where the two O atoms of an O_2 molecule bond to Rh_1 and Ti, respectively. This O=O insertion step is kinetically barrierless, differing from the direct activation of O_2 on an oxygen vacancy of Co_3O_4 and CeO_2, which requires an activation barrier of 0.5–2.0 eV. Clearly, MPO on Rh_1/TiO_2 does not follow an indirect mechanism where complete oxidation of CH_4 occurs first, generating CO_2 and H_2O, and then these products reform CH_4 to generate syngas. As depicted in Figure 6.12, one of the rate-determining steps is the dissociation of CH_3 to CH_2. The activation barrier of this step on Rh_1O_5 of Rh_1/TiO_2 is only half of that on the Ti_1O_5 site of bare TiO_2. The primary reason is that Rh_1 can transfer much more electron density to the C atoms of CH_3, significantly decreasing the energy of the transition state from CH_3 to CH_2 compared to the Ti_1 atom of bare TiO_2 [202].

6.4 Methane Transformation Involving both Heterogeneous and Homogeneous Catalysis

All the aforementioned CH_4 transformations primarily involve heterogeneous catalysis occurring at the solid–gas interface. The observed catalysis is generally explained without considering any radical reactions in the gas phase. The high selectivity for the ideal

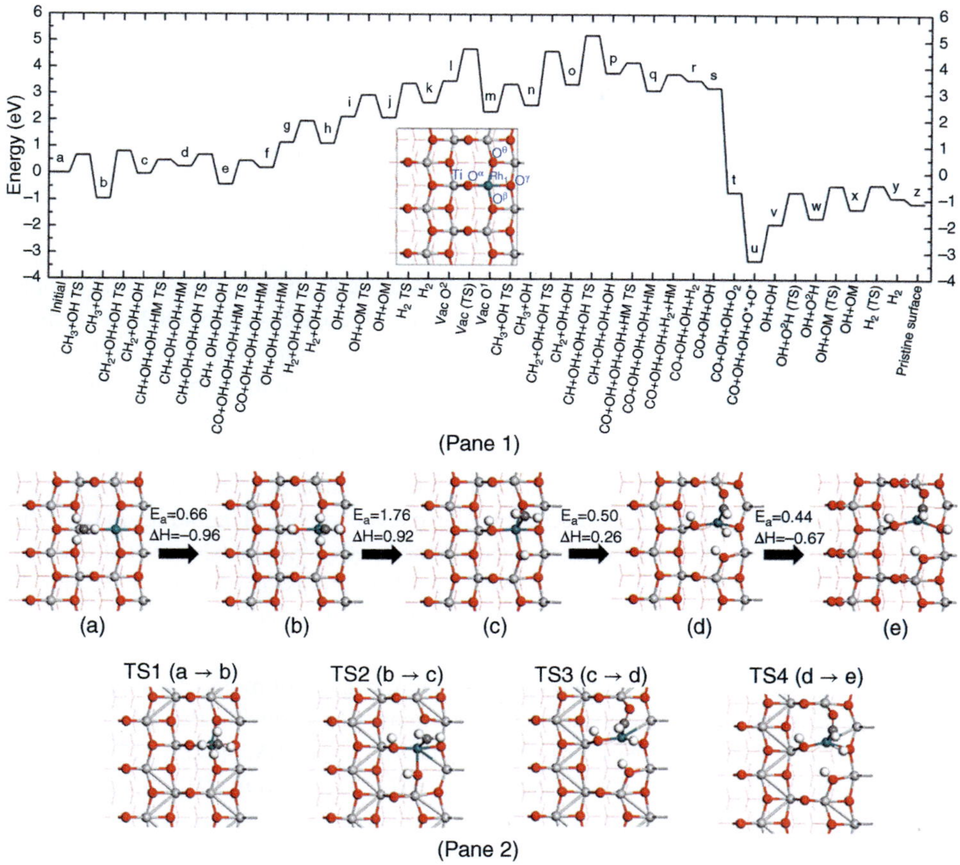

Figure 6.12 Energy profile of the complete pathway of MPO consisting of 26 elementary steps on Rh_1/TiO_2 (panel 1) and optimized structures of intermediates (a–e) of activation of CH_4 on Rh_1O_5 to form CH_3 (a → b), CH_2 (b → c), and CH (c → d) and the transition state 1 (a→ b), transition state 2 (b → c), transition state 3 (c → d), and transition state 4 (d → e) (panel 2). Inset of panel 1 is the clean surface structure of Rh_1/TiO_2. *Source:* Reproduced with permission from [202] / American Chemical Society.

product, syngas, led scientists to exclude the low selectivity paths, radical reactions in the gas phase at high temperatures. Unlike the reforming of CH_4 and partial oxidation of CH_4, oxidative coupling of methane (OCM), methane aromatization, and methane to ethylene are performed at higher temperatures, typically 800 °C or higher. In general, OCM, methane aromatization, and methane to ethylene proceed through a pathway consisting of both heterogeneous catalysis at a gas–solid interface for forming CH_x radicals and then homogeneous catalysis in the gas phase to form products. The following sections will focus on these three types of catalytic reactions. OCM will be discussed first, followed by the nonoxidative transformation of CH_4 to unsaturated hydrocarbons, including methane aromatization and methane to ethylene.

6.5 Oxidative Coupling of Methane

Since the early reports in the 1980s that ethylene can be directly produced from a mixture of CH_4 and O_2, OCM has attracted significant attention from both academic and industrial communities [203–206]. This interest arises from its potential to bypass the complex industrial route, which involves first generating syngas through steam reforming and then producing ethane/ethylene through Fischer–Tropsch synthesis using syngas. A promising production of ethylene from CH_4 through OCM could potentially decrease the cost of ethylene, a crucial intermediate in the chemical industry. Several review articles on this topic have been published, reflecting the sustained interest and research efforts. OCM is exothermic at 800 °C, with a reaction enthalpy of $-139\,kJ\,mol^{-1}$ at this temperature. As of 2011, there have been approximately over two thousand articles published on this topic, covering hundreds of reported catalysts. A statistical analysis suggests that the selectivity for OCM products on a catalyst is high when the conversion is low and vice versa (Figure 6.13). Consequently, there is a limit to the yield of C_2 products, such as ethane and ethylene, at 25–30%. The low selectivity for C_2 molecules at high conversion of CH_4 could result from the activation of the products (C_2 molecules) since ethane or ethylene can be activated readily.

Although the understanding of gas-phase radical reactions at high temperatures is quite limited, the heterogeneous catalysis involved in OCM has been well studied over the last four to five decades. Heterogeneous catalysis is primarily responsible for activating CH_4 to generate CH_3 and molecular O_2 to atomic O. Activation of molecular O_2 to dissociate into atomic O is a common surface reaction extensively studied in the literature. The following paragraphs will discuss four key questions by taking OCM on Li/MgO as a probe catalyst: *What is the evidence for the formation of CH_3 radicals from CH_4 on the surface?*

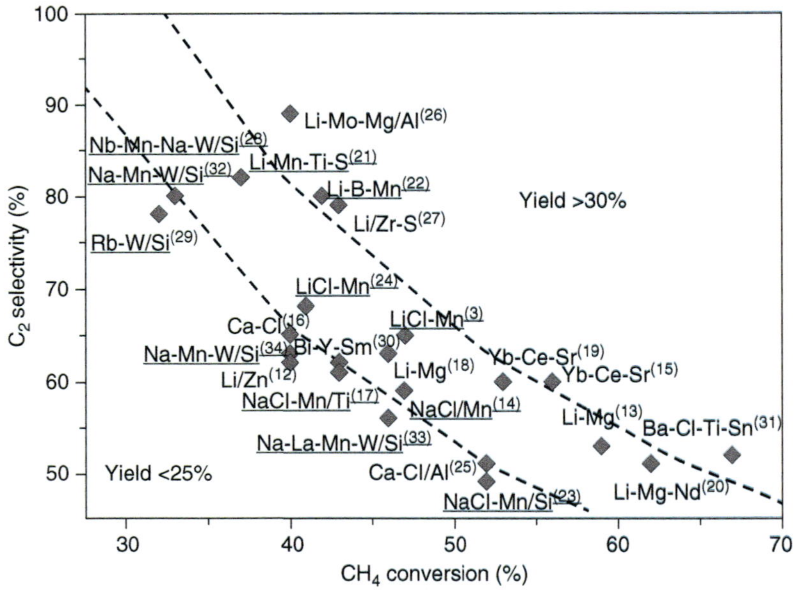

Figure 6.13 Plot of catalytic selectivity for ethane and ethylene as a function of the corresponding conversion of methane for reported catalysts of oxidative coupling of methane. [204] / With permission of John Wiley & Sons.

What is the evidence that C_2H_6 is formed through a step performed in the gas phase? What is the evidence that C_2H_4 is formed through a step performed in the gas phase? What is the reaction mechanism for the formation of CH_3 at a molecular level?

What is the evidence for the formation of CH_3 radicals from CH_4 on the surface?

As early as 1985 [207], a study employing matrix isolation electron spin spectroscopy (MIESR) suggested the existence of $CH_3\cdot$ in the gas phase during OCM on Li/MgO. Two years later, modified MIESR studies indicated that over 45% of the C_2H_6 product molecules could be formed through the coupling of $CH_3\cdot$ in the gas phase [208].

Direct evidence supporting the involvement of $CH_3\cdot$ in OCM comes from isotope-labeled experiments documented in the literature. In this study, a mixture of CH_4 and CD_4 was introduced as a reactant for OCM on Li-promoted MgO (Li/MgO) at 750 °C. Mass spectrometry and high-resolution FTIR, highly sensitive to isotope substitution effects, were utilized. The study found no significant exchange between CH_4 and CD_4, as none of the potential isotope substitutes, including CH_3D, CH_2D_2, CHD_3, and CD_3H, was observed. However, isotope substituted C_2 products, CD_3CH_3 rather than C_2D_6 and C_2H_6, were clearly detected during OCM on Li/MgO. In addition, the absence of CH_2CD_2 or CD_2CD_2 ruled out the possibility of $CH_3\cdot$ further dissociating to $CH_2\cdot$ on the surface and then coupling in the gas phase. These findings indicate that CH_4 dissociates to CH_3 on Li/MgO. Moreover, the measured distributions of isotope-substituted ethane products, CH_3CH_3, and C_2D_6 are highly consistent with the calculated distributions based on the kinetic isotope effect ($k_H/k_D = 1.4$) and a statistical distribution. This consistency demonstrates that the measured distributions of isotope-substituted products reflect the relative concentration of CH_3 and CD_3 [209]. These isotope-labeled studies strongly suggest that $CH_3\cdot$ couples with each other to form ethane [209].

What is the evidence that C_2H_6 is formed through a step performed in the gas phase?

The absence of $CH_2D\cdot$ or $CD_2H\cdot$, deduced from the lack of CH_2DCH_3 or CD_2HCH_3 in the products of OCM on Li/MgO [209], suggests CH_3 does not further dissociate into CH_2 on the surface of Li/MgO at 750 °C. This finding contrasts with literature suggestions that CH_4 and CD_4 exchange on bare MgO surfaces at 600 °C, and CH_4 and H_2 exchange on bare MgO at 300 °C [210, 211]. Moreover, the absence of exchange between $CH_3\cdot$ and $CD_3\cdot$ on Li/MgO during catalysis at 750 °C indicates that $CH_3\cdot$ and $CD_3\cdot$ leave the surface immediately once formed. In other words, this study suggests that $CH_3\cdot$ and $CD_3\cdot$ couple in the gas phase after their formation on Li/MgO [209]. These isotope-labeled experiments suggest: (1) CH_4 dissociates into $CH_3\cdot$ and $H\cdot$ on Li/MgO, (2) $CH_3\cdot$ desorbs from the surface immediately upon formation, and (3) $CH_3\cdot$ radicals couple in the gas phase to form ethane.

What is the evidence that C_2H_4 is formed through a step performed in the gas phase?

In isotope-labeled experiments at 750 °C under OCM conditions, the analysis of labeled ethylene products, including C_2H_4, CD_2CH_2, and C_2D_4, was conducted [209]. As presented in Figure 6.14, the distributions of ethylene products in the gas phase of this nonisotope-labeled experiment are nearly identical to those of ethane products (C_2H_6, CD_3CH_3, and C_2H_6) in the isotope-labeled experiment. These highly consistent distributions, as shown in Table 6.1, can only be explained by the assumption that ethylene is formed from ethane in the gas phase. In other words, ethylene is a successive product following the formation of ethane through OCM in the gas phase [209].

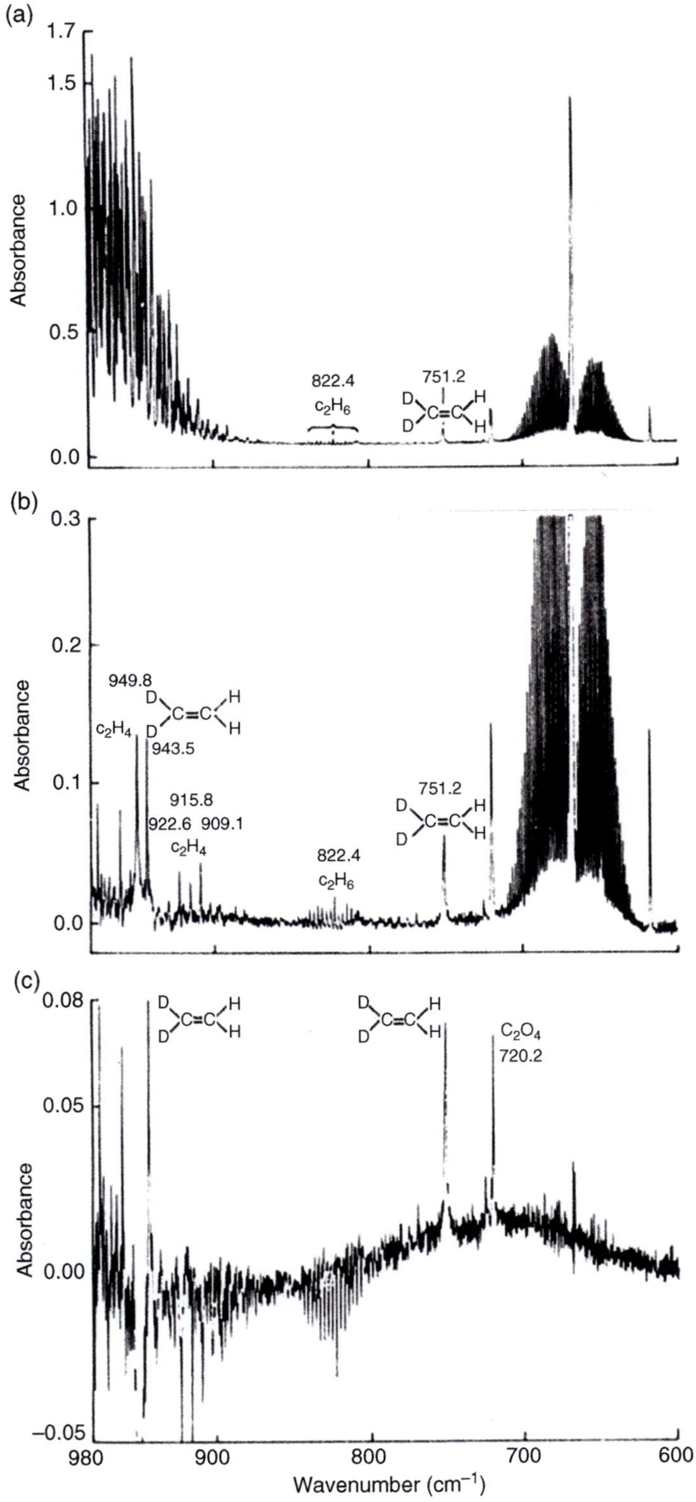

Figure 6.14 FTIR spectra: (a) product gas spectrum, 980–600 cm^{-1}; (b) difference spectrum of (a) minus CD_4; (c) difference spectrum of (b) minus CO_2. *Source:* Reproduced with permission from [209] / American Chemical Society.

Table 6.1 Relative concentration of ethane and ethylene in CH_4/CD_4 experiments (FTIR measurements).

	calcd[a]	obsd		calcd[a]	obsd
C_2H_6	34±2	36 ±4	C_2H_4	34±2	34 ± 2
CD_3CH_3	49±2	(49)[b]	CD_2CH_2	49±2	(51)[b]
$C_2D_6$17 ±2	15±2	C_2D_4	17±2	15±2	

[a] For $k_H/k_D = 1.4 \pm 0.1$.
[b] By difference.
Source: Reproduced with permission from [209] / American Chemical Society.

What is the reaction mechanism for the formation of CH_3 at a molecular level?
Identifying active sites of OCM catalysts during the OCM process is a challenging task due to the high operating temperatures, typically ranging from 800 to 1000 °C. At such elevated temperatures, the catalyst surfaces undergo significant restructuring from their as-synthesized state. Without the ability to identify or reasonably deduce the chemical and coordination environment of catalytic sites during catalysis, proposing a mechanism for the formation of $CH_3\cdot$ on the catalyst surface, such as Li/MgO, becomes a formidable challenge. The substitution of Mg^{2+} by Li^+ introduces an extra electron to the lattice of the MgO cation, contributing to the complexity of the reaction pathway. To balance the charge, this extra electron brought by the Li^+ makes the adjacent oxygen atom lose one electron to form O^-. Thus, a $[Li^+O^-]$ pair is created due to the substitution of Li^+ for Mg^{2+} on MgO surface lattice. In the $[Li^+O^-]$ pair, Li^+ has the same oxidation state as other compounds such as Li_2O. However, its oxygen atom appears as O^-. Thus, the $[Li^+O^-]$ can be considered as an oxygen-centered radical. Literature suggests an oxygen-centered radical on a metal oxide is active for dissociating C–H of CH_4 through a homolytic hydrogen abstraction mechanism [212]. This mechanism can be simply expressed as $MO + H\text{–}CH_3 \rightarrow MO\text{–}H + CH_3\cdot$. Thus, a mechanism for activation of CH_4 on Li/MgO could be expressed as $Li^+O^- + H\text{–}CH_3 \rightarrow Li^+O^{2-}\text{–}H + CH_3$, as schematically shown in Figure 6.15. This proposal is seemingly consistent with a computational study suggesting that Li^+O^- can activate C–H of CH_4 with an activation barrier of only $27 \pm 7\,kJ\,mol^{-1}$ [213, 214]. However, this is quite different from the apparent activation barrier estimated from experimental studies, $89\text{–}160\,kJ\,mol^{-1}$, dependent on preparation details [215, 216]. In fact, experimental studies confirmed that $[Li^+O^-]$ on MgO is not an active site for MPO [214, 217]. Literature confirmed that there are not $[Li^+O^-]$ sites found on a catalyst of MgO powder after Li/MgO was used for OCM although they were found in the as-prepared MgO single crystal doped with Li^+ [218–220]. Interestingly, characterizations of Li/MgO revealed that the added Li^+ induced a significant restructuring of the MgO surface, resulting in stepped surfaces comprising steps and corner sites. This restructuring may have caused the Li^+ ions, being of small size, to either become buried in the bulk or be lost during pretreatment. Consequently, the inconsistency between the experimentally measured apparent activation barrier and the theoretically

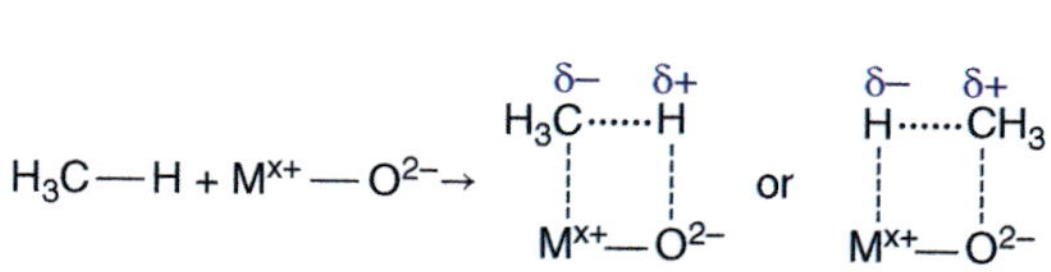

Figure 6.15 Schematic of OCM mechanism. *Source:* Reproduced with permission from [212] / John Wiley & Sons.

calculated activation barrier for the homolytic dissociation of CH_4 on $[Li^+O^-]$, along with the confirmed absence of $[Li^+O^-]$ on the MgO surface, indicates that $[Li^+O^-]$ is not the active site for the formation of $CH_3\cdot$ radicals.

The literature revealed that Li^+ indirectly influences the formation of $CH_3\cdot$ radicals. Li-doped MgO and bare MgO exhibit significantly different catalytic activities in OCM, as observed in their distinct initial phases. While both Li/MgO and bare MgO share the same active sites, the role of doped Li^+ is attributed to an increase in the number of active sites. Multiple characterizations suggest that Li^+ doping results in a MgO surface with a higher concentration of undercoordinated O^{2-} ions located at edges, corners, and other sites compared to bare MgO. In essence, Li^+ acts as a structural promoter rather than an electronic promoter, as the Li^+O^- pair is confirmed not to be an active site for OCM.

A recent report, integrating compositional studies and experimental work, proposes a new mechanism involving the interaction of the C–H bond of CH_4 with an $Mg^{2+}O^{2-}$ ion pair, leading to the formation of $CH_3\cdot$ radicals through homolytic release in the presence of O_2 molecules on the catalyst surface. This proposal is based on computational studies using B3LYP hybrid density function theory, and the structural geometries of the active sites and the transition state, along with spin densities, are presented in Figure 6.16. Experimental findings from both bare MgO and Li^+-doped MgO are the observation of C_2H_6, C_2H_4 and CO at temperatures ≥540 °C. Furthermore, Li^+-doped MgO exhibits significantly higher activity and selectivity for forming C_2 molecules than bare MgO under the same catalytic conditions at temperatures ≥540 °C. The enhanced activity of Li^+-doped MgO in OCM is attributed to the high density of undercoordinated O^{2-} ions present at step edges, corners, or kinks in MgO doped with Li^+. These morphological defects, containing undercoordinated O^{2-} anions, were identified through optical spectroscopy in the literature [214, 217].

The significance of undercoordinated O^{2-} anions at step edges, corners, and kinks in enhancing catalytic activity and selectivity for producing C_2 molecules was supported through a comparative experiment and calculation of the activation barrier for $CH_3\cdot$ radical formation. In the experiment, a bare MgO catalyst underwent OCM for 6 hours and 230 hours, resulting in notable differences in reaction rates, as illustrated in Table 6.2. The reaction rate decreased significantly after 230 hours of OCM catalysis. In addition, TEM images of the catalyst after an OCM catalysis of 230 hours (Figure 6.17b) showed substantial differences compared to those after an OCM catalysis of only 6 hours (Figure 6.17a). Notably, the surface of the catalyst after 230 hours appeared flat. The correlation between the reaction rate and the surface structure of the catalysts suggests that undercoordinated O^{2-} anions at step edges, corners, or kinks play a pivotal role in OCM. To further elucidate the mechanism, computational studies were conducted to calculate the activation barrier for $CH_3\cdot$ radical formation. In this calculation, a Mg_9O_9 cluster with undercoordinated O^{2-} anions was employed, as depicted in Figure 6.18 [214, 217]. Based on the energy diagram in Figure 6.18, the chemisorption of CH_4 to release a $CH_3\cdot$ radical is performed through a heterolytic addition of C–H to the Mg^{2+} and O^{2-}, respectively, as expressed as $[Mg^{2+}O^{2-}]_{MgO} + H\text{–}CH_3 \rightarrow [HO^-(Mg\text{–}CH_3)^+]_{MgO} \rightarrow [HO^-Mg^{\cdot-+}]_{MgO} + {}^{\cdot-}CH_3$. This step requests input of 228 kJ mol^{-1}. The dangling electron in the formed $[HO^-Mg^{\cdot-+}]_{MgO}$ can be used to activate molecular O_2 through the following step: $[HO^-Mg^{\cdot-+}]_{MgO} + O_2 \rightarrow (O_2^{\cdot-})[HO^-Mg^{2+}]_{MgO}$. The reaction enthalpy of this O_2 activation step is –191 kJ mol^{-1}. Thus, the

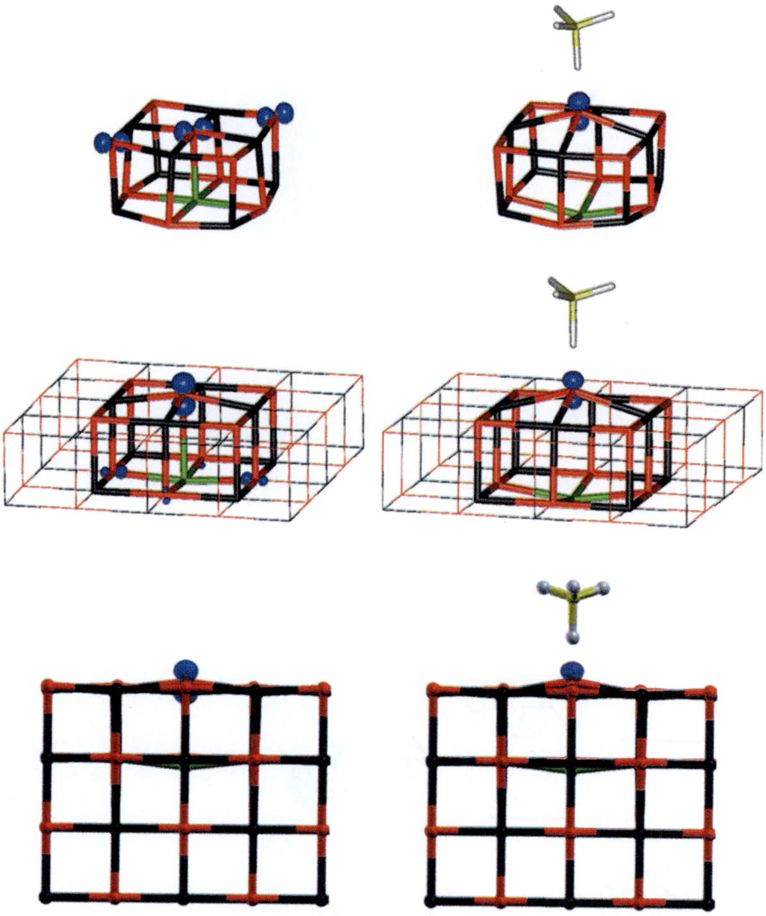

Figure 6.16 Structure models of Li/MgO. From top to bottom: Cluster model, constraint cluster model embedded in a periodic array of point charges, and periodic model of an Li^+O^- site on the MgO(001) terrace. Left: active site structure, right: transition structure for H abstraction from CH_4. O red, Mg black, Li green, C yellow, H gray, spin density blue. *Source:* Reproduced with permission from [214] / Wiley.

Table 6.2 Catalytic performance of the Li-free MgO catalyst in the oxidative coupling of methane.

	Rate ($\mu mol\,s^{-1}\,g_{cat}^{-1}$)	Rate ($\mu mol\,s^{-1}\,m_{cat}^{-2}$)	$X(CH_4)$ (%)[a]	$S(C_2)$ (%)[a]
Initial state	214.4	5.56	26.04	29.84
Final state	8.57	1.26	4.70	13.85

[a] X: conversion, $S(C_2)$: selectivity to ethane and ethylene.

Source: Reproduced with permission from [214] / John Wiley & Sons.

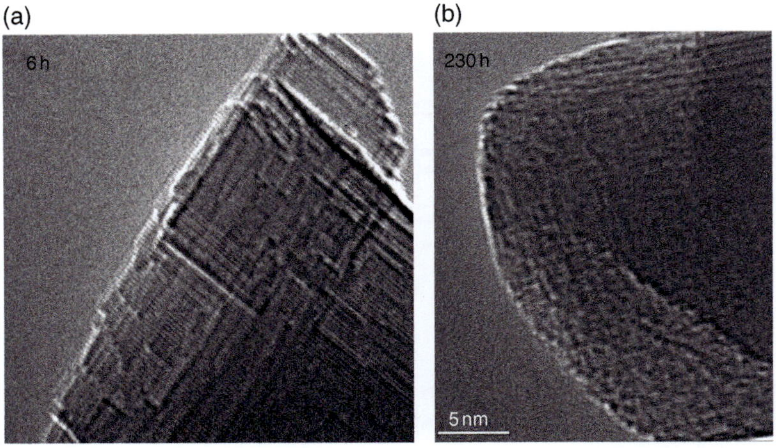

Figure 6.17 TEM images of pure MgO after 6 h (a) and after reaching a steady state at 230 h. (b). *Source:* Reproduced with permission from [214] / Wiley.

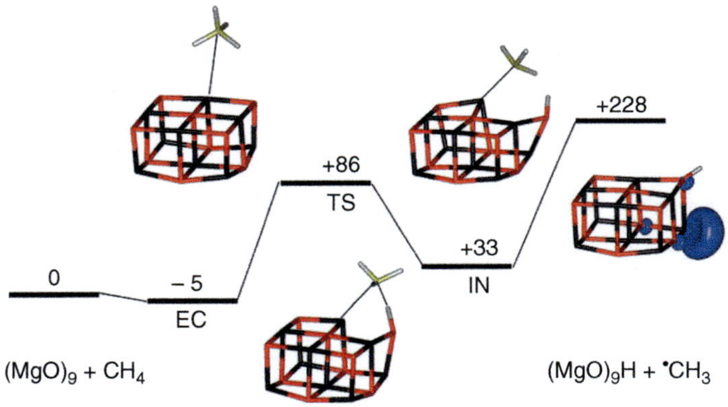

Figure 6.18 Reaction energy diagram for chemisorption of CH_4 onto corner/edge sites of a Mg_9O_9 cluster showing C–H bond addition on an $Mg^{2+}O^{2-}$ pair. B3LYP energies are given in kJ mol^{-1}, for energies including zero-point vibrational contributions see Supporting Information. EC encounter complex, TS transition state, IN intermediate. O red, Mg black, Li green, C yellow, H gray, spin density blue. *Source:* Reproduced with permission from [214] / John Wiley & Sons.

dissociation of CH_4 to form $CH_3\cdot$ along with the activations of molecular O_2 is slightly endothermic, 39 kJ mol^{-1}. The whole reaction performed on MgO surface is $Mg^{2+}O^{2-}]_{MgO} + H\text{–}CH_3 + O_2 \rightarrow (O_2^{\cdot-})[HO^{-}Mg^{2+}]_{MgO} + \cdot CH_3$ [214, 217].

6.6 Aromatization of Methane

6.6.1 Brief

The conversion of CH_4 to aromatics represents a distinct process from syngas routes, where CH_4 reforming with H_2O or CO_2, and partial oxidation of CH_4 produces syngas, subsequently transformed by Fisher–Tropsch Synthesis (FTS) into various value-added

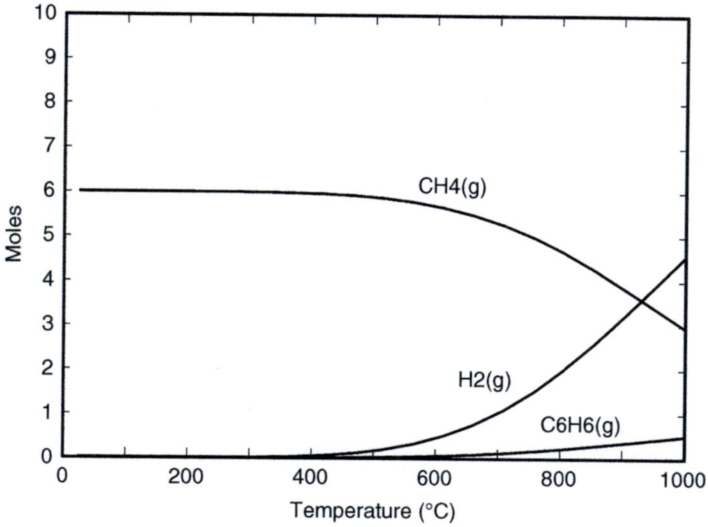

Figure 6.19 Equilibrium amounts (moles), 1 bar, starting with 6 mol CH_4 and allowing only H_2 and benzene as components. Calculated using HSC 7.1 software. *Source:* Reproduced with permission from [222] / Royal Society of Chemistry.

chemicals. Aromatization is a direct transformation of CH_4 and can occur under oxidative or nonoxidative conditions. While oxidative methane aromatization is thermodynamically favorable, introducing O_2 to the system complicates the process, generating numerous byproducts and decreasing selectivity for aromatics. Most literature studies on CH_4 aromatization focus on the nonoxidative approach. An early example of nonoxidative methane aromatization dates back to 1993 [221]. This section exclusively covers nonoxidative CH_4 aromatization.

CH_4 aromatization is highly endothermic: 6 CH_4 (g) → C_6H_6 (g) + 9 H_2 (g) ΔH_r^o =+531 kJ mol^{-1}. This transformation occurs at temperatures of at least 700 °C. Figure 6.19 represents the equilibrium amounts (moles) of CH_4, H_2, and C_6H_6 at different catalysis temperatures while the starting materials are 6 mol CH_4 and the products are limited to H_2 and benzene [222]. It illustrates the calculated CH_4 conversion and C_6H_6 yield across the temperature range of 0–1000 °C, assuming CH_4 aromatization as the sole reaction [222]. From this figure, the equilibrium yield of C_6H_6 at 700 °C is below 15%, reaching approximately 25% at 800 °C, 40% at 900 °C, and 50% at 1000 °C. However, considering carbon formation instead of C_6H_6 as a product, there is minimal benzene formed at 700–1000 °C (Figure 6.20) [222]. The notable thermodynamic discrepancy between aromatization and carbon formation (Figure 6.19 versus 6.20) makes CH_4 aromatization a promising but challenging process for direct CH_4 utilization.

6.6.2 Catalyst Preparation

The preparation of catalysts for CH_4 aromatization appears straightforward; however, as the process is conducted at temperatures ≥800 °C, significant restructuring of the catalyst occurs during the high-temperature catalysis. Little attention has been given to understanding the correlation between the structure of the as-prepared catalyst and the catalyst

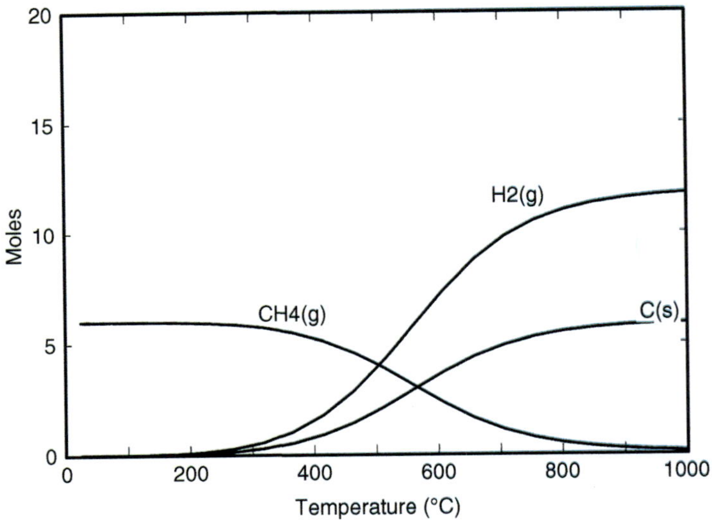

Figure 6.20 Equilibrium amounts (moles), 1 bar, starting with 6 mol CH_4 and allowing C(s), H_2, and benzene as components. Calculated using HSC 7.1 software. *Source:* Reproduced with permission from [222] / Royal Society of Chemistry.

structure during high-temperature catalysis. In this context, "structure" is defined to the chemical and coordination environment of catalytic sites at the atomic scale, as well as the geometric structure of the zeolite framework that encapsulates or supports these sites. Due to the unknown catalyst structure during CH_4 aromatization at temperatures exceeding 800 °C, establishing a correlation between the structure of an as-prepared catalyst and its structure during catalysis remains a challenge. Consequently, literature reports how variations in preparation details lead to catalysts with different catalytic performances.

Most catalysts reported for CH_4 aromatization in the literature involve zeolites with introduced Mo. HZSM-5 impregnated with Fe also exhibits activity in benzene production through CH_4 aromatization, albeit after an induction period [223–226]. Literature notes that HZSM-5 with introduced Fe is effective for CH_4 aromatization, and the addition of a small amount of CO or CO_2 enhances catalyst stability [227]. This section focuses on zeolite-based catalysts with introduced Mo. Various zeolites, including ZSM-5 [221, 228–240], ZSM-8 [241], ZSM-11 [241], MCMC-41 [241, 242], and FSM-16 [234, 243], have been employed for CH_4 aromatization. Typically, these zeolites are in the H-form or NH_4-form, as Bronsted acid sites are essential for the oligomerization of CH_x to form aromatics. Mo can be introduced through wet chemistry or solid-state synthesis methods. Wet chemistry methods include impregnation with excess solution [240, 244] and incipient wetness impregnation [234–237, 245–248]. Typically, $(NH_4)_6Mo_7O_{24}$ serves as the source of Mo atoms in these preparations. Impregnation conditions vary widely, with literature reporting processes conducted at room temperature or elevated temperatures (<100 °C) for durations ranging from 0.5 hour to 24 hours. Post impregnation, pretreatment conditions also vary, involving drying at 80–120 °C for a specific duration followed by calcination in air at temperatures ranging from 500 to 720 °C for periods of 4–30 hours. Most catalysts for CH_4 aromatization reported in the literature are prepared using wet chemistry routes. In solid-state synthesis, Mo is introduced to the zeolite by depositing Mo_2C onto the zeolite or depositing $Mo(CO)_6$ into dehydroxylated ZSM-5 [229, 249].

6.6.3 Catalyst Structure

Understanding the chemical and coordination environment of Mo atoms introduced into zeolite, as well as the location of Mo-based nanoparticles or clusters, is crucial for unraveling the catalytic mechanism. Establishing a correlation between structural and coordination information during catalysis and the resulting catalytic performance is particularly valuable. Given the elevated operating temperature of ≥800 °C for this catalysis, most characterization techniques cannot study the catalyst at such high temperatures in the presence of flowing CH_4. Therefore, the structural and chemical information discussed here pertains to the catalyst before or after CH_4 aromatization.

What is the structure of the Mo-based catalyst? EXAFS studies [243, 250] indicate that Mo in as-prepared Mo/Zeolite catalysts exists in the form of MoO_3 clusters or nanoparticles. The size and location of MoO_3 may vary in different as-prepared catalysts, as MoO_x clusters exhibit mobility during high-temperature pretreatments. Where are the MoO_3 clusters or NPs located? High-resolution transmission electron microscopy (HRTEM) suggests they anchor to the external surface of zeolite particles with sizes ranging from 1 to 5 nm [239, 251]. Larger MoO_3 nanoparticles may coexist with MoO_3 clusters when the Mo loading is higher. Mo atoms are also believed to reside within the channels and cages of zeolite, given the high dispersibility of MoO_x at high temperatures, with MoO_3 even subliming at >600 °C. These Mo atoms likely exist as single-atom sites in the form of Mo_1O_x anchored in the zeolite although electron microscopic techniques may face challenges in detection. As these highly dispersed Mo atoms coexist with MoO_3 clusters or nanoparticles supported on the external surface of zeolite particles, EXAFS alone cannot confirm their presence. In terms of the chemical or coordination environment of Mo atoms, the introduced Mo atoms can be categorized into (1) Mo atoms correlated with a Brønsted acid site and (2) Mo atoms unrelated to a Brønsted acid site. The atomic fraction of Mo correlating with a Brønsted acid site varies depending on preparation parameters, including loading, introduction method, and pretreatment temperature and time.

In terms of obtaining structural information about the Mo/zeolite catalyst during catalysis at high temperatures (≥800 °C) with flowing reactants, few techniques can provide valuable insights into the catalyst structure during a high-temperature aromatization. Technically, EXAFS can be employed to collect data for subsequent analysis when the catalyst is at a high temperature in the presence of reactants. However, the Debye factor poses challenges in fitting an r-space spectrum collected at high temperatures, making it unrealistic to determine the number of oxygen or carbon atoms coordinating to Mo atoms while the Mo/zeolite is functioning as a catalyst. TEM techniques are limited in imaging NPs at temperatures ≥800 °C in a gas phase due to sample thermal drift. In fact, HRTEM has been the primary technique for verifying the formation of coke. EXAFS studies readily reveal the average coordination environment of Mo atoms in a used catalyst at room temperature.

AP–XPS, on the contrary, enables the observation of the catalyst surface during catalysis at temperatures up to 800 °C in the gas phase [252]. As the Mo-containing NPs on the zeolite surface examined by AP–XPS may differ from the Mo atoms anchored next to Brønsted acid sites in the microporous cages or channels, AP–XPS cannot provide chemical information about the Mo atoms encapsulated within the cages and channels of the zeolite, because AP–XPS is a highly surface-sensitive technique. XRD can track the patterns of the catalyst at a wide temperature range, up to 1000 °C or even higher, offering insights into how the Mo-containing phase transforms with changes in catalysis temperature or gas environment. For example, *in situ* XRD characterizations suggest that MoO_3 is reduced to MoO_{3-x} or MoO_2 in the presence of CH_4 and then carbonized to molybdenum oxycarbide (MoO_xC_y) or molybdenum carbide such as β-Mo_2C on zeolite in CH_4, as shown in Figure 6.21 [253–255].

A comparison of acidities between Bronsted acid sites and Lewis acid sites of the parent HZSM-5 and Mo/ZSM-5 with different loadings (2 , 5 , and 8 wt%) was conducted [256]. As depicted in Figure 6.22a, two desorption peaks at 213 and 410 °C were observed on bare HZSM-5. The 410 °C peak, known as the high-temperature peak, is attributed to strong Bronsted acid sites. In comparison to bare HZSM-5, the loading of 5 wt% Mo to

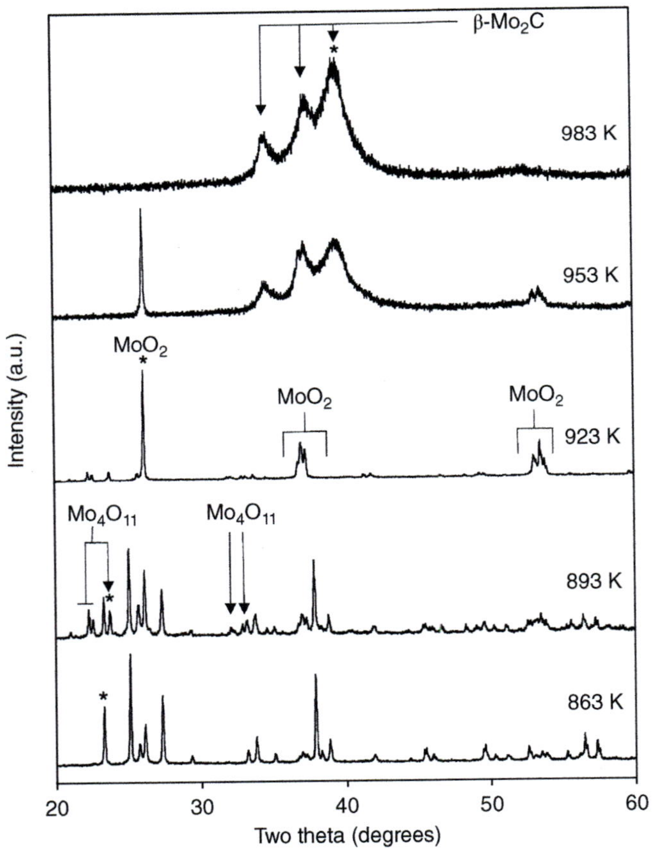

Figure 6.21 *In situ* XRD patterns were recorded during the direct carburization of bulk MoO_3 by methane in the temperature of 863–983 K range. MoO_3 peaks are not indexed. *Source:* Reproduced with permission from [253] / Royal Society of Chemistry.

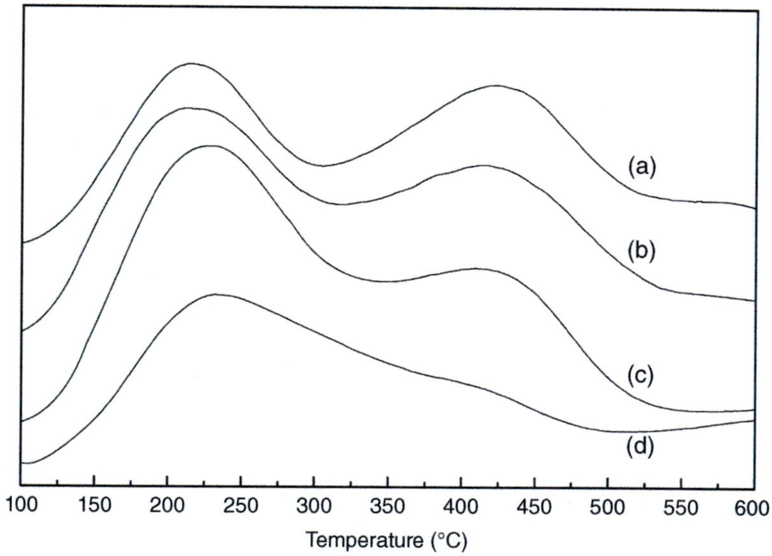

Figure 6.22 NH_3-TPD profiles of the samples: (a) parent HZSM-5; (b) 2Mo/ZSM-5; (c) Mo/ZSM-5 and (d) 8Mo/ZSM-5. *Source:* Reproduced with permission from [256] / Elsevier.

ZSM-5 noticeably weakened the peak of strong Bronsted acid sites. With an increase in Mo loading to 8%, the peak at 410 °C nearly disappeared. The absence of the peak at 410°C suggests the replacement of strong Bronsted acid sites by Mo atoms immigrated from the external surface of ZSM-5. As the intensity of the strong Bronsted acid sites decreases, the peak at 210 °C slightly shifts to 232 °C, and its intensity increases, indicating the presence of Lewis acid sites. The intensity of these Lewis acid sites increases with the rising Mo loading in Figure 6.22. The Pyridine–IR technique measured the number of Bronsted acid sites (B-acid) and Lewis acid sites (L-acid) of bare HZSM-5 and these Mo/ZSM-5 catalysts (Table 6.3). Clearly, compared to bare HZSM-5, the number of Bronsted acid sites significantly decreased due to the replacement of Bronsted acid sites by the Mo species,

Table 6.3 The acidic sites of the Mo/ZSM-5 with different Mo loading from NH_3-TPD and Pyridine-IR.

Sample	Acid sites from NH_3-TPD ($\times 10^{20}\,g^{-1}$)	Acid sites from Py-IR ($\times 10^{20}\,g^{-1}$)	
	Total	B acid	L acid
HZSM-5	6.7	8.0	0.2
2Mo/ZSM-5	4.7	4.7	0.4
5Mo/ZSM-5	4.3	4.2	0.5
8Mo/ZSM-5	3.7	3.2	0.5

Source: Reproduced with permission from [256] / Elsevier.

but the number of Lewis acid sites increased due to the immigration of Mo species from the external surface of HZSM-5 particles. XRD Rietveld structure refinement suggests that $[Mo_5O_{12}]^{6+}$ clusters were formed at the crossing points of zigzag channels and straight channels. Figure 6.23 schematically shows the distribution of $[Mo_5O_{12}]^{6+}$ clusters in ZSM-5 [256, 257]. This as-prepared Mo/ZSM-5 underwent restructuring during the incubation period.

The structural information of used catalysts is not always readily obtained, and the presence of formed coke can pose challenges to characterize the used catalysts. Despite this, the structural information of a used catalyst, while potentially different from that during catalysis, remains valuable. XPS was employed to investigate the surface of 3% Mo/ZSM-5 after O_2 treatment at 700 °C for 1 hour and subsequent reaction with CH_4 at 700 °C for varying durations (10, 20, and 30 minutes) (Figure 6.24) [258]. In addition, as illustrated in Figure 6.25,

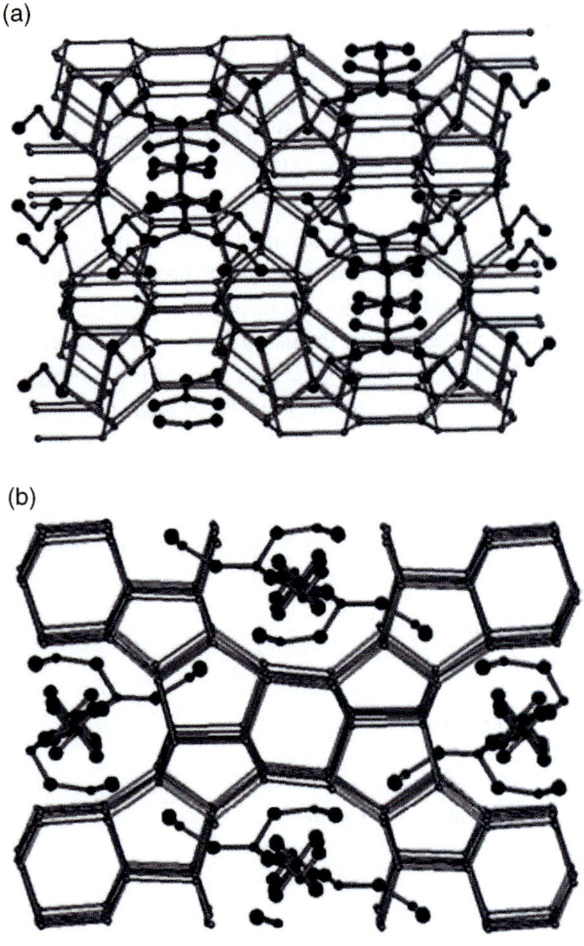

Figure 6.23 Distribution of $[Mo_5O_{12}]^{6+}$ units inside the channels of ZSM-5 in 8Mo/ZSM-5 catalyst projected along the (100) direction of the framework (a) and the (010) direction of the framework (b). *Source:* Reproduced with permission from [256] / Elsevier.

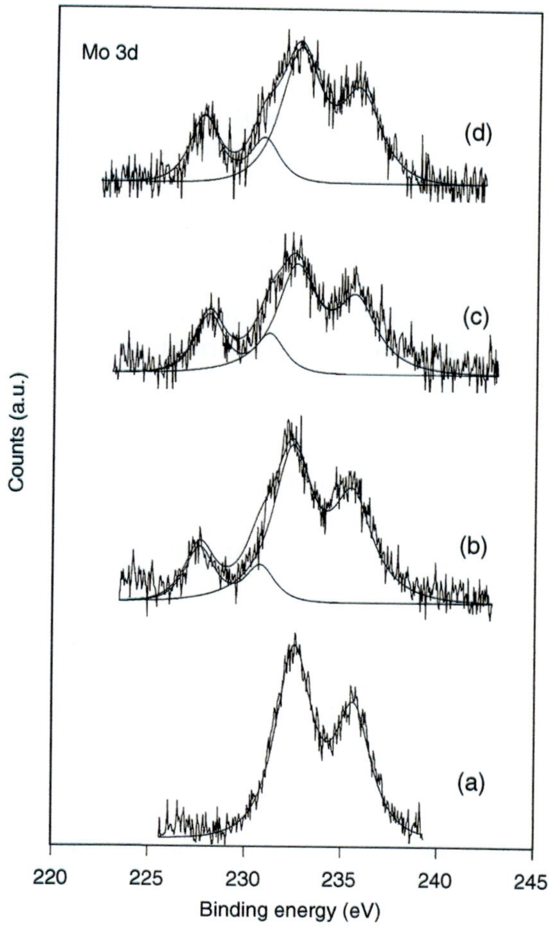

Figure 6.24 Mo 3d XPS spectra obtained from the 3%Mo/ZSM-5 catalysts: (a) after O_2 treatment at 700 °C for 1 h; (b) after reaction with CH_4 at 700 °C for 10 min; (c) reaction for 20 min; (d) reaction for 30 min. *Source:* Reproduced with permission from [258]/World Scientific Publishing.

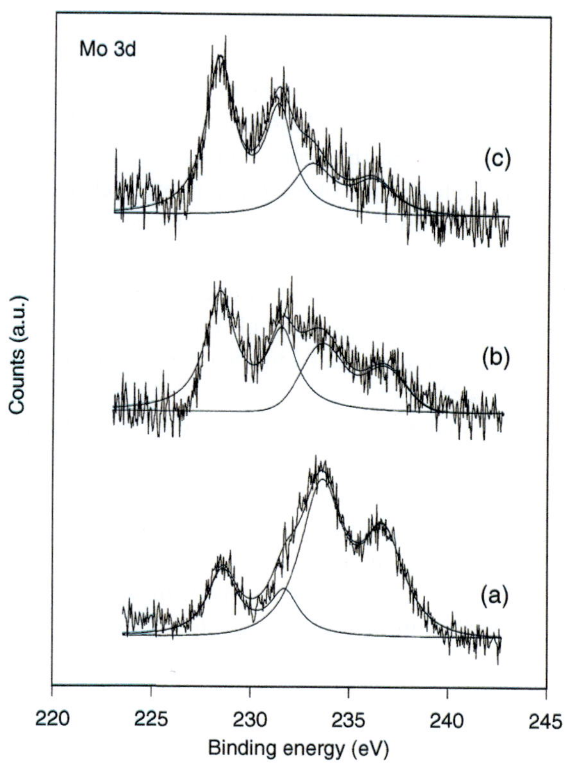

Figure 6.25 Mo3d XPS spectra obtained from 6% Mo/ZSM-5 catalysts after reaction with CH_4 at 700 °C (a) for 10 min; (b) 20 min; (c) 30 min. *Source:* Reproduced with permission from [258] / World Scientific Publishing.

MoO_3 of 6% Mo/ZSM-5 underwent progressive carbonization to Mo_2C, coexisting on the surface after reacting with CH_4 for 30 minutes [258, 259].

6.6.4 Formation of Active Phases

The transformation of the as-prepared catalyst into an active phase may take hours or longer, a phenomenon often referred to as the incubation period. While observing the surface during incubation is challenging, characterizing the catalyst after specific reaction times (10, 20, and 30 minutes) and comparatively analyzing these data can unveil structural evolution to some extent, if not completely. Figure 6.21 clearly illustrates the progressive transformation of the MoO_3 surface to Mo_2C during incubation. Generally, during incubation, the Mo/ZSM-5 catalyst precursor undergoes progressive restructuring to form MoO_xC_y on the external surface of ZSM-5 and $[Mo_5O_aC_b]^{n+}$ within the ZSM-5 [76, 229, 240, 249, 260–268].

Determination of the exact active phases for CH_4 aromatization poses challenging question, as there are limited characterizations aside from in-situ XRD. In general, the literature lacks agreement on the assignment of active phase(s) for Mo/ZSM-5, the most studied catalyst system in recent decades [256, 269, 270]. The active phase(s) formed depend on several factors, including pretreatment temperature and gas, catalysis temperature, stream time, and even gas hourly space velocity. It is reported that MoO_xC_y dispersed on the external surface of ZSM-5 is the active phase responsible for the activity and selectivity of CH_4 aromatization at 700 °C initially. The MoO_3 species on the external surface of ZSM-5 can preferentially carbonize in the early stages of catalysis. As time progresses, more $[Mo_5O_{12}]^{5+}$ species encapsulated in ZSM-5 are carbonized to $[Mo_5O_aC_b]^{n+}$ at 700 °C, making it one of the active phases at this temperature. It is likely the main active phase since coke must form on Mo_2C derived from MoO_xC_y in CH_4 at high temperature. On the other hand, the active phase depends on catalysis temperature. Although MoO_xC_y formed on the external surface of ZSM-5 is considered an active phase for CH_4 aromatization at 700 °C, it morphs into the Mo_2C phase at higher temperatures. Mo_2C on the external surface of ZSM-5 is regarded as one of the active phases at 800 °C or higher before being covered with coke. Indeed, the formation of the Mo_2C phase is closely related to coke information. Once coke forms on the Mo_2C on the external surface of ZSM-5, the Mo-containing species lose their activity, as the Mo_2C covered with coke cannot directly interact with CH_4 molecules. The $[Mo_5O_aC_b]^{n+}$ encapsulated in the channels can survive at 800 °C or higher since it firmly bonds with the framework of ZSM-5. Therefore, it is suggested that the survivable $[Mo_5O_aC_b]^{n+}$ encapsulated in the channels is the sole active phase for CH_4 aromatization at 800 °C or higher.

MoC_{1-x} and Mo_2C were also proposed as active phases of CH_4 aromatization in literature [251, 271], supported by the following experiments. In literature, α-MoC_{1-x} supported on ZSM-5 (α-MoC_{1-x}/ZSM-5) was prepared through carburization of MoO_3 supported on ZSM-5 by two steps: (1) activation at 350 °C in 30 mL min^{-1} H_2 for 24 hours or 30 mL min^{-1} of mixture of H_2 and *n*-butane with a molar ratio of 11:1 for 24 hours, and (2) activation of product of step (1) at 720 °C in 16 mL min^{-1} CH_4. In addition, β-Mo_2C/ZSM-5 was prepared by carburization of MoO_3 supported on ZSM-5 bulk by two steps (1) heating MoO_3 to

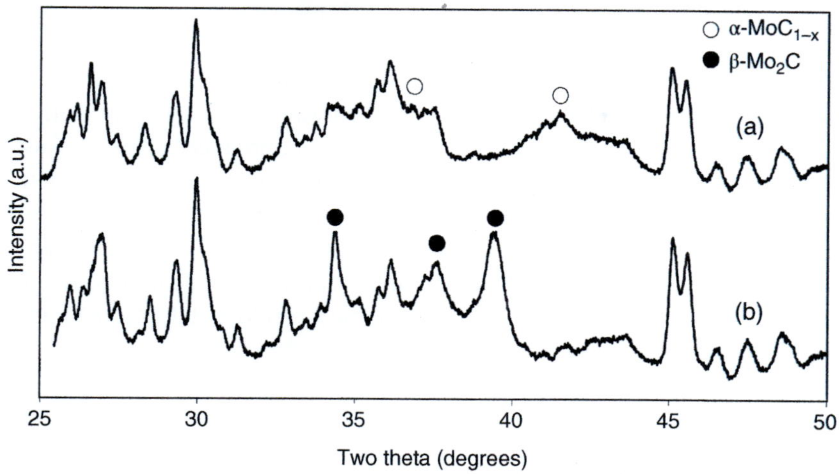

Figure 6.26 XRD patterns of the carburized samples: (a) sample activated under hydrogen/*n*-butane and carburized with methane; (b) sample activated under hydrogen and carburized with methane. HSZM5 peaks are not indexed. *Source:* Reproduced with permission from [224] / Elsevier.

350 °C in 50 mL min^{-1} helium, and (2) heating the product of step (1) to 720 °C in 16 mL min^{-1} CH_4 [224, 272]. Figure 6.26 represents the XRD patterns of α-MoC_{1-x}/ZSM-5 and β-Mo_2C/ZSM-5. Catalytic performances of the prepared α-MoC_{1-x}/ZSM-5 and β-Mo_2C/ZSM-5 catalysts were evaluated. They are active for CH_4 aromatization at 700 °C.

6.6.5 Carburization

Various types of carbonaceous residues formed in CH_4 aromatization have been documented in the literature [228, 232, 234, 236, 239, 273–279]. Given the diverse experimental conditions, including activation in CH_4 or other hydrocarbons, as well as catalysis in the gas CH_4, only specific carbonaceous residues reported in the literature were observed under pretreatment or catalytic conditions. For example, during CH_4 aromatization at 700 °C on 2 wt% Mo/HZSM-5, three types of carbons were identified. These include graphitic-like carbon reflected at 284.6 eV in the C 1s spectrum, carbide-like carbon at 282.7 eV, and hydrogen-poor sp-type carbon at 283.2 eV, as revealed in XPS studies of a used catalyst [267, 280].

6.6.6 Generally Agreed Reaction Intermediate

In general, the goal of fundamental studies of catalytic reactions is to understand a catalytic mechanism at a molecular level. Broad consensus suggests that CH_4 aromatization follows a bifunctional mechanism, involving both Mo sites and nearby Bronsted acid sites in the formation of aromatics [229, 232, 234, 243, 281–286]. Specifically, the process begins with the formation of CH_x species on Mo-based sites, which then undergo coupling to generate C_2H_y as a primary intermediate. Subsequently, C_2H_y species oligomerize on Bronsted acid sites, leading to the formation of aromatics. This mechanism is supported by a series of comparative experiments [287].

In one such experiment, unsupported Mo_2C was prepared through carburization and tested under conditions similar to those for Mo/ZSM-5. However, only ethylene, with a small amount of ethane, was observed, and no aromatics were produced. This contrast underscores the necessity of ZSM-5 as a support for aromatic production. In addition, Mo_2C supported on nonacidic SiO_2 exhibited higher CH_4 conversion than unsupported Mo_2C, yet the predominant product on Mo_2C/nonacidic SiO_2 was carbon, and the selectivity for benzene was much lower than that of Mo_2C/ZSM-5. This implies the indispensability of acidic sites for aromatic formation. Importantly, bare HZSM-5 did not yield any aromatics under CH_4 aromatization conditions, emphasizing the essential role of acidic sites in H-ZSM-5.

These studies collectively affirm that both Mo-based sites and acidic sites are vital for completing a catalytic cycle leading to aromatic production. Moreover, the proximity of the Mo-based site and the acidic site is crucial, as demonstrated by comparative experiments. For instance, a 1:1 physical mixture of 2 wt% Mo/SiO_2 exhibited lower selectivity for aromatics compared to 2 wt% Mo/ZSM-5, and the catalytic performance of a 1:1 physical mixture of 2 wt% Mo/SiO_2 and HZSM-5 was nearly identical to that of 2 wt% Mo/SiO_2. This underscores the necessity of proximate acidic sites in HZSM-5 to the Mo-based sites. Another comparison experiment involving HZSM-5 and Mo/ZSM-5, using ethylene as a reactant, revealed similar product distributions on both catalysts. This similarity suggests that Mo-based sites are not essential for the oligomerization of ethylene to benzene, indicating that this oligomerization process primarily occurs on the acidic sites of ZSM-5, while Mo-based sites are mainly responsible for the activation of CH_4.

6.6.7 Facing Challenge and Promising Topic

Catalyst stability remains a ubiquitous challenge in the context of CH_4 aromatization, primarily due to carbon formation. Addressing this challenge necessitates the development of innovative approaches to prevent coke formation, demanding further research to enhance our understanding of the carbon formation mechanism. Another formidable challenge lies in conducting *in situ*/operando studies of the catalyst, particularly focusing on the chemical and coordination environments of Mo-based sites during catalysis. In comparison to other high-temperature processes for CH_4 transformation, computational studies on CH_4 aromatization are relatively scarce. This scarcity can be attributed, in part, to the ongoing debates surrounding the active phase(s) of catalysts during catalysis. Furthermore, the addition of various metals, such as Ru [288, 289], Zr [248, 290], Co [234, 291, 292], or W [248, 293], to Mo/ZSM-5 has been shown to enhance catalytic activity and selectivity towards aromatics. Additionally, the introduction of Zn [235, 294] or Pt [295] to Mo/ZSM-5 has proven effective in increasing the stability of Mo/ZSM-5 by reducing the concentration of carbon formed during catalysis. However, a fundamental understanding of the underlying mechanisms behind these promotional effects of secondary metals is lacking. Exploring and unraveling these effects represent intriguing and significant areas for further investigation.

6.7 Direct Activation of Methane on Single Sites of Fe to Synthesize Ethylene and Aromatics

As discussed in Section 6.7, similar to Mo, Fe was introduced into HZSM-5 for the nonoxidative aromatization of CH_4 [223–226, 296–299]. In Fe/ZSM-5, Fe cations are dispersed within the zeolite structure and exhibit activity for CH_4 aromatization. However, like many CH_4 aromatization catalysts, Fe/ZSM-5 faces challenges associated with coke formation. A notable departure from this trend is observed with a catalyst featuring single-atom Fe sites anchored on a unique support as explained in the next paragraph, demonstrating high activity for nonoxidative CH_4 activation with minimal coke formation [300].

The preparation of the catalyst featuring single Fe sites involves intricate steps. SiO_2 ($<1\,m^2\,g^{-1}$) is mixed with Fe_2SiO_4, followed by ball milling under an Ar atmosphere (99.999%) to ensure thorough mixing. The mixture undergoes fusion at 1700 °C in air, allowing iron to diffuse fully and distribute uniformly. Subsequent removal of iron oxide particles on SiO_2 is achieved through treatment with HNO_3 ($0.5\,mol\,l^{-1}$). The resulting material is then dried at 80 °C for 12 hours to eliminate excess HNO_3 solution. A subsequent activation step at 900 °C for 2 hours in a 90% CH_4/10% N_2 atmosphere leads to carbonatization, resulting in singly dispersed Fe atoms bonded with two carbon atoms and one Si atom. Additional details can be found in the literature. Under the reported conditions, the CH_4 conversion at 1090 °C reached 48.1%, with a selectivity of 48.4% for ethylene and a total hydrocarbon selectivity of nearly 100%. This exceptional catalytic performance has been corroborated by other studies. In contrast, a catalyst with the same Fe loading (0.5 wt%) where Fe exists as NPs of 2–5 nm (0.5 wt% Fe/SiO_2) exhibits a selectivity for carbon exceeding 98% [300].

Based on the literature [300], the primary function of singly dispersed Fe sites is to cleave the first C–H bond of CH_4, forming $CH_3\cdot$ radicals and H radicals. The coupling of two $CH_3\cdot$ radicals then forms a C_2H_6 molecule. Subsequently, C_2H_6 can dissociate to C_2H_5 and further to C_2H_4 with the assistance of H radicals. Figure 6.27 illustrates the thermodynamics of the direct activation of CH_4 to C_2H_4, C_6H_6, and $C_{10}H_8$, CH_4 dissociation to C and H_2, OCM, and CH_4 partial oxidation. According to Figure 6.27, nonoxidative conditions are thermodynamically unfavorable over a wide temperature range. Only at extremely high temperatures is nonoxidative CH_4 activation thermodynamically feasible ($\Delta G_r^\circ<0$). Carbon formation (the brown line) is consistently thermodynamically favorable compared to the reactions leading to ethylene (green-blue line) and benzene formation (pink line). Therefore, the main challenge in designing a catalyst for the high-temperature formation of ethylene and benzene is understanding and controlling carbon formation from the direct dissociation of CH_4. The unique feature of the catalyst with fused silica and anchored single Fe sites largely prevents coke formation.

As discussed, both OCM and CH_4 aromatization involve the activation of the first C–H bond of CH_4, with the formed $CH_3\cdot$ radicals desorbing to the gas phase to couple and form C_2H_6. In the case of OCM, this process includes CH_4 activation on Mo-based species to form CH_3, the generation of C_2H_x species, and even oligomerization to produce aromatics on acidic sites. Continuous activation of the second C–H bond is likely performed on acidic sites, leading to the formation of CH_2, the intermediate for oligomerization on the acidic

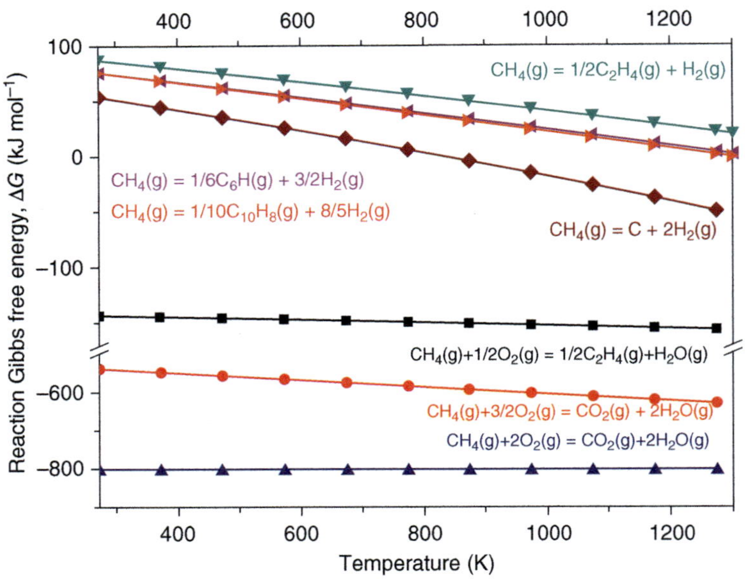

Figure 6.27 Thermodynamics of direct conversion of methane to hydrocarbons both oxidatively and nonoxidatively. *Source:* [331] / with permission of American Chemical Society.

sites. Controlling the ending species of continuous C–H activation is challenging, and the formation of atomic carbon through CH→C + H· on acidic sites competes with the oligomerization of CH·. The subsequent coupling of carbon atoms to form a carbon chain and carbon layer, and ultimately graphitic carbon, leads to coke formation. The distinctive characteristic of the catalyst composed of fused silica with anchored single-atom Fe sites is the absence of proximate Fe sites, effectively preventing the coupling of atomic carbon and, consequently, inhibiting the formation of a carbon chain.

6.8 Transformation of Methane to Form Hydrogen and Carbon

From a molecular perspective, hydrogen stands out as an energy carrier owing to its robust H–H bond. However, the low density of gaseous hydrogen poses a significant challenge to safe storage. Although hydrogen-fueled transportation systems remain challenging, ensuring safe storage on mobile systems remains a formidable task. Beyond its potential applications in transportation, hydrogen serves as a crucial chemical intermediate for the production of a myriad of compounds and the manufacturing of essential materials for our daily lives, such as fertilizers. Generally, CH_4 serves as a primary source of hydrogen in manufacturing industries.

6.8.1 Noncatalytic Approaches

In contrast to the steam reforming processes prevalent in chemical industries, the dissociation of CH_4 into carbon and hydrogen can be achieved through alternative physical or catalytic methods. The resulting carbon materials holds substantial commercial value as the U.S. carbon black market size in 2023 is US$19.2B. For instance, this carbon material finds applications as a reinforcing agent in tire manufacturing, as well as in ink and pigment production. One such approach to carbon production from CH_4 involves the use of an electric arc [301]. The electric arc generates plasma to decompose CH_4 into carbon and hydrogen. However, this process may not be deemed cost-effective, given the considerable amount of electricity required.

6.8.2 Catalysis by Supported Fe, Co or Ni

In comparison to steam reforming, catalytic pyrolysis of CH_4 at high temperatures provides an alternative method for hydrogen generation from CH_4. The direct decomposition of CH_4 into carbon and hydrogen, $CH_4 \rightarrow C + 2H_2$, is an endothermic process with $\Delta H_r(1000\,°C)$ $(CH_4 \rightarrow C + 2H_2) = 91\,kJ\,mol^{-1}$. This reaction occurs at temperatures of 700 °C or higher. Early transition metals such as Fe, Co, and Ni supported on MgO, SiO_2, $LiAlO_2$, and ZrO_2 are well-known catalysts for this reaction. Deactivation is a significant challenge for these catalysts, although regeneration using O_2 or CO_2 could extend the lifespan of these supported metal catalysts [302–304]. However, the catalytic performance of a regenerated catalyst is highly dependent on the dispersion and morphology of the metal phase. The direct decomposition of CH_4 to generate hydrogen is an oxygen-free catalytic pyrolysis, producing CO_x-free H_2, as this process does not involve any oxygen-containing molecules. CO_x-containing H_2 exhibits a limited application due to the potential poisoning effect of CO. The direct decomposition of CH_4 is a unique feature of the production of non-CO_x H_2, as many other H_2 production processes such as CH_4 steam reforming with a followed water gas shift produce H_2 always containing a certain amount of CO or CO_2.

6.8.3 Catalysis by Melted Metal

Regarding hydrogen generation at the solid–gas interface through catalytic pyrolysis, another approach involves the use of liquid metal catalysts, often referred to as a molten metal reactor [305–309]. Early studies demonstrated that molten Sn or Cu at 900 °C in a bubbling bath reactor can catalyze the decomposition of CH_4 to hydrogen and carbon [310]. Unlike the carbon generated from an electric-field-based plasma [301], the carbon produced in a molten reactor can be readily collected from the surface of the liquid alloy catalysts, as it is formed into a phase separated from the liquid phase of the catalyst [311]. Using a bubble column reactor loaded with liquid tin, a yield of 75% for H_2 was achieved at 1175 °C.

6.8.4 Catalysis by Melted Alloy

Recently, a liquid alloy composed of Group VIII metals dissolved in low-melting-temperature metals (In, Ga, Sn, and Pb) has been identified as a robust and active alloy catalyst for producing pure H_2 and removable carbon. This dilute melted alloy serves dual functions. First, it has a low-melting temperature that enables decarbonization at the liquid–gas interface. Second, the liquid state of the catalyst allows for rapid carbon diffusion and easy segregation to the surface of the liquid metal, facilitating the removal of carbon black. In this bubble reactor, each bubble can be considered a small batch reactor [312]. In this unique batch reactor, the concave catalyst (liquid) -reactant (gas) interface in a gas bubble in liquid chuck is continuously refreshed, given the high temperature of the liquid catalyst, promoting fast atomic diffusion. The significant advantage of this bubble column reactor with liquid metal lies in the fact that the carbon formed at the interface of the concave liquid alloy surface and the gas of gas bubble in the liquid metal can rise up to the surface of the liquid alloy chuck along with the movement of gas bubbles and accumulate at the top of liquid alloy chuck. The carbon formed at the liquid–gas interface and then dissolved in the liquid precipitates once saturated, resulting in the segregation of carbon species at the topmost surface of the liquid column. The carbon materials formed through the catalytic CH_4 pyrolysis remains as a separated solid phase on the liquid chuck and thus can be readily removed using physical methods.

Another function of this liquid alloy is to serve as a unique catalyst hosting guest metal atoms of Ni, Pd, or Pt. Figure 6.28 schematically illustrates the concept of using a Pt–Bi alloy for the catalysis of CH_4 decarbonization [313]. A differential reactor, as depicted in Figure 6.29a, was employed to evaluate 20 pure metals or alloys [313]. Over 20 liquid metals, either pure or alloyed, were screened under the same conditions. Table 6.4 provides the rates of hydrogen production at 1000 or 950 °C. These screenings revealed that alloys such as $Pt_{0.17}Sn_{0.83}$ consistently exhibit higher reactivity than pure Sn, indicating a promotional effect of Group VIII metals (Ni, Pd, Pt) on low-melting point metals (In, Sn, Ga, Pb). Among these liquid alloys, $Ni_{0.17}Bi_{0.83}$ demonstrates the highest activity, being over 1000 times faster than pure liquid Bi. In terms of CH_4 conversion, 98% can be achieved at 1065 °C using $Ni_{0.27}Bi_{0.73}$ in the reported bubble column reactor containing molten metal (Figure 6.29) which is one of these reactors used for melted metals schematically shown in Figure 6.29.

Kinetics were examined for this high-temperature catalyst [313]. By measuring the rate of hydrogen production at various CH_4 pressures, the reaction order of CH_4 was determined to be first order. Apparent activation barriers for different catalysts were evaluated from Arrhenius plots. The activation energies for the decarbonization of CH_4 in the gas phase without any catalyst, with liquid Bi, and with liquid $Ni_{0.27}Bi_{0.73}$ alloy are 422, 310, and 208 kJ mol^{-1}, respectively [313]. Clearly, liquid Bi can catalyze this reaction. Notably, the liquid $Ni_{0.27}Bi_{0.73}$ alloy significantly reduces the activation barrier compared to pure Bi. In other words, the guest metal Ni plays a crucial role, as does Pt in the liquid Pt–Bi alloy.

Computational studies have delved into the electronic states of the introduced guest metals, predicting the chemical and coordination environments within the metal solvent. The average Bader charges of VIII metals in liquid metals (Sn, Bi, or Pb) were calculated, revealing negative charges on all guest metal atoms in the alloy liquid compared to their gas phase counterparts. Table 6.5 illustrates this phenomenon, where the negative charge originates from the solvent metal (Sn, Bi, or Pb) as the guest metal atoms disperse within it [313]. The charge transfer from Sn to Pt was substantiated by the calculated density of states, as depicted in Figure 6.30c [313]. The density of states of Sn atoms neighboring Pt atoms in a Pt–Sn liquid alloy within the energy window of −4 to 0 eV is noticeably lower than that of Sn atoms in pure liquid Sn metal. In the Pt–Sn liquid alloy, the average charge of Pt atoms is −0.91 electrons, and the surrounding Sn atoms exhibit a slightly more positive Bader charge compared to pure liquid Sn metal. This negative charge on Pt atoms,

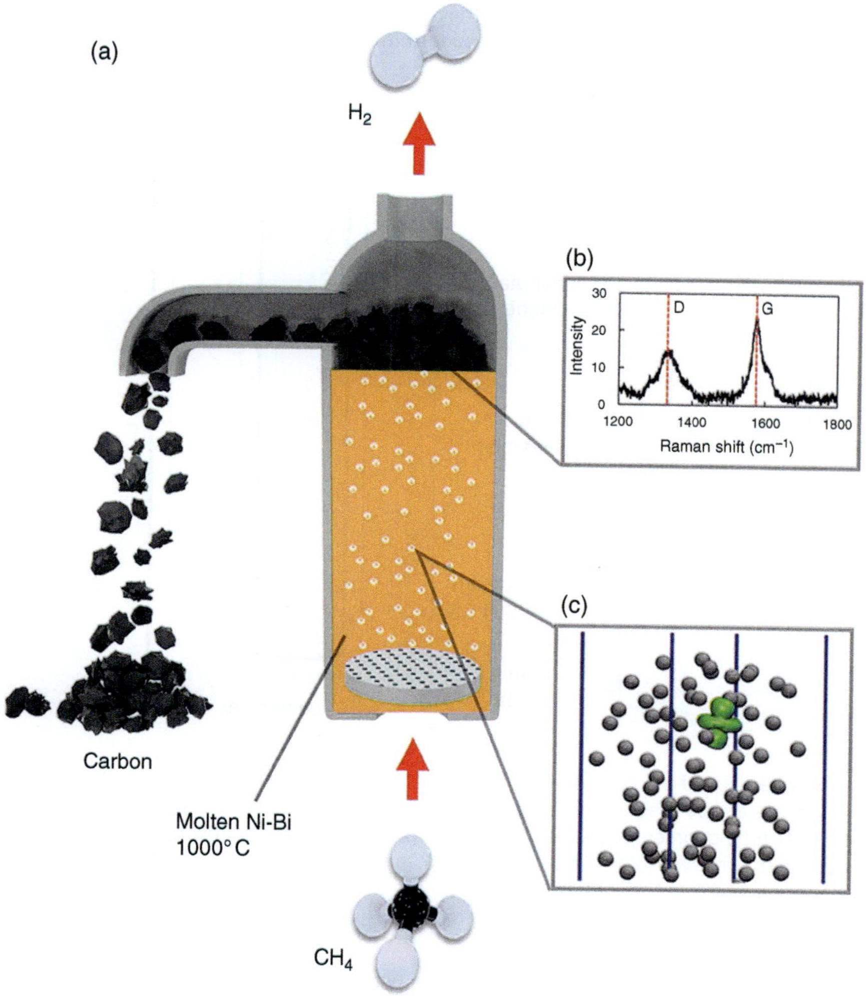

Figure 6.28 Hydrogen production with a Ni-Bi molten catalyst. (a) Reactor for CH_4 conversion to H_2 and carbon in a molten-metal bubble column with continuous carbon removal. (b) Raman spectrum of surface carbon. The dashed line labeled "D" is at 1350 cm^{-1}, and the dashed line labeled "G" is at 1582 cm^{-1}. (c) Ab initio molecular dynamics simulation showing an orbital (green) of a Pt atom dissolved in molten Bi (gray) alloy. *Source:* [313] / American Association for the Advancement of Science - AAAS. / CC BY 4.0.

relative to a neutral Pt atom, weakens their binding to carbon atoms. Computational studies revealed that Pt_2 or Pt_8 clusters, stable in an inert gas phase, become unstable in liquid Sn due to the negative charge, as the valence electrons of Sn occupy the unoccupied antibonding orbitals of the Pt_n cluster, thereby weakening Pt–Pt bonds. Consequently, Pt atoms in the Pt–Sn liquid alloy exist as singly dispersed entities within the liquid Sn [313]. Similarly, computational studies also identified negatively charged Ni atoms dispersed in Bi liquid metal. Figure 6.30c illustrates the substantial difference in catalytic performance between singly dispersed Ni atoms in Bi liquid and solid Ni wire.

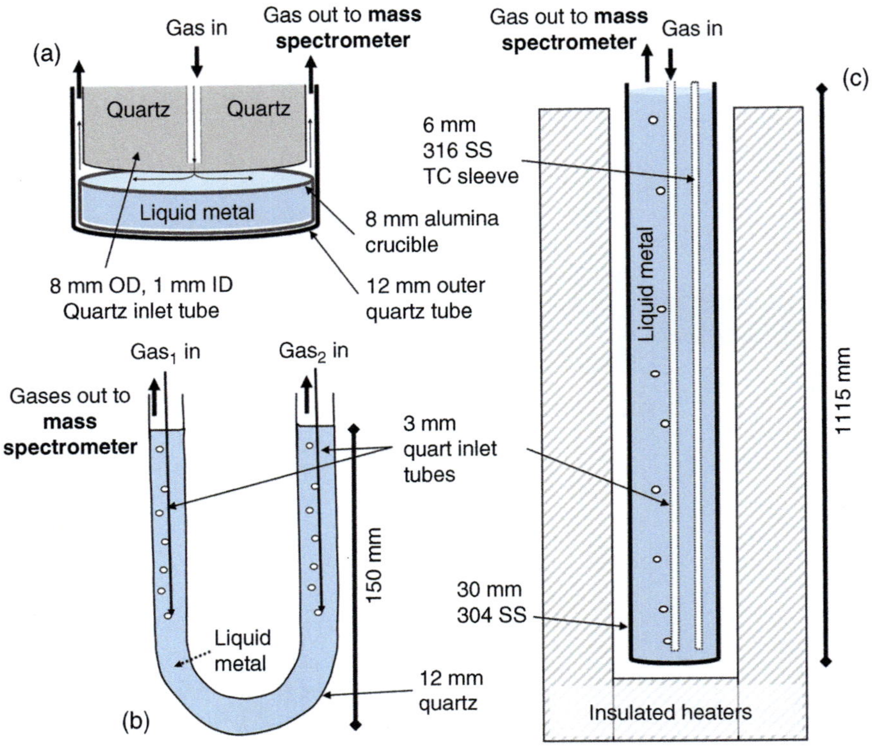

Figure 6.29 (a) Schematic of the screening reactor used to determine liquid metal activity. 38.5 mm^2 of liquid metal was used and filled the alumina crucible. (b) Schematic of quartz bubble column reactor in U-tube geometry. (c) Schematic of 1.1-m stainless steel bubble column reactor. All reactors were in thermally insulated electric heaters. *Source:* [313] / American Association for the Advancement of Science - AAAS. / CC BY 4.0.

Another alloy liquid metal, Cu–Bi, has recently been reported by the same research group that investigated the Ni–Bi and Pt–Sn systems [312, 314]. In contrast to the expensive Pt–Sn and Ni–Bi alloys, the Cu–Bi alloy catalyst is cost-effective and environmentally friendly. The Cu–Bi alloy liquid metal stands out as the most active catalyst for CH_4 decomposition to produce H_2 and carbon, even though pure liquid Cu or Bi is nearly inactive. The notable catalytic activity of the Cu–Bi alloy liquid aligns with an earlier reported correlation between the surface tension of an alloy liquid and its catalytic activity [315, 316]. Precise measurements of the surface tension of Cu, Bi, and Cu–Bi alloys indicate that dissolved Bi atoms in liquid Cu act as a surfactant, significantly lowering the surface tension of liquid Cu. In other words, Bi atoms are enriched on the surface of the Cu–Bi alloy liquid.

Computational studies on the catalytic pyrolysis of CH_4 on Cu–Bi revealed that the observed high activity of the Cu–Bi alloy liquid is associated with the charge redistribution of Bi and Cu on the surface. In the bulk of the Cu–Bi alloy liquid, Cu and Bi have slightly negative and slightly positive charges, respectively. This charge redistribution varies from the bulk to the surface region. In the surface region, Bi is less positive, and Cu is more

Table 6.4 Comparison of activity for methane pyrolysis at 1000 °C when CH_4 is flowed over 38.5 mm^2 of molten metal.

Liquid catalyst	Rate of hydrogen production (mol H_2 produced × cm^{-2} s^{-1})
In	8.2×10^{-11}
Bi	8.2×10^{-11}
Sn	8.5×10^{-10}
Ga	3.2×10^{-9}
Pb	3.3×10^{-9}
Ag	4.3×10^{-9}
Pb vapor	2.1×10^{-9}
17% Cu–Sn*	3.1×10^{-9}
17% Pt–Sn	1.6×10^{-9}
17% Pt–Bi	4.2×10^{-9}
62% Pt–Bi*	6.5×10^{-9}
17% Ni–In	4.7×10^{-9}
17% Ni–Sn	5.6×10^{-9}
73% Ni–In*	6.4×10^{-9}
17% Ni–Ga	7.9×10^{-9}
17% Ni–Pb	8.3×10^{-9}
17% Ni–Bi	9.0×10^{-8}
27% Ni–Au*	1.2×10^{-8}
27% Ni–Bi*	1.7×10^{-8}
27% Ni–Bi* (replicate)*	1.7×10^{-8}

The same reactor volume was used in all cases, including for Pb vapor. All compositions are molar percent. An asterisk (*) indicates that alloy is at the solubility limit of the dissolved active metal at 950 °C
Source: [313] / American Association for the Advancement of Science - AAAS. / CC BY 4.0.

Table 6.5 Average Bader charges of active metals (Pt, Ni, and Pd) in molten "inert" metals (Sn, Bi, Pb).

Atom	Bader charge		
	Sn-liquid	Bi-liquid	Pb-liquid
Ni	-0.44 ± 0.009	-0.28 ± 0.006	-0.30 ± 0.004
Pd	-0.66 ± 0.007	-0.45 ± 0.004	-0.53 ± 0.005
Pt	-1.01 ± 0.007	-0.82 ± 0.004	-0.87 ± 0.005

The averages are computed for 5 ps with data collected at every 100 fs.
Source: [313] / American Association for the Advancement of Science - AAAS. / CC BY 4.0.

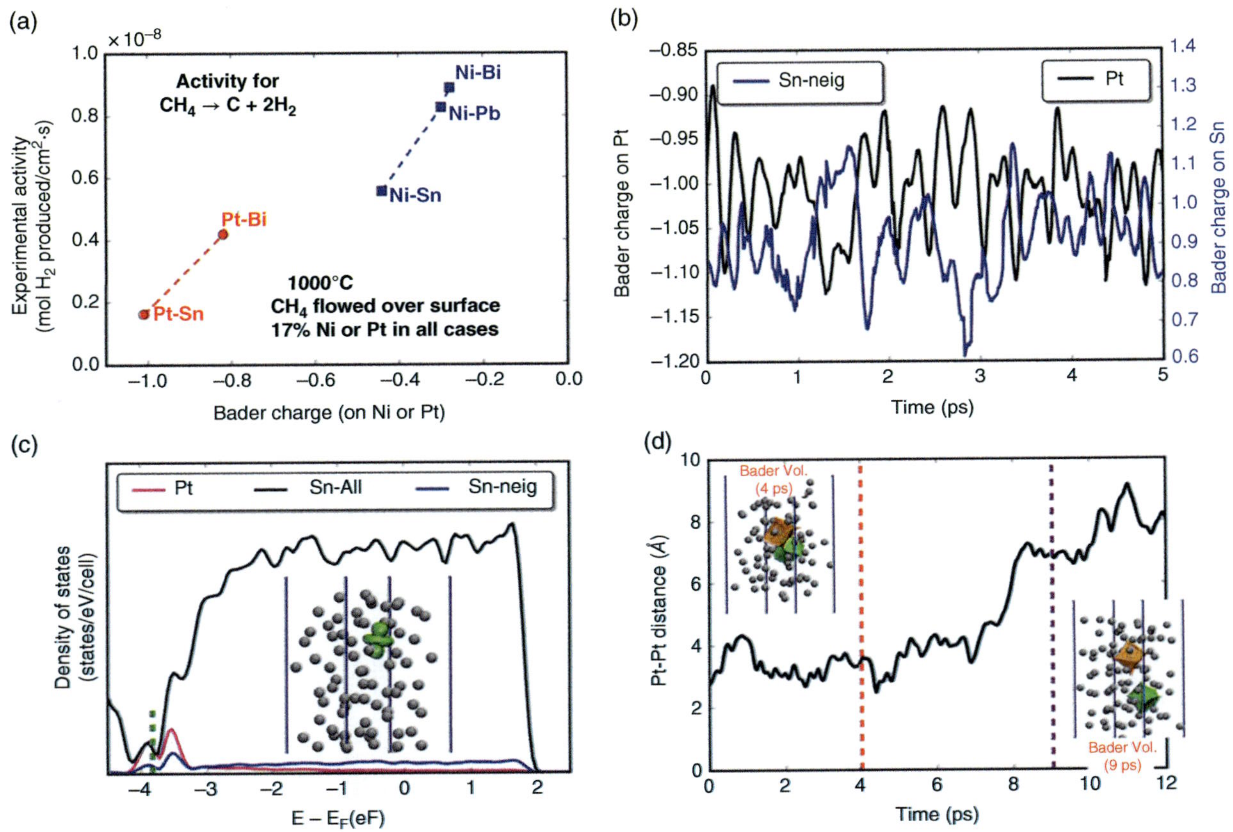

Figure 6.30 Theoretical results. (a) Activity for pyrolysis at 1000 °C plotted versus calculated Bader charge on the active element. In all experiments, 17 mol% of platinum or nickel was used. The calculations used 1.2 mol% Pt or Ni, at 627 °C. (b) The time evolution of the Bader charge on a Pt atom dissolved in molten Sn, and the sum of Bader charges on nearest-neighboring Sn atoms during ab initio molecular dynamics (AIMD) run. (c) Projected density of states of a Pt atom, of the Sn atoms neighboring the Pt atom, and of all Sn atoms. The graph also shows, in green, the shape of one Pt orbital. The boundaries of the simulation box are shown in blue. (d) The graph shows the distance between Pt atoms in a Pt_2 cluster in molten Sn during an AIMD run. The two pictures show the charges on the Pt atoms of a Pt_2 cluster at two different times. *Source:* [313] / American Association for the Advancement of Science - AAAS. / CC BY 4.0.

positive compared to the bulk. The electron-deficient Bi atoms on the surface promote the dissociation of C–H bonds in CH_4. Computational studies have identified that CH_4 is adsorbed on Bi atoms enriched on the surface of Bi–Cu and subsequently dissociates into H and CH_3; the H and CH_3 then bind to Cu and Bi atoms, respectively. The binding of H atoms to Cu is consistent with the higher stability of copper hydride than bismuth hydride [312, 317]. The rate-determining step of CH_4 pyrolysis, in terms of the $CH_4 \rightarrow CH_3 + H$ step, was simulated on both the Bi-rich surface of the Cu–Bi alloy liquid and the surface of pure Bi-metal liquid. It was found that the activation barriers for this step on a $Cu_{45}Bi_{55}$ liquid cluster, 2.53 or 2.60 eV, are lower than the 2.80 or 2.90 eV on a pure Bi surface, respectively [317]. The evaluated activation barrier from the simulation is consistent with the experimentally measured apparent activation energy of 2.3 eV [317].

6.9 Methane Oxidation to Formaldehyde

CH_4 can be transformed to formaldehyde (CH_2O). The reported yield of CH_2O from CH_4 achieved in academic laboratories is too low for industrial applications. Consequently, formaldehyde is currently produced from the oxidation of methanol in chemical industries.

According to computational simulations, the noncatalytic oxidation of CH_4 by O_2 at high temperatures involves at least eight elementary steps. The most challenging step is the formation of the $CH_3\cdot$ radical through a bimolecular process in which CH_4 and O_2 react to form $CH_3\cdot$ and $HO_2\cdot$, $CH_4 + O_2 \rightarrow CH_3\cdot + HO_2\cdot$. This step requires crossing an activation barrier of 230–235 kJ mol^{-1} [318, 319]. When a catalyst is introduced, it is generally accepted that the mechanism of CH_4 oxidation to CH_2O involves both heterogeneous catalysis for forming the $CH_3\cdot$ radical and homogeneous catalysis. The proposed formation of $CH_3\cdot$ is supported by the observation that the addition of C_2H_6 to the reactant mixture of CH_4 and O_2 can increase selectivity to CH_2O. This is deduced from the fact that CH_3CH_3 must have been dissociated into $CH_3\cdot$, and then $CH_3\cdot$ is transformed into $CH_3O\cdot$ by coupling with an O atom. Thus, the addition of CH_3CH_3 increases selectivity to CH_2O. If CH_2O were not formed from $CH_3\cdot$, the addition of CH_3CH_3 to the catalysis system could not increase the selectivity for producing CH_2O [318, 319].

Scientists have been dedicated to gaining a fundamental understanding of CH_4 oxidation to CH_2O for decades, with MoO_3/SiO_2 and V_2O_5/SiO_2 emerging as the two most extensively studied catalysts. However, the yield of CH_2O from CH_4 has not surpassed 3–4%. A comprehensive review article has discussed tens of catalysts for CH_4 oxidation to CH_2O reported in the literature [319], shedding light on the challenges and opportunities in this transformation. Most of these catalysts are nonprecious metals, and the catalysis temperature typically ranges from 550 to 750 °C. Within this temperature range, CH_4 conversion remains below 10%, and selectivity for CH_2O spans from 10% to 90% [319, 320].

The main challenge in CH_4 oxidation lies in the intricate relationship between conversion and selectivity. Literature indicates that high selectivity for CH_2O can be achieved at low conversion and the selectivity for CH_2O decreases along the increase of CH_4 conversion. This interlocking relationship resembles OCM and poses limitations on the potential industrial applications of both OCM and CH_4 oxidation to CH_2O. The decline in CH_2O selectivity at high conversion is linked to multiple reaction channels.

The catalytic oxidation of CH_4 to CH_2O initiates with the formation of a $CH_3\cdot$ radical after the chemisorption of CH_4 onto surface lattice oxygen atoms, as observed in V2O5 supported on silica: $CH_3\cdot + M^{n+}O^{2-} \rightarrow M^{n+}(OCH_3)^-$ [321]. The methoxy group, CH_3O can lose one H atom upon coupling with a surface oxygen atom, forming CH_2O and HO_{surf}. Alternatively, the methoxy group, OCH_3 can also gain a H atom from H_2O, forming a CH_3OH. An HO is formed upon the H_2O molecule loses a H atom. Both processes produce an HO. It will couple with an H atom to form H_2O. The overall reactions of forming CH_2O and CH_3OH can be written as: $2CH_4 + O_2 \rightarrow 2CH_2O + H_2O$ and $2CH_4 + O_2 \rightarrow 2CH_3OH$. Both CH_2O and CH_3OH take O atoms away from the surface, creating surface vacancies. These surface vacancies are refilled with atomic oxygen available through the dissociation of molecular O_2. Although literature raised the possibility that CH_2O is formed from CH_3OH [321, 322], there is no direct evidence supporting that CH_3OH is the intermediate for forming CH_2O. It is concluded that CH_3OH is a byproduct and $O_{surf}CH_3\cdot$ is the common intermediate of CH_2O and CH_3OH. Thus, the $O_{surf}CH_3\cdot$-based mechanism for forming CH_2O is confirmed. This mechanism was indirectly supported by other studies in which $CH_3\cdot$ radicals were assumed to be the intermediate. For instance, other than CH_3OH, C_2H_6, and C_2H_4 are byproducts formed through intermediates $CH_3\cdot$ radicals as well [323].

A crucial aspect of the oxidation of CH_4 to CH_2O ($2\ CH_4 + O_2 \rightarrow 2\ CH_2O + H_2O$) involves identifying the source of oxygen atoms for the products CH_2O and H_2O. Primarily, the molecular O_2 serves as the original source. The preceding paragraph implies that the oxygen atoms in CH_2O and H_2O products directly originate from surface lattice oxygen, ultimately derived from molecular O_2. This holds true for certain catalysts involved in oxidizing CH_4 to produce CH_2O and H_2O. For example, it has been observed that oxygen atoms in Mo–O species of MoO_3/SiO_2 and V_2O_5/SiO_2 readily exchange with oxygen atoms of CH_2O molecules during the oxidation of CH_4 to CH_2O [324, 325]. However, isotope-labeled experiments indicate that $^{18}O_2$ does not undergo exchange with surface lattice oxygen of the Fe-Nb-O catalyst, suggesting a distinct reaction mechanism for the oxidation of CH_4 to form CH_2O [326].

The two most extensively studied catalysts for converting CH_4 to CH_2O are MoO_3/SiO_2 and V_2O_5/SiO_2. The selectivity for CH_2O varies over the time-on-stream and differs concerning the catalysis temperature and Mo concentration. Methanol formation was observed in some studies on MoO_3/SiO_2, while other studies did not detect it [327], likely due to variations in catalyst preparation methods. A distinctive feature of catalysis on MoO_3/SiO_2 is its concentration-dependent catalytic selectivity. For loadings below 5%, the selectivity for CH_2O ranges from 75% to 86%, but it sharply decreases to 30% at a Mo loading of 8% under the same catalytic conditions [328]. No CH_2O is formed at a 15% loading, but ethane and ethylene were observed under this loading.

The active site for CH_4 oxidation to CH_2O using NO_2 was investigated with EPR in the literature [327, 329]. The Mo atoms in an as-prepared catalyst are Mo^{VI}. Upon reduction, Mo^V with 2×10^{18} spins g^{-1} is formed. Upon exposing the Mo^V species to 100 Torr N_2O at 150 °C for 4 hours, the EPR signal of Mo^V completely disappeared, while an O^- spectrum corresponding to 1.4×10^{18} spins g^{-1} was observed. Subsequently, CH_4 was introduced to the catalyst surface consisting of O^- anions at 196 °C. The EPR signal of O^- nearly disappeared [329]. The formation and disappearance of O upon exposure to CH_4 suggests that O^- clearly reacts with CH_4. The formation of methoxy CH_3O was confirmed by IR spectra collected under the same conditions as the EPR experimental setup. In the IR spectrum, in addition to the two sharp peaks at 2918 cm^{-1} and 2857 cm^{-1}, a shoulder at 2959 cm^{-1} was observed, assigned to the formed methoxy group. Thus, the formation of methoxy was confirmed. Regarding the formed methoxy, its role as an intermediate leading to the formation of CH_2O was confirmed by the disappearance of the vibrational signature of the methoxy group when the catalyst surface was exposed to H_2O or heated in a vacuum. The disappearance of the shoulder at 2959 cm^{-1} suggests that CH_3O reacted with water or desorbed from the surface. The literature proposed that O^- instead of O^{2-} participated in the activation of CH_4 to form $CH_3\cdot$ through the following elementary steps [327]. Based on this proposal, the active center is Mo^V.

$$Mo^V + N_2O \rightarrow Mo^{VI}O^- + N_2 \text{ (a)}$$
$$Mo^{VI}O^- + CH_4 \rightarrow Mo^{VI}OH^- + CH_3\cdot \text{ (b)}$$
$$Mo^{VI}O^{2-} + CH_3\cdot \rightarrow Mo^VOCH_3^- \text{ (c)}$$
$$Mo^VOCH_3^- + H_2O \rightarrow Mo^{VI}OH^- + CH_3OH \text{ (d)}$$
$$Mo^{VI}OH^- + Mo^VOH^- \rightarrow Mo^V + Mo^{VI}O^{2-} + H_2O \text{ (e)}$$

CH_2O could be formed through either the loss of an H atom from Mo^VOCH_3 after step (c) or further oxidation of CH_3OH after step (e). The loss of H to form a CH_2O can be expressed with $Mo^VOCH_3^- + Mo^{VI}O^{2-} \rightarrow Mo^{VI} + Mo^VOH^- + CH_2O$. It is suggested that active Mo atoms are highly dispersed or even singly dispersed on silica since MoO_x species are highly diffused, particularly at a high temperature. The high or even single dispersion of Mo atoms on SiO_2 is supported by the observation that MoO_3/SiO_2 with a low loading of Mo exhibits a higher selectivity for CH_2O than MoO_3/SiO_2 with a high loading while the conversions of CH_4 on the two catalysts are similar. [330] Figure 6.31 is a representation of the catalytic cycle of oxidation of CH_4 to form CH_2O on $Mo^{VI}O°$.

To some extent, oxidation mechanism of CH_4 to form CH_2O on V_2O_5/SiO_2 is similar to MoO_3/SiO_2. The active center of a catalytic site is V^{IV}. Different from MoO_3/SiO_2, the oxidation state of V^{IV} atoms does not change.

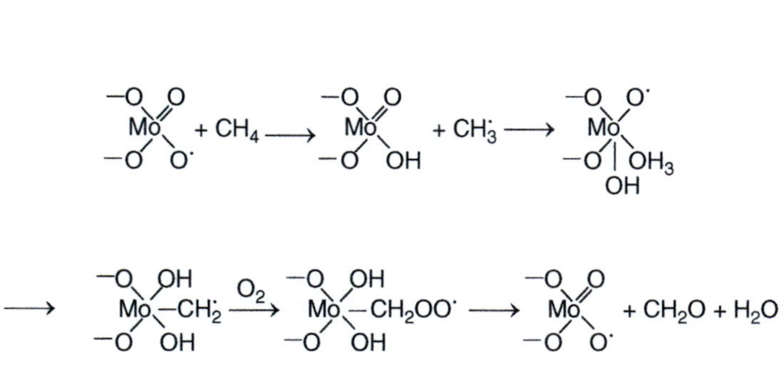

Figure 6.31 Representation of the scheme of the catalytic cycle of partial oxidation of CH_4 to form formaldehyde (CH_2O) on MoO_3. [To be drawn for representing the scheme on p. 10 of Ref 320. It should be shown as a cycle]. *Source:* Reproduced with permission from de [319] / Springer Nature.

References

1 Rostrup-Nielsen, J. R. 1973. "Activity of nickel catalysts for steam reforming of hydrocarbons." *J. Catal.* 31, 173–199.

2 Yamaguchi, A., Iglesia, E. 2010. "Catalytic activation and reforming of methane on supported palladium clusters." *J. Catal.* 274, 52–63.

3 Jones, G., Jakobsen, J. G., Shim, S. S., Kleis, J., Andersson, M. P., Rossmeisl, J., Abildedersen, F., Bligaard, T., Helveg, S., Hinnemann, B., Rostrup-Nielsen, J. R., Chorkendorff, I., Sehested, J., Nørskov, J. K. 2008. "First principles calculations and experimental insight into methane steam reforming over transition metal catalysts." *J. Catal.* 259, 147–160.

4 Bengaard, H. S., Nørskov, J. K., Sehested, J., Clausen, B. S., Nielsen, L. P., Molenbroek, A. M., Rostrup-Nielsen, J. 2002. "Steam reforming and graphite formation on Ni catalysts" *J. Catal.* 209, 365.

5 Rostrup-Nielsen, J., Dybkjaer, I., Christiansen, L. J. 1993. "Steam Reforming Opportunities and Limits of the Technology." *Chemical Reactor Technology for Environmentally Safe Reactors and Products*. Kluwer Academic Publishers. 225, 249–281.

6 Aasberg-Petersen, K., Dybkjær, I., Ovesen, C. V., Schjødt, N. C., Sehested, J., Thomsen, S. G. 2011. "Natural gas to synthesis gas – Catalysts and catalytic processes." *J. Nat. Gas Sci. Eng.* 3, 423–459.

7 Rostrup-Nielsen, J. R. 1984. In: Anderson, J. R., Boudart, M. (Eds.), *Catalysis, Science and Technology*. Berlin, Springer-Verlag. 5, 1–117.

8 Yarlagadda, O. S., Morton, L. A., Hunter, N. R., Gessar, H. D. 1990. "Temperature oscillations during the high-pressure partial oxidation of methane in a tubular flow reactor." *Combust. Flame* 79, 216–218.

9 Crabtree, R. H. 1994. "Current ideas and future prospects in metal-catalyzed methane conversion." *Stud. Surf. Sci. Catal.* 81, 85–92.

10 Crabtree, R. H. 1995. "Aspects of methane chemistry." *Chem. Rev.* 95, 987–1007.

11 Lange, J. P. 1997. "Perspectives for manufacturing methanol at fuel value." *Ind. Eng. Chem. Res.* 36, 4282–4290.

12 Rostrup-Nielsen, J. R. 2002. "Syngas in perspective." *Catal. Today* 71, 243–247.

13 Sehested, J. 2003. "Sintering of nickel steam-reforming catalysts." *J. Catal.* 217, 417–426.

14 Ouyang, R., Liu, J. X., Li, W. X. 2013. "Atomistic theory of Ostwald ripening and disintegration of supported metal particles under reaction conditions." *J. Am. Chem. Soc.* 135, 1760–1771.

15 Chen, Y., deGlee, B., Tang, Y., Wang, Z., Wei, Y., Zhang, L., Yoo, S., Zhao, B., Hu. P., Tao, F., Liu, M. 2018. "A robust low-cost fuel cell operated on 'dry' methane at 500°C." *Nat. Energy.* 3, 1042–1050.

16 Tang, Y., Wei, Y., Wang, Z., Zhang, S., Li, Y., Nguyen, L., Li, Y., Zhou, Y., Shen, W., Hu, P., Tao, F. 2019. "Synergy effect of two sets of single-atom sites (Ni_1 and Ru_1) on catalyst surface for reforming CH_4." *J. Am. Chem. Soc.* 141, 7283–7293.

17 Seets, D. C., Reeves, C. T., Ferguson, B. A., Wheeler, M. C., Mullins, C. B. 1997. "Dissociative chemisorption of methane on Ir(111): Evidence for direct and trapping-mediated mechanisms." *J. Chem. Phys.* 107, 10229–10241.

18 German, E. D., Sheintuch, M. 2013. "Predicting CH4 dissociation kinetics on metals: Trends, sticking coefficients, H tunneling, and kinetic isotope effect." *J. Phys. Chem. C* 117, 22811–22826.

19 Hu, Y. H., Ruckenstein, E. 2004. "Catalytic conversion of methane to synthesis gas by partial oxidation and CO2 reforming." *Adv. Catal.* 48, 297–345.

20 Rostrup-Nielsen, J. R., Sehested, J., Norskov, J. K. 2002. "Hydrogen and synthesis gas by steam- and CO2 reforming." *Adv. Catal.* 47, 65–139.

21 Snoeck, J.-W., Froment, G. F., Fowles, M. 2002. "Steam/CO_2 reforming of methane. Carbon filament formation by the Boudouard reaction and gasification by CO_2, by H_2, and by steam: Kinetic study." *Ind. Eng. Chem. Res.* 41, 4252–4265.

22 Bradford, M. C. J., Vannice, M. A. 1999. "CO_2 reforming of CH_4." *Catal. Rev. Sci. Eng.* 41, 1–42.

23 Bradford, M. C. J., Vannice, M. A. 1999. "The role of metal–support interactions in CO_2 reforming of CH_4." *Catal. Today.* 50, 87–96.

24 Bradford, M. C. J., Vannice, M. A. 1996. "Catalytic reforming of methane with carbon dioxide over nickel catalysts I. Catalyst characterization and activity." *Appl. Catal. A* 142, 73–96.

25 Aartun, I., Gjervan, T., Venvik, H., Gorke, O., Pfeifer, P., Fathi, M., Holmen, A., Schubert, K. 2004. "Catalytic conversion of propane to hydrogen in microstructured reactors." *Chem. Eng. J.* 101, 93–99

26 Nielsen, B. O., Luntz, A. C., Holmblad, P. M., Chorkendorff, I. 1995. "Activated dissociative chemisorption of methane on Ni(100): A direct mechanism under thermal conditions?" *Catal. Lett.* 32, 15–30.

27 Horiuchi, T., Sakuma, K., Fukui, T., Kubo, Y., Osaki, T., Mori, T. 1996. "Suppression of carbon deposition in the CO_2-reforming of CH_4 by adding basic metal oxides to a Ni/Al2O3 catalyst." *Appl. Catal. A* 144, 111–120.

28 Ahmed, K., Foger, K. 2000. "Kinetics of internal steam reforming of methane on Ni/YSZ-based anodes for solid oxide fuel cells." *Catal. Today* 63, 479–487.

29 Beebe, T. P., Goodman, D. W., Kay, B. D., Yates, J. T. 1987. "Kinetics of the activated dissociative adsorption of methane on the low index planes of nickel single crystal surfaces." *J. Chem. Phys.* 87, 2305–2315.

30 Yang, H., Whitten, J. L.1992. "Dissociative chemisorption of CH_4 on Ni(111)." *J. Chem. Phys.* 96, 5529–5537.

31 Chorkendorff, I., Alstruo, I., Ullmann, S. 1990. "XPS study of chemisorption of CH_4 on Ni(100)." *Surf. Sci.* 227, 291–206.

32 Jansen, A. P. J., Burghgraef, H. 1995. "MCTDH study of CH_4 dissociation on Ni(111)." *Surf. Sci.* 344, 149–158.

33 Freund, H.-J., Roberts, M. W. 1996. "Surface chemistry of carbon dioxide." *Surf. Sci. Rep.* 25, 225–273.

34 Wang, S. G., Liao. X. Y., Hu, J., Cao, D. B., Li, Y. W., Wang, J., Jiao, H. 2007. "Kinetic aspect of CO_2 reforming of CH4 on Ni(111): A density functional theory calculation." *Surf. Sci.* 601, 1271–1284.

35 Padovani, C., Franchetti, P. 1933. "Incomplete oxidation of methane with oxygen and air." *Giorn. Chem. Ind. Appl.* 15, 429.

36 Olsbye, U., Tangstad, E., Dahl, I. M. 1994. "Partial oxidation of methane to synthesis gas in a fluidized bed reactor." *Stud. Surf. Sci. Catal.* 81, 303–308.

37 Marschall, K. -J., Mleczko, L. 1999. "Short-contact-time reactor for catalytic partial oxidation of methane." *Ind. Eng. Chem. Res.* 38, 1813–1821.

38 Ji, Y., Li, W., Xu, H., Chen, Y. 2001. "Catalytic partial oxidation of methane to synthesis gas over Ni/γ-Al_2O_3 catalyst in a fluidized-bed." *Appl. Catal. A* 213 (2001), 25–31.

39 Ji, Y., Li, W., Xu, H., Chen, Y. 2001. "A study of carbon deposition on catalysts during the catalytic partial oxidation of methane to syngas in a fluidized bed." *React. Kinet. Catal. Lett.* 73, 27–32.

40 Bharadwaj, S. S., Schmidt, L. D. 1994. "Synthesis gas formation by catalytic oxidation of methane in fluidized bed reactors." *J. Catal.* 146, 11–21.

41 Choudhary, V. R., Uphade, B. S., Mamman, A. S. 1995. "Large enhancement in methane-to-syngas conversion activity of supported Ni catalysts due to precoating of catalyst supports with MgO, CaO or rare-earth oxide." *Catal. Lett.* 32, 387–390.

42 Ayabe, S., Omoto, H., Utaka, T., Kikuchi, R., Sasaki, K., Teraoka, Y., Eguchi, K. 2003. "Catalytic autothermal reforming of methane and propane over supported metal catalysts." *Appl. Catal. A* 241, 261–269.

43 Li, B., Maruyama, K., Nurunnabi, M., Kunimori, K., Tomishige, K. 2004. "Temperature profiles of alumina-supported noble metal catalysts in autothermal reforming of methane." *Appl. Catal. A* 275, 157–172.

44 Hagh, B. F. 2003. "Optimization of autothermal reactor for maximum hydrogen production." *Int. J. Hydrogen Energy* 28,1369–1377.

45 Chan, S. H., Wang, H. M. 2000. "Effect of natural gas composition on autothermal fuel reforming products." *Fuel Process. Technol.* 64, 221–239.

46 Glockler, B., Kolios, G., Eigenberger, G. 2003. "Analysis of a novel reverse-flow reactor concept for autothermal methane steam reforming." *Chem. Eng. Sci.* 58, 593–601.

47 Lutz, A. E., Bradshaw, R. W., Bromberg, L., Rabinovich, A. 2004. "Thermodynamic analysis of hydrogen production by partial oxidation reforming." *Int. J. Hydrogen Energy* 29, 809–816.

48 Souzaa, M. M. V. M., Schmal, M. 2005. "Autothermal reforming of methane over Pt/ZrO_2/Al_2O_3 catalysts." *Appl. Catal. A* 281, 19–24.

49 Lee, S. H. D., Applegate, D. V., Ahmed, S., Calderone, S. G., Harvey, T. L. 2005. "Hydrogen from natural gas: Part I—Autothermal reforming in an integrated fuel processor." *Int. J. Hydrogen Energy* 30, 829–842.

50 Li, B., Watanabe, R., Maruyama, K., Kunimori, K., Tomishige, K. 2005. "Thermographical observation of catalyst bed temperature in oxidative steam reforming of methane over Ni supported on α-alumina granules: Effect of Ni precursors." *Catal. Today* 104, 7–17.

51 Mukainakano, Y., Li, B., Kado, S., Miyazawa, T., Okumura, K., Miyao, T., Naito, S., Kunimori, K., Tomishige, K. 2007. "Surface modification of Ni catalysts with trace Pd and Rh for oxidative steam reforming of methane." *Appl. Catal. A* 318, 252–264.

52 Roh, H. S., Jun, K. W., Dong, W. S., Park, S. E., Baek, Y. S. 2001. "Highly stable Ni catalyst supported on Ce–ZrO2 for oxy-steam reforming of methane." *Catal. Lett.* 74, 31–36.

53 Puolakka, K. J., Krause, A. O. I. 2007. "Combined CO_2 reforming and partial oxidation of fuel compounds." *Catal. Lett.* 116, 87–93.

54 Ruckenstein, E., Hu, Y. H. 1998. "Combination of CO_2 reforming and partial oxidation of methane over NiO/MgO solid solution catalysts." *Ind. Eng. Chem. Res.* 37, 1744–1747.

55 Wang, H. Y., Ruckenstein, E. 2001. "Conversions of methane to synthesis gas over Co/γ-Al_2O_3 by CO2 and/or O2." *Catal. Lett.* 75, 13–18.

56 Reyes, S. C., Sinfelt, J. H., Feeley, J. S. 2003. "Evolution of processes for synthesis gas production: Recent developments in an old technology." *Ind. Eng. Chem. Res.* 42, 1588–1597.

57 Dissanayake, D., Rosynek, M. P., Kharas, K. C. C., Lunsford, J. H. 1991. "Partial oxidation of methane to carbon monoxide and hydrogen over a Ni/Al_2O_3 catalyst." *J. Catal.* 132, 117–127.

58 Vermeiren, W. J. M., Blomsma, E., Jacobs, P. A. 1992. "Catalytic and thermodynamic approach of the oxyreforming reaction of methane." *Catal. Today* 13, 427–436.

59 Claridge, J. B., Green, M. L. H., Tsang, S. C., York, A. P. E., Ashcroft, A. T., Battle, P. D. 1993. "A study of carbon deposition on catalysts during the partial oxidation of methane to synthesis gas." *Catal. Lett.* 22, 299–305.

60 Dias, J. A. C., Assaf, J. M. 2004. "Autothermal reforming of methane over Ni/γ- Al_2O_3 catalysts: The enhancement effect of small quantities of noble metals." *J. Power Sources* 130, 106–110.

61 Marnasidou, K. G., Voutetakis, S. S., Tjatjopoulos, G. J., Vasalos, I. A. 1999. "Catalytic partial oxidation of methane to synthesis gas in a pilot-plant-scale spouted-bed reactor." *Chem. Eng. Sci.* 54, 3691–3699.

62 Ruckenstein, E., Wang, H. Y. 2000. "Carbon dioxide reforming of methane to synthesis gas over supported cobalt catalysts." *Appl. Catal. A.* 204, 257–263.

63 Mo, L., Zheng, X., Chen, Y., Fei, J. 2003. "Combination of CO_2 reforming and partial oxidation of CH4 over Ni/Al_2O_3 catalysts using fluidized bed reactor." *React. Kinet. Catal. Lett.* 78, 233–242.

64 Puolakka, K. J., Juutilainen, S., Krause, A. O. I. 2006. "Combined CO_2 reforming and partial oxidation of n-heptane on noble metal zirconia catalysts." *Catal. Today* 115, 217–211.

65 Slagtern, A., Olsbye, A. 1994. "Partial oxidation of methane to synthesis gas using La-M-O catalysts." *Appl. Catal. A* 110, 99–108.

66 Marschall, K. -J., Mleczko, L. 1999. "Experimental investigations of catalytic partial oxidation of methane to synthesis gas in various types of fluidized-bed reactors." *Chem. Eng. Technol.* 21, 31–37.

67 Jin, R., Chen, Y., Li, W., Cui, W., Cui, W., Ji, Y., Yu, C., Jiang, Y. 2013. "Mechanism for catalytic partial oxidation of methane to syngas over a Ni/Al_2O_3 catalyst." *Appl. Catal. A* 201, 71–80.

68 Zhu, T., Flytzani-Stephanopoulos, M. 2001. "Catalytic partial oxidation of methane to synthesis gas over Ni–CeO_2." *Appl. Catal. A* 208, 403–417.

69 Hu, Y. H., Ruckenstein, E. 1998. "Catalyst temperature oscillations during partial oxidation of methane." *Ind. Eng. Chem. Res.* 37, 2333–2335.

70 Feng, W., Ji, P., Zheng, D., Tan, T., de Swaan Arons, J. 2005. "Simulation and thermodynamic analysis of an integrated process with a two-membrane catalytic partial oxidation (CPO) reactor for producing pure hydrogen." *Ind. Eng. Chem. Res.* 44, 9191–9198.

71 Enger, B. C., Lødeng, R., Holmen, A. 2007. "Catalytic partial oxidation of CH_4 to H_2 over cobalt catalysts at moderate temperatures." *Appl. Catal. A.* 333, 11–23.

72 Ostrowski, T., Giroir-Fendler, A., Mirodatos, C., Mleczko, L. 1998. "Comparative study of the catalytic partial oxidation of methane to synthesis gas in fixed-bed and fluidized-bed membrane reactors: Part I: A modeling approach." *Catal. Today* 40, 181–190.

73 Swaan, H. M., Rouanet, R., Widyananda, P., Mirodatos, C. 1997. "Partial oxidation of methane over nickel- and cobalt-based catalysts." *Stud. Surf. Sci. Catal.* 107, 447–453.

74 Nishimoto, H., Nakagawa, K., Ikenaga, K., Nishitani-Gamo, M., Ando, T., Suzuki, T. 2004. "Partial oxidation of methane to synthesis gas over oxidized diamond catalysts." *Appl. Catal. A* 264, 65–72

75 Koh, A. C. W., Chen, L., Leong, W. K., Johnson, B. F. G., Khimyak, T., Lin, J. 2007. "Hydrogen or synthesis gas production via the partial oxidation of methane over supported nickel–cobalt catalysts." *Int. J. Hydrogen Energy* 32, 725–730.

76 Steghuis, A. G., van Ommen, J. G., Seshan, K., Lercher, J. A. 1997. "New highly active catalysts in direct partial oxidation of methane to synthesis gas." *Stud. Surf. Sci. Catal.* 107, 403–408.

77 Pompeo, F., Nichio, N. N., Ferritti, O. A., Resasco, D. 2005. "Study of Ni catalysts on different supports to obtain synthesis gas." *Int. J. Hydrogen Energy* 30, 1399–1405.

78 Choudhary, V. R., Rajput, A. M., Prabhakar, B. 1992. "Low temperature oxidative conversion of methane to syngas over NiO-CaO catalyst." *Catal. Lett.* 15, 363–370.

79 Basile, F., Fornasari, G., Trifiro, F., Vaccari, A. 2001. "Partial oxidation of methane: Effect of reaction parameters and catalyst composition on the thermal profile and heat distribution." *Catal. Today* 64, 21–30.

80 Takehira, K., Shishido, T., Shoro, D., Murakami, K., Honda, M., Kawabata, T., Takaki, K. 2004. "Preparation of egg-shell type Ni-loaded catalyst by adopting "Memory Effect" of Mg–Al hydrotalcite and its application for CH4 reforming." *Catal. Commun.* 5, 209–213.

81 Takehira, K., Shishido, T., Wang, P., Kosaka, T., Takaki, K. 2004. "Autothermal reforming of CH_4 over supported Ni catalysts prepared from Mg–Al hydrotalcite-like anionic clay." *J. Catal.* 221, 43–54.

82 Morioka, H., Shimizu, Y., Sukenobu, M., Ito, K., Tanabe, E., Shishido, T., Takehira, K. 2001. "Partial oxidation of methane to synthesis gas over supported Ni catalysts prepared from Ni–Ca/Al-layered double hydroxide." *Appl. Catal. A* 215, 11–19.

83 Lee, K. M., Lee, W. Y. 2002. "Partial oxidation of methane to syngas over calcined Ni–Mg/Al layered double hydroxides." *Catal. Lett.* 83, 65–70.

84 Shishido, T., Sukenobu, M., Morioka, H., Kondo, M., Wang, Y., Takaki, K., Takehira, K. 2002. "Partial oxidation of methane over Ni/Mg-Al oxide catalysts prepared by solid phase crystallization method from Mg-Al hydrotalcite-like precursors." *Appl. Catal. A* 223, 35–42.

85 Arpentinier, P., Basile, F., del Gallo, G., Fornasari, P., Gary, D., Rosetti, V., Vaccari, A. 2006. "New catalysts with low amounts of active phase for CPO processes." *Catal. Today* 117, 462–467.

86 Tsipouriari, V. A., Zhang, Z., Verykios, X. E. 1998. "Catalytic partial oxidation of methane to synthesis gas over Ni-based catalysts: I. Catalyst performance characteristics." *J. Catal.* 179, 283–291.

87 Tsipouriari, V. A., Verykios, X. E., 1998. "Catalytic partial oxidation of methane to synthesis gas over Ni-based catalysts: II. Transient, FTIR, and XRD measurements." *J. Catal.* 179, 292–299.

88 Tang, S., Lin, J., Tan, K. L. 1998. " Partial oxidation of methane to syngas over Ni/MgO, Ni/CaO and Ni/CeO2." *Catal. Lett.* 51, 169–175.

89 Choudhary, V. R., Mamman, A. S. 1999. "Simultaneous oxidative conversion and CO2 or steam reforming of methane to syngas over CoO–NiO–MgO catalyst." *J. Chem. Technol. Biotechnol.* 73, 345–350.

90 Shamsi, A. 2006. "Partial oxidation and dry reforming of methane over Ca/Ni/K(Na) catalysts." *Catal. Lett.* 109, 189–193.

91 Choudhary, V. R., Rajput, A. M., Mamman, A. S. 1998. "NiO-Alkaline earth oxide catalysts for oxidative methane-to-syngas conversion: Influence of alkaline earth oxide on the surface properties and temperature-programmed reduction/reaction by H2 and methane." *J. Catal.* 178, 576–585.

92 Tsyganok, A. I., Inaba, M., Tsunoda, T., Suzuki, K., Takehira, K., Hayakawa, T. 2004. "Combined partial oxidation and dry reforming of methane to synthesis gas over noble metals supported on Mg–Al mixed oxide." *Appl. Catal. A* 275, 149–155.

93 O'Connor, A. M., Ross, J. R. H. 1998. "The effect of O_2 addition on the carbon dioxide reforming of methane over Pt/ZrO_2 catalysts." *Catal. Today* 46, 203–210.

94 Choudhary, V. R., Rane, V. H., Rajput, A. M. 1993. "Selective oxidation of methane to CO and H_2 over unreduced NiO-rare earth oxide catalysts." *Catal. Lett.* 22, 289–297.

95 Steghuis, A.G., van Ommen, J.G., Lercher, J.A. 1998. "On the reaction mechanism for methane partial oxidation over yttria/zirconia." *Catal. Today* 46, 91–97.

96 Bi, X.-j., Hong, P.-j., Xie, X.-g., Dai, S.-s. 1999. "Microwave effect on partial oxidation of methane to syngas." *React. Kinet. Catal. Lett.* 66, 381–386.

97 Tsipouriari, V. A., Verykios, X. E.1999. "Carbon and oxygen reaction pathways of CO_2 reforming of methane over Ni/La_2O_3 and Ni/Al_2O_3 catalysts studied by isotopic tracing techniques." *J. Catal.* 187, 85–94.

98 Au, C. T., Hu, Y. H., Wan, H. L. 1996. "Methane activation over unsupported and La_2O_3-supported copper and nickel catalysts." *Catal. Lett.* 36, 159–163.

99 Hu, Y. H., Ruckenstein, E. 1995. "Pulse-MS study of the partial oxidation of methane over Ni/La_2O_3 catalyst." *Catal. Lett.* 34, 41–50.

100 Hu, Y. H., Ruckenstein, E. 1996. "Transient kinetic studies of partial oxidation of CH_4." *J. Catal.* 158, 260–266.

101 Ji, Y., Li, W., Xu, H., Chen, Y. 2001. "A study on the ignition process for the catalytic partial oxidation of methane to synthesis gas by MS-TPSR technique." *Catal. Lett.* 71, 45–48.

102 Choudhary, V. R., Prabhakar, B., Rajput, A. M. 1995. "Beneficial effects of noble metal addition to Ni/Al_2O_3 catalyst for oxidative methane-to-syngas conversion." *J. Catal.* 157, 752–754.

103 Dias, J. A. C., Assaf, J. M. 2008. "Autothermal reforming of methane over Ni/γ-Al_2O_3 promoted with Pd: The effect of the Pd source in activity, temperature profile of reactor and in ignition." *Appl. Catal. A* 334, 243–250.

104 Li, B., Kado, S., Mukainakano, Y., Nurunnabi, M., Miyao, T., Naito, S., Kunimori, K., Tomishige, K. 2006. "Temperature profile of catalyst bed during oxidative steam reforming of methane over Pt-Ni bimetallic catalysts." *Appl. Catal. A* 304, 62–71.

105 Rodrigues, L. M. T. S., Silva, R. B., Rocha, M. G. C., Bargiela, P., Noronha, F. B., Brandão, S. T. 2012. "Partial oxidation of methane on Ni and Pd catalysts: Influence of active phase and CeO_2 modification." *Catal. Today* 197, 137–142.

106 Choudhary, V.R., Prabhakar, B., Rajput, A.M., Mamman, A.S. 1998. "Oxidative conversion of methane to CO and H_2 over Pt or Pd containing alkaline and rare earth oxide catalysts." *Fuel* 77, 1477–1481.

107 Au, C. T., Liao, M. S., Ng, C. F. 1998. "A detailed theoretical treatment of the partial oxidation of methane to syngas on transition and coinage metal (M) catalysts (M = Ni, Pd, Pt, Cu)." *J. Phys. Chem. A* 102, 3959–3969.

108 Gonzalez, M., Nichio, N. N., Moraweck, B., Martin, G. 2000. "Role of chromium in the stability of Ni/Al_2O_3 catalysts for natural gas reforming." *Mater. Lett.* 45, 15–18

109 Lanza, R., Järås, S. G., Canu, P. 2007. "Partial oxidation of methane over supported ruthenium catalysts." *Appl. Catal. A: Gen.* 325, 57–67.

110 Liu, X.-z., Wang, J.-g., Liu, C.-j., He, F., Eliasson, B. 2003. "Partial oxidation of methane to syngas over $Ni\text{-}Fe/Al_2O_3$ catalyst with plasma enhanced activity." *React. Kinet. Catal. Lett.* 79, 69–76.

111 Basile, F., Benito, P., Fornasari, G., Monti, M., Scavetta, E., Tonelli, D., Vaccari, A. 2010. "Novel Rh-based structured catalysts for the catalytic partial oxidation of methane." *Catal. Today* 157, 183–190.

112 Basile, F., Fornasari, G., Trifiro, F., Vaccari, A. 2002. "Rh–Ni synergy in the catalytic partial oxidation of methane: surface phenomena and catalyst stability." *Catal. Today* 77, 215–223.

113 Yan, Q. G., Wu, T. H.,Weng, W. Z., Toghiani, H., Toghiani, R. K., Wan, H. L., Pittman Jr., C. U. 2004. "Partial oxidation of methane to H_2 and CO over Rh/SiO_2 and Ru/SiO_2 catalysts." *J. Catal.* 226, 247–259.

114 Tsang, S. C., Claridge, J. B., Green, M. L. H. 1995. "Recent advances in the conversion of methane to synthesis gas." *Catal. Today* 23, 3–15.

115 Dissanayake, D., Rosynek, M. P., Lunsford, J. H. 1993. "Are the equilibrium concentrations of carbon monoxide and hydrogen exceeded during the oxidation of methane over a nickel/ytterbium oxide catalyst?" *J. Phys. Chem.* 97, 3644–3646.

116 Pantaleo, G., La Parola, V., Deganello, F., Singha, R. K., Bal, R., Venezia, A. M. 2016. "Ni/CeO_2 catalysts for methane partial oxidation: Synthesis driven structural and catalytic effects." *Appl. Catal. B* 189, 233–241.

117 Santos, A., Menendez, M., Santamarı, J. 1994. "Partial oxidation of methane to carbon monoxide and hydrogen in a fluidized bed reactor." *Catal. Today* 21, 481–488.

118 Dantas, S. C., Escritori, J. C., Soares, R. R., Hori, C. E. 2010. "Effect of different promoters on $Ni/CeZrO_2$ catalyst for autothermal reforming and partial oxidation of methane." *Chem. Eng. J.* 156, 380–387.

119 Requies, J., Cabrero, M. A., Barrio, V. L., Güemez, M. B., Cambra, J. F., Arias, P. L., Pérez-Alonso, F. J., Ojeda, M., Peña, M. A., Fierro, J. L. G. 2005. "Partial oxidation of methane to syngas over Ni/MgO and Ni/La_2O_3 catalysts." *Appl. Catal. A* 289, 214–223.

120 Wu, T., Yan, Q., Wan, H. 2005. "Partial oxidation of methane to hydrogen and carbon monoxide over a Ni/TiO_2 catalyst." *J. Mol. Catal. A* 226, 41–48.

121 Choudhary, V. R., Mamman, A. S. 1999. "Oxidative conversion of methane to syngas over NiO/MgO solid solution supported on low surface area catalyst carrier." *Fuel Process. Technol.* 60, 203–211.

122 Ruckenstein, E., Hu, Y. H. 1999. "Methane partial oxidation over NiO/MgO solid solution catalysts." *Appl. Catal. A* 183, 85–92.

123 Nichio, N., Casella, M., Ferretti, O., Gonza´lez, M., Nicot, C., Moraweck, B., Frety, R. 1996. "Partial oxidation of methane to synthesis gas. Behaviour of different Ni supported catalysts." *Catal. Lett.* 42, 65–72.

124 Zhang, Y., Xiong, G., Yang, W., Sheng, S. 2000. "Partial oxidation of methane to syngas over $NiO/\gamma\text{-}Al_2O_3$ catalysts prepared by the sol-gel method." *Stud. Surf. Sci. Catal.* 136, 21–26.

125 Zhang, Y., Xiong, G., Sheng, S., Yang, W. 2000. "Deactivation studies over NiO/γ-Al_2O_3 catalysts for partial oxidation of methane to syngas." *Catal. Today* 63, 517–522.

126 Zhang, Y., Xiong, G., Kou, Y., Yang, W., Sheng, S. 2000. "Influence of the sol-gel method on a NiO/Al_2O_3 catalyst for CH_4/O_2 to syngas reaction." *React. Kinet. Catal. Lett.* 69, 325–329.

127 Basile, F., Basini, L., Fornasari, G., Guarinoni, A., Trifiro, F., Vaccari, A.1998. "Catalytic behaviour of Ni-and Rh-containing catalysts in the partial oxidation of methane at short residence times." *Stud. Surf. Sci. Catal.* 119, 693–698.

128 Cao, L., Chen, Y., Li, W. 1997. "Effect of La_2O_3 added to NiO/Al_2O_3 catalyst on partial oxidation of methane to syngas." *Stud. Surf. Sci. Catal.* 107, 467–471.

129 Cesar, D. V., Baldanza, M. A. S., Henriques, C. A., Pompeo, F., Santori, G., Múnera, J., Lombardo, E., Schmal, M., Cornaglia, L., Nichio, N. 2013. "Stability of Ni and Rh–Ni catalysts derived from hydrotalcite-like precursors for the partial oxidation of methane." *Int. J. Hydrogen Energy* 38, 5616–5626.

130 Bshish, A., Yaakob, Z., Ebshish, A., Alhasan, F. H. 2013. "Hydrogen production via ethanol steam reforming over Ni/Al_2O_3 catalysts: Effect of Ni loading." *J. Energy Source Technol.* 136, 012601.

131 Hu, Y. H., Ruckenstein, E. 1998. "Isotopic GCMS study of the mechanism of methane partial oxidation to synthesis gas." *J. Phys. Chem. A* 102,10568–10571.

132 Hu, Y. H., Ruckenstein, E. 1998. "Broadened pulse-step change–isotopic sharp pulse analysis of the mechanism of methane partial oxidation to synthesis gas." *J. Phys. Chem. B* 102, 230–233.

133 Choudhary, V. R., Rajput, A. M., Rane, V. H. 1992. "Low temperature oxidative conversion of methane to synthesis gas over Co/rare earth oxide catalysts." *Catal. Lett.* 16, 269–272.

134 Choudhary, V. R., Sansare, S. D., Mamman, A. S. 1992. "Low-temperature selective oxidation of methane to carbon monoxide and hydrogen over cobalt-MgO catalysts." *Appl. Catal. A* 90, L1–L5.

135 Choudhary, V. R., Mondal, K. C., Choudhary, T. V. 2007. "Oxy-methane reforming over high temperature stable NiCoMgCeOx and NiCoMgOx supported on zirconia–haffnia catalysts: Accelerated sulfur deactivation and regeneration." *Catal. Commun.* 8, 561–564.

136 Choudhary, V. R., Rajput, A. M., Prabhakar, B., Mamman, A. S. 1998. "Partial oxidation of methane to CO and H2 over nickel and/or cobalt containing ZrO_2, ThO_2, UO_2, TiO_2 and SiO_2 catalysts." *Fuel* 77, 1803–1807.

137 Sato, K., Fujita, S., Suzuki, K., Mori, T. 2007. "High performance of Ni-substituted calcium aluminosilicate for partial oxidation of methane into syngas." *Catal. Commun.* 8 (2007), 1735.

138 Slagtern, A., Swaan, H. M., Olsbye, U., Dahl, I. M., Mirodatos, C.1998. "Catalytic partial oxidation of methane over Ni-, Co- and Fe-based catalysts." *Catal. Today* 46, 107–115.

139 Weng, T., Wolf, E. E.1993. "Partial oxidation of methane on Mo/Sn/P silica supported catalysts." *Appl. Catal. A.* 96, 383–396.

140 Sokolovskii, V. D., Coville, N. J., Parmaliana, A., Eskendirov, I., Makoa, M. 1998. "Methane partial oxidation. Challenge and perspective." *Catal. Today* 42, 191–195.

141 Lødeng, R., Bjørgum, E., Enger, B. C., Eilertsen, J. L., Holmen, A., Krogh, B., Rønnekleiv, M., Rytter, E. 2007. "Catalytic partial oxidation of CH_4 to H_2 over cobalt catalysts at moderate temperatures." *Appl. Catal. A* 233 (2007), 11.

142 Yang, S., Kondo, J. N., Hayashi, K., Hirano, M., Domen, K., Hosono, H. 2004. "Partial oxidation of methane to syngas over promoted C12A7." *Appl. Catal. A* 277, 239–246.

143 Sokolovskii, V. D., Jeannot, J. C., Coville, N. J., Glasser, D., Hildebrandt, D., Makao, M. 1997. "High yield syngas formation by partial oxidation of methane over Co-alumina catalysts." *Stud. Surf. Sci. Catal.* 107, 461–467.

144 Figen, H. E., Baykara, S. Z. 2015. "Hydrogen production by partial oxidation of methane over Co based, Ni and Ru monolithic catalysts." *Int. J. Hydrogen Energy* 40, 7439–7451.

145 Tang, S., Lin, J., Tan, K. L. 1999. "Partial oxidation of methane to synthesis gas over α-Al_2O_3-supported bimetallic Pt–Co catalysts." *Catal. Lett.* 59,129–135.

146 Mo, L., Zheng, X., Huang, C., Fei, J. 2002. "A novel catalyst $Pt/CoAl_2O_4/Al_2O_3$ for combination CO_2 reforming and partial oxidation of CH_4." *Catal. Lett.* 80, 165–169.

147 Liu, R., Yang, M., Huang, C., Weng, W., Wan, H. 2013. "Partial oxidation of methane to syngas over mesoporous Co-Al_2O_3 catalysts." *Chinese J. Catal.* 34, 146–151.

148 Moral, A., Reyero, I., Llorca, J., Bimbela, F., Gandía, L. M. 2019. "Partial oxidation of methane to syngas using Co/Mg and Co/Mg-Al oxide supported catalysts." *Catal. Today* 333, 259–267.

149 Iwasaki, N.-o., Miyake, T., Yagasaki, E., Suzuki, T. 2006. "Partial oxidation of ethane to synthesis gas over Co-loaded catalysts." *Catal. Today* 111, 391–397.

150 Zhu, J., Rahuman, M. S. M. M., van Ommen, J. G., Lefferts, L. 2004. "Dual catalyst bed concept for catalytic partial oxidation of methane to synthesis gas." *Appl. Catal. A* 259, 95–100.

151 Ruckenstein, E., Wang, H. Y. 2001. "Combined catalytic partial oxidation and CO_2 reforming of methane over supported cobalt catalysts." *Catal. Lett.* 73, 99–105.

152 Choudhary, V. R., Mondal, K. C., Mamman, A. S. 2005. "High-temperature stable and highly active/selective supported NiCoMgCeOx catalyst suitable for autothermal reforming of methane to syngas." *J. Catal.* 233, 36–40.

153 Naito, S., Tanaka, H., Kado, S., Miyao, T., Naito, S., Okumura, K., Kunimori, K., Tomishige, K. 2008. "Promoting effect of Co addition on the catalytic partial oxidation of methane at short contact time over a Rh/MgO catalyst." *J. Catal.* 259, 138–146.

154 Wang, H. Y., Ruckenstein, E. 2001. "Partial oxidation of methane to synthesis gas over alkaline Earth metal oxide supported cobalt catalysts." *J. Catal.* 199, 309–317.

155 Chuenjai, M., Wongsakulphasatch, S., Yong, N., Maneeprakorn, W., Sudsakorn, K., Tongnan, V., Hartley, U. W., Ratchahat, S., Kiatkittipong, W., Assabumrungrat, A. 2022. "Synthesis of $NiO/MgO/ZrO_2$ catalyst for syngas production from partial oxidation and dry reforming of biogas." *Int. J. Hydrogen Energy* 47, 41386–41396.

156 Veselov, G. B., Stoyanovskii, V. O., Cherepanova, S. V., Vedyagin, A. A. 2023. "Sol–gel prepared Co–Mg–O oxide systems: Redox behavior, thermal stability and catalytic performance in CO oxidation." *React. Kinet. Mech. Catal.* 136, 233–250.

157 Makoa, M. P., Coville, N. J., Sokolovskii, V. D. 1999. "Characterisation of $Co/Mg/Al_2O_3$ high pressure partial oxidation catalysts." *Catal. Today* 49, 11–16.

158 Nyathi, T. M., Fischer, N., York, A. P. E., Morgan, D. J., Hutchings, G. J., Gibson, E. K., Wells, P. P., Catlow, C. R. A., Claeys, M. 2019. "Impact of nanoparticle–support interactions in Co_3O_4/Al_2O_3 catalysts for the preferential oxidation of carbon monoxide." *ACS Catal.* 9, 7166–7178.

159 Yu, C. L., Zhou, X. C., Weng, W. Z., Hu, J. B., Chen, X. R., Wei, L. F. 2012. "Effects of alkaline-earth strontium on the performance of Co/Al_2O_3 catalyst for methane partial oxidation." *J. Fuel Chem. Technol.* 40, 1222–1229

160 Yu, C., Weng, W., Shu, Q., Meng, X., Zhang, B., Chen, X., Zhou, X. 2011. "Additive effects of alkaline-earth metals and nickel on the performance of Co/γ-Al_2O_3 in methane catalytic partial oxidation." *J. Nat. Gas Chem.* 20, 135–139.

161 Hong, J. H., Ha, H. J., Han, J. D. 2012. "The promotion effects on partial oxidation of methane for hydrogen production over Co/Al_2O_3 and Ni/Al_2O_3 catalysts." *Clean Technol.* 18, 95–101.

162 Zavyalova, U., Scholz, P., Ondruschka, B. 2007. "Influence of cobalt precursor and fuels on the performance of combustion synthesized Co3O4/γ-Al2O3 catalysts for total oxidation of methane." *Appl. Catal. A* 323, 226–233.

163 Stefanov, P., Todorov, S., Naydenov, A., Tzaneva, B., Kolev, H., Atanasova, G., Stoyanova, D., Karakirova, Y., Aleksieva, K. 2015. On the development of active and stable Pd-Co/γ-Al_2O_3 catalyst for complete oxidation of methane. *Chem. Eng. J.* 266, 329–338

164 Babakouhi, R., Alavi, S. M., Rezaei, M., Akbari, E., Varbar, M. 2023. "Combined CO_2 reforming and partial oxidation of methane over mesoporous nanostructured Ni/M-Al_2O_3 catalyst: Effect of various support promoters and nickel loadings." *J. CO2 Utilization* 70, 102427.

165 Dossumov, K., Yergazyieva, G. Y., Myltykbayeva, L. K., Suyunbayev, U., Asanov, N. A., Gyulmaliev, A. M. 2015. "Oxidation of methane over polyoxide catalysts." *Catal. Lett.* 58, 178–183.

166 Zhang, N. W., Huang, C. J., Zhu, X. Q., Xu, J. D., Weng, W. Z., Wan, H. L. 2012. "Effect of calcination temperature and pretreatment with reaction gas on properties of Co/γ-Al_2O_3 catalysts for partial oxidation of methane." *Chem. A Asian J.* 7, 1895–1901.

167 Sang-Sik, L., Ju-Hwan, H., Ho-Jung, H., Byung-Kwan, K., Jong-Dae, H. 2010. "Partial oxidation of methane for hydrogen production over Co and Ni catalysts." *Korean Chem. Eng. Res.* 48, 776–783.

168 Jozwiak, W. K., Szubiakiewicz, E., Góralski, J., Klonkowski, A., Paryjczak, T. 2004. "Physico-chemical and catalytic study of the Co/SiO_2 catalysts." *Kinetics Catal.* 45, 247–255.

169 Yang, C., Zhang, J., Wang, J., Li, D., Li, K., Zhu, X. 2023. "Oxygen vacancies enriched Ni-Co/SiO_2@CeO_2 redox catalyst for cycling methane partial oxidation and CO_2 splitting." *Chinese J. Chem. Eng.* 63, 235–245.

170 Dedov, A. G., Loktev, A. S., Komissarenko, D. A., Mazo, G. N., Shlyakhtin, O. A., Parkhomenko, K. V., Kiennemann, A. A., Roger, A. C., Ishmurzin, A. V., Moiseev, I. I. 2015. "Partial oxidation of methane to produce syngas over a neodymium–calcium cobaltate-based catalyst." *Appl. Catal. A.* 489, 140–146.

171 Hayakawa, T., Andersen, A. G., Shimizu, M., Suzuki, K., Takehira, K. 1993. "Partial oxidation of methane to synthesis gas over some titanates-based perovskite oxides." *Catal. Lett.* 22, 307–317.

172 Khoja, A. H., Javed, A. H., Naqvi, S. R., Shakir, S., Din, I. U., Arshad, Z., Rashid, U., Qazi, U. Y., Naeem, N. 2022. "Partial oxidation of methane over CaO decorated TiO_2 nanocatalyst for syngas production in a fixed bed reactor." *Catalysts* 12, 1089.

173 Takanabe, K., Nagaoka, K., Nariai, K., Aika, K. 2005. "Influence of reduction temperature on the catalytic behavior of Co/TiO_2 catalysts for CH_4/CO_2 reforming and its relation with titania bulk crystal structure." *J. Catal.* 230, 75–85.

174 Jing, Q., Lou, H., Fei, J., Hou, Z., Zheng, X. 2004. "Syngas production from reforming of methane with CO_2 and O_2 over Ni/SrO–SiO_2 catalysts in a fluidized bed reactor." *Int. J. Hydrogen Energy* 29, 1245–1251.

175 Milt, V. G., Ulla, M. A., Lombardo, E. A. 2000. "Cobalt-containing catalysts for the high-temperature combustion of methane." *Catal. Lett.* 65, 65–73

176 Lago, R., Bini, G., Peña, M. A., Fierro, J. L. G. 1997. "Partial oxidation of methane to synthesis gas using $LnCoO_3$ perovskites as catalyst precursors." *J. Catal.* 167, 198–209.

177 Nishimoto, H., Ikenaga, N., Nakagawa, K., Konishi, T., Suzuki, T. 2005. "Partial oxidation of methane to synthesis gas over oxidized diamond supported catalysts—Catalytic behavior of nickel and cobalt species." *J. Jpn. Petrol. Inst.* 48, 290–300.

178 Alarcón, A., Palma, O., Morales, E. M., Biset-Peiró, M., Andreu, T., Guilera, J. 2023. "Bimetallic cobalt catalysts promoted by La_2O_3 for the production of high-calorie synthetic gas." *Fuel* 341, 127726.

179 Yu, C., Hu, J., Weng, W., Zhou, X., Chen, X. 2012. "Preparation of Co/$Ce_{0.5}Zr_{0.5}O_2$ catalysts and their catalytic performance in methane partial oxidation to produce synthesis gas." *J. Fuel Chem. Technol.* 40, 418–423.

180 Bouarab, R., Akdim, O., Auroux, A., Cherifi, O., Mirodatos, C. 2004. "Effect of MgO additive on catalytic properties of Co/SiO_2 in the dry reforming of methane." *Appl. Catal. A* 264, 161–168.

181 Chang, Y. -F., Heinemann, H. 1993. "Partial oxidation of methane to syngas over Co/MgO catalysts. Is it low temperature?" *Catal. Lett.* 21, 215–244.

182 Freni, S., Calogero, G., Cavallaro, S. 2000. "Hydrogen production from methane through catalytic partial oxidation reactions." *J. Power Sources* 87, 28–38

183 Mo, L., Fei, J., Huang, C., Zheng, X. 2003. "Reforming of methane with oxygen and carbon dioxide to produce syngas over a novel Pt/$CoAl_2O_4$/Al_2O_3 catalyst." *J. Mol. Catal. A* 193, 177–184.

184 Chen, L., Huang, Q., Wang, Y., Xiao, H., Liu, W., Zhang, D., Yang, T. 2019. "Tailoring performance of Co–Pt/MgO–Al_2O_3 bimetallic aerogel catalyst for methane oxidative carbon dioxide reforming: Effect of Pt/Co ratio." *Int. J. Hydrogen Energy* 44, 19878–19889.

185 Enger, B. C., Lødeng, R., Holmen, A. 2008. "A review of catalytic partial oxidation of methane to synthesis gas with emphasis on reaction mechanisms over transition metal catalysts." *Appl. Catal. A* 346, 1–27

186 Tang, S., Lin, J., Tan, K. L. 1998. "Pulse-MS studies on CH4/CD4 isotope effect in the partial oxidation of methane to syngas over Pt/α-Al_2O_3." *Catal. Lett.* 55, 83–86.

187 Yan, Q. -G., Chu, W., Gao, L. -Z., Yu, Z. -L., Yuan, S. -Y. 1998. "Reactivity of Pt/Al_2O_3 and Pt/CeO_2/Al_2O_3 catalysts for partial oxidation of methane to syngas." *Stud. Surf. Sci. Catal.* 119, 855–860.

188 Pantu, P., Gavalas, G. R. 2002. "A methane partial oxidation on Pt/CeO_2 and Pt/Al_2O_3 catalysts." *Appl. Catal. A* 223, 253–260.

189 Otsuka, K., Wang, Y., Sunada, E., Yamanaka, I. 1998. "Direct partial oxidation of methane to synthesis gas by cerium oxide." *J. Catal.* 175, 152–160.

190 Mattos, L. V., de Oliveira, E. R., Resende, P. D., Noronha, F. B., Passos, F. B. 2002. "Partial oxidation of methane on Pt/Ce–ZrO_2 catalysts." *Catal. Today* 77, 245–256.

191 Buyevskaya, O.V., Wolf, D., Baerns, M. 1994. "Rhodium-catalyzed partial oxidation of methane to CO and H_2. Transient studies on its mechanism." *Catal. Lett.* 29, 249–260.

192 Buyevskaya, O. V., Walter, K., Wolf, D., Baerns, M. 1996. "Primary reaction steps and active surface sites in the rhodium-catalyzed partial oxidation of methane to CO and H_2." *Catal. Lett.* 38, 81–88.

193 Tian, Z., Dewaele, O., Marin, G. B. 1999. "The state of Rh during the partial oxidation of methane into synthesis gas." *Catal. Lett.* 57, 9–17.

194 Wang, H. Y., Ruckenstein, E. 1999. "CH_4/CD_4 isotope effect and the mechanism of partial oxidation of methane to synthesis gas over Rh/γ-Al_2O_3 catalyst." *J. Phys. Chem. B* 103, 11327–11331.

195 Wang, H.Y., Ruckenstein, E. 1999. "Catalytic partial oxidation of methane to synthesis gas over γ-Al_2O_3-supported rhodium catalysts." *Catal. Lett.* 59, 121–127.

196 Nakagawa, K., Ikenaga, N., Teng, Y., Kobayashi, T., Suzuki, T. 1999. "Transient response of catalyst bed temperature in the pulsed reaction of partial oxidation of methane to synthesis gas over supported rhodium and iridium catalysts." *J. Catal.* 186, 405–413.

197 Mallens, E. P. J., Hoebink, J. H. B. J., Marin, G. B. 1997. "The reaction mechanism of the partial oxidation of methane to synthesis gas: A transient kinetic study over rhodium and a comparison with platinum." *J. Catal.* 167, 43–56.

198 Balient, I., Miyazaki, A., Aika, K.-ie. 2003. "The relevance of Ru nanoparticles morphology and oxidation state to the partial oxidation of methane." *J. Catal.* 220, 74–83.

199 Rabe, S., Nachtegaal, M., Vogel, F. 2007. "Catalytic partial oxidation of methane to synthesis gas over a ruthenium catalyst: The role of the oxidation state." *Phys. Chem. Chem. Phys.* 9, 1461–1468.

200 Elmasides, C., Ioannides, T., Verykios, X. E. 1998. "Kinetic behaviour of the Ru/TiO_2 catalyst in the reaction of partial oxidation of methane." *Stud. Surf. Sci. Catal.* 119, 801–806.

201 Elmasides, C., Kondarides, D. I., Neophytides, S. G., Verykios, X. E. 2001. "Partial oxidation of methane to synthesis gas over Ru/TiO_2 catalysts: Effects of modification of the support on oxidation state and catalytic performance." *J. Catal.* 198, 195–207.

202 Tang, Y., Fung, V., Zhang, X., Li, Y., Nguyen, L., Sakata, T., Higashi, K., Jiang, D., Tao, F. 2021. "Single-atom high-temperature catalysis on a Rh_1O_5 cluster for production of syngas from methane." *J. Am. Chem. Soc.* 143, 16566–16579.

203 Keller, G. E., Bhasin, M. M. 1982. "Synthesis of ethylene via oxidative coupling of methane: I. Determination of active catalysts." *J. Catal.* 73, 9–19.

204 Zavyalova, Z. Holena, M. Schlogl, R. Baerns, M. 2011. "Statistical analysis of past catalytic data on oxidative methane coupling for new insights into the composition of high-performance catalysts." *ChemCatChem* 3, 1935–1947.

205 Somorjai, G. A., Li, Y. 2010. *Introduction to Surface Chemistry and Catalysis*. Wiley.

206 Ertl, G., Knözinger, H., Schüth, F., Weitkamp, J. 2008. *Handbook of Heterogeneous Catalysis*. Wiley.

207 Driscoll, D. J., Martir, W. m., Wang, J. X., Lunsford, J. H. 1985. "Formation of gas-phase methyl radicals over magnesium oxide." *J. Am. Chem. Soc.* 107, 58–63.

208 Campbell, K. D., Morales, E., Lunsford, J. H. 1987. "Gas-phase coupling of methyl radicals during the catalytic partial oxidation of methane." *J. Am. Chem. Soc. 109*, 7900–7901.

209 Nelson, P. F., Lukey, C. A., Cant, N. W. 1988. "Isotopic evidence for direct methyl coupling and ethane to ethylene conversion during partial oxidation of methane over lithium/magnesium oxide." *J. Phys. Chem.* 92, 6176–6179.

210 Roos, J. A., Korf, S. J., Veehof, R. H. J., van Ommen, J. G., Ross, J. R. H. 1989. "Kinetic and mechanistic aspects of the oxidative coupling of methane over a Li/MgO catalyst". *Appl. Catal.* 52, 131–145.

211 Amenomiya, Y., Birss, V. I., Goledzinowski, M., Galuszka, J., Sanger, A. R. 1990. "Conversion of methane by oxidative coupling." *Catal. Rev.* 32, 163–227

212 Schwarz, H. 2011. "Chemistry with methane: Concepts rather than recipes." *Angew. Chem. Int. Ed.* 50, 10096–10115.

213 Lunsford, J. H. 1995. "The catalytic oxidative coupling of methane." *Angew. Chem. Int. Ed.* 34, 970–980.

214 Kwapien, K., Paier, J., Sauer, J., Geske, M., Zavyalova, U., Horn, R., Schwach, P., Trunschke, A., Schlogl, R. 2014. "Sites for methane activation on lithium-doped magnesium oxide surfaces." *Angew. Chem. Int. Ed.* 53, 8774–8778.

215 Xu, M., Shi, C., Yang, X., Rosynek, M. P., Lunsford, J. H. 1992. "Effect of carbon dioxide on the activation energy for methyl radical generation over lithium/magnesia catalysts." *J. Phys. Chem.* 96, 6395–6398.

216 Arndt, S., Simon, U., Heitz, S., Berthold, A., Beck, B., Gçrke, O., Epping, J. D., Otremba, T., Aksu, Y., Irran, E., Laugel, G., Driess, M., Schubert, H., Schomcker, R. 2011. "Li-doped MgO from different preparative routes for the oxidative coupling of methane." *Top. Catal.* 54, 1266.

217 Qian, K., You, R., Guan, Y., Wen, W., Tian, Y., Pan, Y., Huang, W. 2020. "Single-site catalysis of Li-MgO catalysts for oxidative coupling of methane reaction." *ACS Catal.* 10, 15142–15148.

218 Myrach, P., Nilius, N., Levchenko, S. V., Gonchar, A., Risse, T., Dinse, K. P., Boatner, L. A., Frandsen, W., Horn, R., Freund, H. J., Schlçgl, R., Scheffler, M. 2010. "Temperature-dependent morphology, magnetic and optical properties of Li-doped MgO." *ChemCatChem* 2, 854–862.

219 Zavyalova, U., Weinberg, G., Frandsen, W., Girgsdies, F., Risse, T., Dinse, K. P., Schloegl, R., Horn, R. 2011. "Lithium as a modifier for morphology and defect structure of porous magnesium oxide materials prepared by gel combustion synthesis." *ChemCatChem* 3, 1779–1788.

220 Zavyalova, U., Geske, M., Horn, R., Weinberg, G., Frandsen, W., Schuster, M., Schlogl, R. 2011. "Morphology and microstructure of Li/MgO catalysts for the oxidative coupling of methane." *ChemCatChem* 3, 949–959.

221 Wang, L., Tao, L., Xie, M., Xu, G., Huang, J., Xu, Y. 1993. "Dehydrogenation and aromatization of methane under non-oxidizing conditions." *Catal. Lett.* 21, 35–41.

222 Spivey, J. J., Hutchings, G. 2014. "Catalytic aromatization of methane." *Chem. Soc. Rev.* 43, 792–803.

223 Weckhuysen, B. M., Wang, D., Rosynek, M. P., Lunsford, J. H. 1998. "Conversion of methane to benzene over transition metal Ion ZSM-5 zeolites." *J. Catal.* 175, 347–351.

224 Tan, P. 2016. "Active phase, catalytic activity, and induction period of Fe/zeolite material in nonoxidative aromatization of methane." *J. Catal.* 338, 21–29.

225 Weckhuysen, B. M., Wang, D., Rosynek, M. P., Lunsford, J. H. 1997. "Catalytic conversion of methane into aromatic hydrocarbons over iron oxide loaded ZSM-5 zeolites." *Angew. Chem. Int. Ed.* 36, 2374–2376.

226 Lai, Y., Veser, G. 2016. "The nature of the selective species in Fe/HZSM-5 for non-oxidative methane dehydroaromatization." *Catal. Sci. Technol.* 6, 5440–5452.

227 Wang, L., Ohnishi, R., Ichikawa, M. 2000. "Selective dehydroaromatization of methane toward benzene on Re/HZSM-5 catalysts and effects of CO/CO_2 addition." *J. Catal.* 190, 276–283.

228 Liu, W., Xu, Y., Wong, S. -T., Wang, L., Qui, J., Yang, N. 1997. "Methane dehydrogenation and aromatization in the absence of oxygen on: A study on the interaction between Mo species and HZSM-5 by using 27Al and 29Si MAS NMR." *J. Mol. Catal. A* 120, 257–265.

229 Chen, L., Lin, L., Xu, Z. S., Li, X. S., Zhang, T. 1995. "Dehydro-oligomerization of methane to ethylene and aromatics over molybdenum/HZSM-5 catalyst." *J. Catal.* 157, 190–200.

230 Solymosi, F., Cserenyi, A., Szoke, A., Bansagi, T., Oszko, A. 1997. "Aromatization of methane over supported and unsupported Mo-based catalysts." *J. Catal.* 165, 150–161.

231 López-Martín, A., Sini, M. F., Cutrufello, M. G., Caballero, A., Colón, G. 2022. "Characterization of Re-Mo/ZSM-5 catalysts: How Re improves the performance of Mo in the methane dehydroaromatization reaction." *Appl. Catal. B.* 304, 120960.

232 Weckhuysen, B. M., Wang, D., Rosynek, M. P., Lunsford, J. H. 1998. "Conversion of methane to benzene over transition metal Ion ZSM-5 zeolites: I. Catalytic characterization." *J. Catal.* 175, 338–346.

233 Xu, Y., Liu, S., Guo, X., Wang, L., Xie, M. 1994. "Methane activation without using oxidants over Mo/HZSM-5 zeolite catalysts." *Catal. Lett.* 30, 135–149.

234 Liu, S., Wang, L., Ohnishi, R., Ichikawa, M. 2000. "Bifunctional catalysis of Mo/HZSM-5 in the dehydroaromatization of methane with CO/CO_2 to benzene and naphthalene." *Kinet. Catal.* 41, 132–144.

235 Jiang, H., Wang, L., Cui, W., Xu, Y. 1999. "Study on the induction period of methane aromatization over Mo/HZSM-5: Partial reduction of Mo species and formation of carbonaceous deposit." *Catal. Lett.* 57, 95–102.

236 Matus, E. V., Ismagilov, I. Z., Sukhova, O. B., Zaikovskii, V. I., Tsikoza, L. T., Ismagilov, Z. R., Moulijn, J. A. 2007. "Study of methane dehydroaromatization on impregnated Mo/ZSM-5 catalysts and characterization of nanostructured molybdenum phases and carbonaceous deposits." *Ind. Eng. Chem. Res.* 46, 4063–4074.

237 Ma, D., Zhang, W., Shu, Y., Xu, Y., Bao, X. 2000. "MAS NMR, ESR and TPD studies of Mo/HZSM-5 catalysts: Evidence for the migration of molybdenum species into the zeolitic channels." *Catal. Lett.* 66, 155–160.

238 Zeng, J. L., Xiong, Z. T., Zhang, H. B., Lin, G. D., Tsai, K. R. 1998. "Nonoxidative dehydrogenation and aromatization of methane over W/HZSM-5-based catalysts" *Catal. Lett.* 53, 119–124.

239 Xu, Y., Shu, Y., Liu, S., Huang, J., Guo, X. 1995. "Interaction between ammonium heptamolybdate and NH4ZSM-5 zeolite: The location of Mo species and the acidity of Mo/HZSM-5." *Catal. Lett.* 35, 233–243.

240 Wang, D., Lunsford, J. H., Rosynek, M. P. "Characterization of a Mo/ZSM-5 catalyst for the conversion of methane to benzene." *J. Catal.* 169, 347–358.

241 Zhang, C.-L., Li, S., Yuan, Y., Zhang, W.-X., Wu, T.-H., Lin, L.-W. 1998. "Aromatization of methane in the absence of oxygen over Mo-based catalysts supported on different types of zeolites." *Catal. Lett.* 56, 207–213.

242 Li, S., Zhang, C.L., Kan, Q. B., Yu, J. F., Yuan, Y., Wu, T. H. 2000. "H-ZSM-8 supported mo-based catalysts for methane conversion under non-oxidative condition." *Stud. Surf. Sci. Catal.* 130, 3609–3613.

243 Liu, S., Wang, L., Ohnishi, R., Ichikawa, M. 1999. "Bifunctional catalysis of Mo/HZSM-5 in the dehydroaromatization of methane to benzene and naphthalene XAFS/TG/DTA/MASS/FTIR characterization and supporting effects." *J. Catal.* 181, 175–188.

244 Smiešková, A., Hudec, P., Kumar, N., Salmi, T., Murzin, Y., Jorík, V. 2010. "Aromatization of methane on Mo modified zeolites: Influence of the surface and structural properties of the carriers." *Appl. Catal. A* 377, 83–91.

245 Lu, Y., Hu, J., Han, J., Yu, F. 2016. "Synthesis of gasoline-range hydrocarbons from nitrogen-rich syngas over a Mo/HZSM-5 bi-functional catalyst." *J. Energy Inst.* 89, 782–792.

246 Xu, Y., Wang, J., Ma, G., Bai, J., Du, Y. 2020. "Direct synthesis of aromatics from syngas over Mo-modified Fe/HZSM-5 bifunctional catalyst." *Appl. Catal. A* 598, 117589.

247 Tempelman, C. H. L., Zhu, X., Hensen, E. J. M. 2015. "Activation of Mo/HZSM-5 for methane aromatization." *Chinese J. Catal.* 36, 829–837.

248 Wang, L., Xu, Y., Wong, S. T., Cui, W., Guo, X. 1997. "Activity and stability enhancement of catalysts for methane non-oxidative transformation to aromatics and C_2 hydrocarbons: Effect of additives and pretreatment conditions." *Appl. Catal. A* 152, 173–182.

249 Solymosi, F., Erdöhelyi, A., Szöke. A. 1995. "Dehydrogenation of methane on supported molybdenum oxides. Formation of benzene from methan." *Catal. Lett.* 32, 43–53.

250 Rahman, M., Infantes-Molina, A., Hoffman, A. S., Bare, S. R., Emerson, K. L., Khatib, S. J. 2020. "Effect of Si/Al ratio of ZSM-5 support on structure and activity of Mo species in methane dehydroaromatization." *Fuel* 278, 118290.

251 Ismagilov, Z. R., Matus, E. V., Tsikoza, L. T. 2008. "Direct conversion of methane on Mo/ZSM-5 catalysts to produce benzene and hydrogen: achievements and perspectives." *Energy Environ. Sci.* 1, 526–541.

252 Takanabe, K., Khan, A. M., Tang, Y., Nguyen, L., Ziani, A., Jacobs, B. W., Elbaz, A. M., Sarathy, S. M., Tao, F. "integrated in situ characterization of a molten salt catalyst surface: Evidence of sodium peroxide and hydroxyl radical formation." *Angew. Chem. Int. Ed.* 56, 10403–10407.

253 Bouchy, C., Schmidt, I., Anderson, J. R., Jacobsen, C. J. H., Derouane, E. G., Derouane-Abd Hamid, S. B. 2000. "Metastable fcc α-MoC1−x supported on HZSM5: preparation and catalytic performance for the non-oxidative conversion of methane to aromatic compounds." *J. Mol. Catal. A* 163, 283–296.

254 Derouane-Abd Hamid, S. B., Anderson, J. R., Schmidt, I., Bouchy, C., Jacobsen, C. J. H., Derouane, E. G. 2000. "Effect of the activation procedure on the performance of Mo/H-MFI catalysts for the non-oxidative conversion of methane to aromatics." *Catal. Today* 63, 461–469.

255 Yuan, S., Hamid, S. B. D. A., Li, Y., Ying, P., Xin, Q., Derouane, E. G., Li, C. 2002. "Preparation of Mo_2C/HZSM-5 and its catalytic performance for the conversion of n-butane into aromatics." *J. Mol. Catal. A* 184, 257–266.

256 Li, B., Li, S., Li, N., Chen, H., Zhang, W., Bao, X., Lin, B. 2006. "Structure and acidity of Mo/ZSM-5 synthesized by solid state reaction for methane dehydrogenation and aromatization." *Microporous Mesoporous Mater.* 88, 244–253.

257 Tessonnier, J. P., Louis, B., Walspurger, S., Sommer, J., Ledoux, M. J., Pham-Huu, C. 2006. "quantitative measurement of the brönsted acid sites in solid acids: Toward a single-site design of Mo-modified ZSM-5 zeolite." *J. Phys. Chem. C* 110,10390–10395.

258 Chen, H. Y., Tang, S., Zhong, Z. Y., Lin, J., Tan, K. L. 2001. "XPS and FTIR studies of Mo/ZSM-5 catalysts for nonoxidative conversion of methane to aromatic." *Surf. Rev. Lett.* 8, 627–632.

259 López-Martín, A., Platero, F., Colóna, G., Caballero, A. 2021. "Elucidating the nature of Mo species on ZSM-5 and its role in the methane aromatization reaction." *React. Chem. Eng.* 6, 1265–1276.

260 Li, W., Meitzner, G. D., Borry, R. W., Iglesia, E. 2000. "Raman and X-ray absorption studies of Mo species in Mo/H-ZSM5 catalysts for non-oxidative CH4 reactions." *J. Catal.* 191, 373–383.

261 Borry, R. W., Kim, Y. H., Huffsmith, A., Reimer, J. A., Iglesia, E. 1999. "Structure and density of Mo and acid sites in Mo-exchanged H-ZSM5 catalysts for nonoxidative methane conversion." *J. Phys. Chem. B* 103, 5787–5796.

262 Lacheen, H. S., Iglesia, E. 2005. "Stability, structure, and oxidation state of Mo/H-ZSM-5 catalysts during reactions of CH_4 and CH_4–CO_2 mixtures." *J. Catal.* 230, 173–185.

263 Kosinov, N., Coumans, F. J. A. G., Uslamin, E. A., Wijpkema, A. S. G., Mezari, B., Hensen, E. J. M. 2017. "Methane dehydroaromatization by Mo/HZSM-5: Mono- or bifunctional catalysis?" *ACS Catal.* 7, 520–529.

264 Solymosi, F., Szöke, A., Cserényi, J. 1996. "Conversion of methane to benzene over Mo_2C and Mo_2C/ZSM-5 catalysts." *Catal. Lett.* 39, 157–161.

265 Szoke, A., Solymosi, F. 1996. "Selective oxidation of methane to benzene over K_2MoO_4/ZSM-5 catalysts." *Appl. Catal.* 142, 361–374.

266 Liu, H., Yang, S., Hu, J., Shang, F., Li, Z., Xu, C., Guan, J., Kan, Q. 2012. "A comparison study of mesoporous Mo/H-ZSM-5 and conventional Mo/H-ZSM-5 catalysts in methane non-oxidative aromatization." *Fuel Process. Technol.* 96, 195–202.

267 Weckhuysen, B. M., Rosynek, M. P., Lunsford, J. H. 1998. "Characterization of surface carbon formed during the conversion of methane to benzene over Mo/H-ZSM-5 catalysts." *Catal. Lett.* 52, 31–36.

268 Budde, P. K., Singh, A. K., Upadhyayula, S. 2018. "Non-oxidative methane dehydroaromatization reaction over highly active α—MoC1-x-ZSM-5 derived from pretreatment". *J. Chem. Sci.* 130, 27.

269 Zhou, D., Ma, D., Liu, X., Bao, X. 2001. "Study with density functional theory method on methane dehydro-aromatization over Mo/HZSM-5 catalysts I: Optimization of active Mo species bonded to ZSM-5 zeolite." *J. Chem. Phys.* 114, 9125–9129.

270 Yuan, S., Li, J., Hao, Z., Feng, Z., Xin, Q., Ying, P., Li, C. 1999. "The effect of oxygen on the aromatization of methane over the Mo/HZSM-5 catalyst." *Catal. Lett.* 63, 73–77.

271 Matus, E. V., Sukhova, O. B., Ismagilov, I. Z., Tsikoza, L. T., Ismagilov, Z. R. 2000. "Peculiarities of dehydroaromatization of CH_4–C_2H_6 and CH_4 over Mo/ZSM-5 catalysts." *Reaction Kinetics Catal. Lett.* 98, 59–67.

272 Denardin, F., Perez-Lopez, O. W. 2019. "Tuning the acidity and reducibility of Fe/ZSM-5 catalysts for methane dehydroaromatization." *Fuel* 236, 1293–1300.

273 Ismagilov, Z. R., Tsikoza, L. T., Matus, E. V., Litvak, G. S., Ismagilov, I. Z., Sukhova, O. B. 2005. "Carbonization and regeneration of Mo/ZSM-5 catalysts for methane dehydroaromatization." *Eurasian Chem.-Technol. J.* 7, 115–121.

274 Karakaya, C., Morejudo, S. H., Zhu, H., Kee, R. J. 2016. "Catalytic chemistry for methane dehydroaromatization (MDA) on a bifunctional Mo/HZSM-5 catalyst in a packed bed." *Ind. Eng. Chem. Res.* 55, 9895–9906.

275 Tempelman, C. H. L., Hensen, E. J. M. 2015. "On the deactivation of Mo/HZSM-5 in the methane dehydroaromatization reaction." *Appl. Catal. B* 176-177, 731–739.

276 Kosinov, N., Uslamin, E. A., Coumans, F. J. A. G, Wijpkema, A. S. G., Rohling, R. Y., Hensen, E. J. M. 2018. "Structure and evolution of confined carbon species during methane dehydroaromatization over Mo/ZSM-5." *ACS Catal.* 8, 8459–8467.

277 Ma, D., Shu, Y., Bao, X., Xu, Y. 2000. "Methane dehydro-aromatization under nonoxidative conditions over Mo/HZSM-5 catalysts: EPR study of the Mo species on/in the HZSM-5 zeolite." *J. Catal.* 189, 314–325.

278 Rahman, M., Sridhar, A., Khatib, S. J. 2018. "Impact of the presence of Mo carbide species prepared ex situ in Mo/HZSM-5 on the catalytic properties in methane aromatization." *Appl. Catal. A* 558, 67–80.

279 Tan, P., Wong, K. W., Au, C. T., Lai, S. Y. 2003. "Effects of Co-fed O_2 and CO_2 on the deactivation of Mo/HZSM-5 for methane aromatization." *Appl. Catal. A* 253, 3–5–316.

280 Pasupulety, N., Al-Zahrani, A. A., Daous, M. A., Driss, H., Petrov, L. A. 2021. "Methane aromatization study on M-Mo_2C/HZSM-5 (M = Ce or Pd or Nb) nano materials." *J. Mater. Res. Technol.* 14, 363–373.

281 Shu, J., Adnot, A., Grandjean, B. P. A. 1999. "Bifunctional behavior of Mo/HZSM-5 catalysts in methane aromatization." *Ind. Eng. Chem. Res.* 38, 3860–3867.

282 Ding, W., Li, S., Meitzner, G. D., Iglesia, E. 2001. "Methane conversion to aromatics on Mo/H-ZSM5: Structure of molybdenum species in working catalysts." *J. Phys. Chem. B* 105, 506–513.

283 Cheng, M., Cruchade, H., Pinard, L., Dib, E., Liu, H., Wang, J., Liu, X., Yan, Z. F., Qin, Z., Mintova, S. 2023. "Coke relocation and Mo immobilization in donut-shaped Mo/HZSM-5 catalysts for methane dehydroaromatization." *J. Mater. Chem. A* 11, 24991–24998.

284 Yang, Z., Kumar, A., Apblett, A. W., Moneeb, A. M. 2017. "Co-pyrolysis of torrefied biomass and methane over molybdenum modified bimetallic HZSM-5 catalyst for hydrocarbons production." *Green Chem.* 19, 757–768.

285 Gao, R., Li, Xin, Wang, Z., Liu, Y., Sun, J., Zhu, Y., Yao, S. 2022. "Reaction regeneration cycle of Mo/HZSM-5 catalyst in methane dehydroaromatization with the addition of oxygen-containing components." *Appl. Catal. A* 647, 118916.

286 Gu, Y., Chen, P., Wang, X., Lyu, Y., Tian, Y., Liu, W., Yan, Z., Liu, X. 2021. "Coking mechanism of Mo/ZSM-5 catalyst in methane dehydroaromatization." *Appl. Catal. A* 613, 118019.

287 Julian, I., Roedern, M. B., Hueso, J. L., Irusta, S., Baden, A. K., Mallada, R., Davis, Z., Santamaria, J. 2020. "Supercritical solvothermal synthesis under reducing conditions to increase stability and durability of Mo/ZSM-5 catalysts in methane dehydroaromatization." *Appl. Catal. B* 263, 118360.

288 Sily, P. D., Noronha, F. B., Passos, F. B. 2006. "Methane direct conversion on Mo/ZSM-5 catalysts modified by Pd and Ru." *J. Nat. Gas Chem.* 15, 82–86.

289 Shu, Y., Xu, Y., Wong, S. T., Wang, L., Guo, X. 1997. "Promotional effect of Ru on the dehydrogenation and aromatization of methane in the absence of oxygen over Mo/HZSM-5 catalysts." *J. Catal.* 170, 11–19.

290 Zhang, Y., Wang, D., Fei, J., Zheng, X. 2002. "Effect of Cr addition on the methane aromatization performance of the Mo/HZSM-5 catalyst." *Australian J. Chem.* 55, 531–534

291 Burns, S., Hargreaves, J. S. J., Pal, P., Parida, K. M., Parija, S. 2006. "The effect of dopants on the activity of MoO_3/ZSM-5 catalysts for the dehydroaromatisation of methane." *Catal. Today* 114, 383–387.

292 Sridhar, A., Rahman, M., Infantes-Molina, A., Wylie, B. J., Borcik, C. G., Khatib, S. J. 2020. "Bimetallic Mo-Co/ZSM-5 and Mo-Ni/ZSM-5 catalysts for methane dehydroaromatization: A study of the effect of pretreatment and metal loadings on the catalytic behavior." *Appl. Catal. A* 589, 117247.

293 Zarei, T.; Davoodbeygi, Y., Islamimanesh, S., Zarei, Z. 2022. "Methane dehydroaromatization over Mo and W catalysts supported on ZSM-5." *Iran. J. Chem. Chem. Eng.* 41, 2278–2287.

294 Abdelsayed, V., Shekhawat, D., Smith, M. W. 2015. "Effect of Fe and Zn promoters on Mo/HZSM-5 catalyst for methane dehydroaromatization." *Fuel* 139, 401–410.

295 Chen, L., Lin, L., Xu, Z., Zhang, T., Li, X. 1996. "Promotional effect of Pt on non-oxidative methane transformation over Mo-HZSM-5 catalyst." *Catal. Lett.* 39, 169–172

296 Liu, Y., Bolshakov, A., Ćoza, M., Drozhzhin, V., Hensen, E.J. M., Kosinov, N. 2023. "Isomorphously substituted [Fe,Al]ZSM-5 catalysts for methane dehydroaromatization." *ACS Catal.* 13, 8128–8138.

297 Xu, Y., Yuan, X., Chen, M., Dong, A., Liu, B., Jiang, F., Yang, S., Liu, X. 2021. "Identification of atomically dispersed Fe-oxo species as new active sites in HZSM-5 for efficient non-oxidative methane dehydroaromatization." *J. Catal.* 396, 224–241.

298 Gu, Y., Chen, P., Wang, X., Lyu, Y., Liu, W., Liu, X., Yan, Z. 2021. "Active sites and induction period of Fe/ZSM-5 catalyst in methane dehydroaromatization." *ACS Catal.* 11, 6771–6786.

299 Deng, Y., Veser, G. 2023. "Toward accelerated activation of Fe-ZSM-5 in methane dehydroaromatization." *Energy Fuels* 37, 13282–13295.

300 Guo, X., Fang, G., Li, G., Ma, H., Fan, H., Yu, L., Ma, C., Wu, X., Deng, D., Wei, M., Tan, D., Zhang, S., Li, J., Sun, L., Tang, Z., Pan, X., Bao, X. 2014. "Direct, nonoxidative conversion of methane to ethylene, aromatics, and hydrogen." *Science* 344, 616–619

301 Kheirollahivash, M., Rashidi, F., Moshrefi, M. M. 2019. "Hydrogen production from methane decomposition using a mobile and elongating Arc plasma reactor." *Plasma Chem. Plasma Process.* 39, 445–459.

302 Ermakova, M. A., Ermakov, Y. D., Chuvilin, A. L., Kuvshinov, G. G. 2001. "Decomposition of methane over iron catalysts at the range of moderate temperatures: The influence of structure of the catalytic systems and the reaction conditions on the yield of carbon and morphology of carbon filaments." *J. Catal.* 201, 183–197.

303 Avdeeva, L. B., Reshetenko, T. V., Ismagilov, Z. R., Likholobov, V. A. 2002. "Iron-containing catalysts of methane decomposition: Accumulation of filamentous carbon." *Appl. Catal., A* 228, 53–63.

304 Bonura, G., Blasi, O. D., Spadaro, L., Arena, F., Frusteri, F. 2006. "A basic assessment of the reactivity of Ni catalysts in the decomposition of methane for the production of "CO_x-free" hydrogen for fuel cells application." *Catal. Today* 116, 298–303.

305 Serban, M., Lewis, M. A., Marshall, C. L., Doctor, R. D. 2003. "Hydrogen production by direct contact pyrolysis of natural gas." *Energy Fuels* 17, 705–713.

306 Schultz, I., Agar, D. W. 2015. "Decarbonisation of fossil energy via methane pyrolysis using two reactor concepts: Fluid wall flow reactor and molten metal capillary reactor." *Int. J. Hydrogen Energy* 40, 11422–11427.

307 Geißler, T., Plevan, M., Abánades, A., Heinzel, A., Mehravaran, K., Rathnam, R. K., Rubbia, C., Salmieri, D., Stoppel, L., Stückrad, S., Weisenburger, A., Wenninger, H., Wetzel, Th. 2015. "Experimental investigation and thermo-chemical modeling of methane pyrolysis in a liquid metal bubble column reactor with a packed bed." *Int. J. Hydrogen Energy* 40, 14134–14146.

308 Abánades, A., Ruiz, E., Ferruelo, E. M., Hernández, F., Cabanillas, A., Martínez-Val, J. M., Rubio, J. A., López, C., Gavela, R., Barrera, G., Rubbia, C., Salmieri, D., Rodilla, E., Gutiérrez, D. 2011. "Experimental analysis of direct thermal methane cracking." *Int. J. Hydrogen Energy* 36, 12877–12886.

309 Plevan, M., Geißler, T., Abánades, A., Mehravaran, K., Rathnam, R. K., Rubbia, C., Salmieri, D., Stoppel, L., Stückrad, S., Wetzel, Th. 2015. "Thermal cracking of methane in a liquid metal bubble column reactor: Experiments and kinetic analysis." *Int. J. Hydrogen Energy* 40, 8020–8033.

310 Steinberg, M.1999. "Fossil fuel decarbonization technology for mitigating global warming." *Int. J. Hydrogen Energy* 24, 771–777.

311 Abánades, A., Rathnam, R. K., Geißler, T., Heinzel, A., Mehravaran, K., Müller, G., Plevan, M., Rubbia, C., Salmieri, D., Stoppel, L., Stückrad, S., Weisenburger, A., Wenninger, H., Wetzel, T. 2016. "Development of methane decarbonization based on liquid metal technology for CO_2-free production of hydrogen." *Int. J. Hydrogen Energy* 41, 815–8167.

312 Palmer, C., Tarazkar, M., Kristoffersen, H. H., Gelinas, J., Gordon, M. J., McFarland, E. W., Metiu, H. 2019. "Methane pyrolysis with a molten Cu–Bi alloy catalyst." *ACS Catal.* 9, 8337–8345.

313 Upham, D. C., Agarwal, V., Khechfe, A., Snodgrass, Z. R., Gordon, M. J., Metiu, H., McFarland, E. W. 2017. "Catalytic molten metals for the direct conversion of methane to hydrogen and separable carbon", *Science* 358, 917–921.

314 Farmer, T. C., McFarland, E. W., Doherty, M. F. 2019. "Membrane bubble column reactor model for the production of hydrogen by methane pyrolysis." *Int. J. Hydrogen Energy* 44, 14721–14731.

315 Ogino, Y. 2007. "Catalysis by molten metals and molten alloys." *Catal. Rev.* 23, 505–551.

316 Ogino, Y. 1987. "*Catalysis and Surface Properties of Liquid Metals and Alloys*." New York, Marcel Dekker.

317 Rahimi, N., Kang, D., Gelinas, J., Menon, A., Gordon, M. J., Metiu, H., McFarland, E. W. 2019. "Solid carbon production and recovery from high temperature methane pyrolysis in bubble columns containing molten metals and molten salts." *Carbon* 151, 181–191.

318 Karmilova, L. V., Enikolopyan, N. S., Nalbandyan, A. B., Semenov, N. N. 1960, "Mechanism of the Oxidation of Methane with the Aid of Labeled Atoms." *Zh. Fiz. Khim.*, 34, 1176.

319 de Vekki, A. V., Marakaev, S. T. 2009. "Catalytic partial oxidation of methane to formaldehyde." *Russ J. Appl. Chem.* 82, 521–536.

320 de Vekki, A. V., Marakaev, S. T. 2008. "Kinetics of oxidation of methane to formaldehyde on Na4[PFeMo11O40]/SiO2." *Russ J. Appl. Chem.* 81, 1568–1572.

321 Pak, S., Rosynek, M. P., Lunsford, J. H. 1997. "Conversion of methyl radicals to methanol and formaldehyde over vanadium oxide catalysts." *J. Catal.*, 165, 73–79.

322 Pak, S., Rosynek, M. P., Lunsford, J. H. 1994. "Conversion of methyl radicals to methanol and formaldehyde over molybdenum oxide catalysts." *J. Phys. Chem.* 98, 11786–11790.

323 Otsuka, K., Wang, Y., Yamanaka, I., Marikawa, A. 1993. "Kinetic study of the partial oxidation of methane over $Fe_2(MoO_4)_3$ catalyst." *J. Chem. Soc., Faraday Trans.* 89, 4225–4230.

324 Banares, M. A., Spenscer, N. D., Jones, M. D., Wachs, I. E. 1994. "Effect of alkali metal cations on the structure of catalysts and its relevance to the selective oxidation of methane and methanol." *J. Catal.* 146, 204–210.

325 Smith, M.R., Zhang, L., Driscol, S.A., Ozkan, U. S. 1993. "Effect of surface species on activity and selectivity of MoO_3/SiO_2 catalysts in partial oxidation of methane to formaldehyde." *Catal. Lett.* 19, 1–15.

326 Dang, T. T. H., Seeburg, D., Radnik, J., Kreyenschulte, C., Atia, H., Vu, T. T. H., Wohlrab, S. 2018. "Influence of V-sources on the catalytic performance of VMCM-41 in the selective oxidation of methane to formaldehyde." *Catal. Comm.* 103, 56–59.

327 Liu, H. F., Liu, R.S., Liew, K.Y., Johnson, R. E., Lunsford, J. H.1984. "Partial oxidation of methane by nitrous oxide over molybdenum on silica." *J. Am. Chem. Soc.* 106, 4117–4121.

328 Barbaux, Y., Elamrani, A. R., Payen, E., Gengembre, L., Bonnelle, J. P., Grzybowska, B. 1988. "Silica supported molybdena catalysts: Characterization and methane oxidation." *Appl. Catal.* 44, 117–132.

329 Takemoto, T., Tabata, K., Teng, Y., Dai, L. X., Suzuki, E. 2001. "Selective oxidation of methane in CH_4–O_2–NO_2 over MoO_3." *Catal. Today* 71, 47–53.

330 Liu, R.-S. Iwamoto, M. Lunsford, J. H. 1982. "Partial oxidation of methane by nitrous oxide over molybdenum oxide supported on silica." *J. Chem. Soc. Chem. Commun.* 1982, 78–79.

331 Schwach, P. Pan, X., Bao, X. 2017. "Direct conversion of methane to value-added chemicals over heterogeneous catalysts: challenges and prospects." *Chem. Rev.* 117(13), 8497–8520.

7

Electrochemical Conversion of Chemical Energy of CH_4 to Electrical Energy at Intermediate Temperature

There are two methods for converting the chemical energy stored in methane (CH_4) into electricity. The first method involves electricity generation in power plants, which follows two main steps. Initially, CH_4 undergoes combustion to produce CO_2 and H_2O, constituting a highly exothermic reaction that converts chemical energy into heat [1–3]. Subsequently, the generated heat is used to vaporize water into steam, which then drives an electric generator to produce electricity.

According to the United States Energy Information Administration report, methane combustion-based power plants supply approximately 40% of the electricity generated in the United States in 2022. Compared to coal power plants, methane power plants are significantly cleaner since coal often contains sulfur, nitrogen, and phosphorus, leading to the formation of acidic oxides during combustion. These oxides form inorganic acids in the atmosphere that are either poisonous or corrosive. In addition, residual coal combustion dust contributes to environmental issues. Consequently, methane power plants have largely replaced traditional coal power plants in North America and Europe. However, a remaining concern with methane power plants is the release of unburnt methane into the atmosphere, in addition to the greenhouse effect resulting from the released CO_2. Since CH_4 combustion typically occurs at temperatures exceeding 1000 °C, transforming unburnt methane to CO_2 requires an active catalyst for complete CH_4 oxidation at temperatures as high as 1300 °C. Barium hexaaluminate nanoparticles have been reported as effective catalysts for this purpose [4].

Apart from power plants, methane can be electrochemically oxidized on an anode in a high-temperature fuel cell at 1000 °C or higher temperatures to generate electricity. However, the high-temperature process presents various technical challenges, including the need for special materials that can withstand high temperatures, leading to increased manufacturing and maintenance costs of a high-temperature fuel cell. In contrast, an intermediate-temperature fuel cell process has shown promise for converting chemical energy from CH_4 to electricity at lower temperatures [5–12].

One example of an intermediate-temperature fuel cell process integrates thermal and electrochemical catalysis into the same system. The kinetics of electrooxidation of CH_4 at 300–500 °C are typically sluggish. However, compared to CH_4, the kinetics of electrooxidation of H_2 are much faster. Therefore, the intermediate-temperature fuel cell facilitates

Low-temperature Activation and Catalytic Transformation of Methane to Non-CO_2 Products,
First Edition. Franklin Tao.

both the production of H_2 from the steam reforming of CH_4 ($CH_4 + H_2O \rightarrow CO + 3H_2$) and the generation of electricity through the electrochemical oxidation of H_2/CO to H_2O/CO_2 since the kinetics of electrooxidation of H_2 are much faster than CH_4. A thermal catalyst capable of transforming CH_4 to H_2 is utilized, and a highly active catalyst is essential for fast kinetics at intermediate temperatures. A unique catalyst consisting of two sets of single-atom sites, Ru_1 and Ni_1, has been developed for this purpose. Figure 7.1a schematically illustrates the structure of this intermediate-temperature fuel cell [5]. It comprises a cathode, electrolyte, and anode. Fuel molecules such as H_2 are oxidized on the anode to form protons, generating electrons and protons. O_2 is reduced on the cathode, producing oxygen anions that transfer from the cathode to the anode through the electrolyte. The received O^{2-} on the anode couples with protons, forming H_2O molecules. The flowing electrons from the anode to a load generate electrical current. The cell potential and electrical current determine the power of the fuel cell. Multiple fuel cells can be arranged in parallel or series for various applications, including driving transportation vehicles and providing power for household electrical facilities and various machines of public facilities.

The layer closest to the electrolyte consists of the fine Ni-$BaZr_{0.1}Ce_{0.7}Y_{0.1}Yb_{0.1}O_{3-\delta}$ (Ni-BZCYYb) anode functional layer (AFL), strategically designed to maximize the phase boundary of three phases: the AFL, anode supporting layer (ASL), and the gas phase. The surface of the Ni-BZCYYb anode supporting layer is decorated with $Ce_{0.8}Sm_{0.2}O_2$ (SDC) nanoparticles, which inhibit coke formation [5, 13]. Below the ASL lies an anode reforming layer (ARL) directly exposed to the flow of raw fuel, which in this case is CH_4. The ARL serves as a thermal catalyst, converting raw CH_4 fuel to H_2 through thermal catalysis. This thermal catalyst comprises a unique single-atom catalyst with a CeO_2 support and two sets of single-atom sites, Ru_1 and Ni_1, anchored on CeO_2. The synergistic effect of the single-atom sites Ru_1 and Ni_1 was described in the section on steam reforming of CH_4.

The cathode backbone consists of crushed $PrBa_{0.5}Sr_{0.5}Co_{1.5}Fe_{0.5}O_{5+\delta}$ (PBSCF) coated with nanoparticles of PrO_x, known for its high activity in oxygen reduction. The electrolyte is a $Ce_{0.8}Sm_{0.2}O_2$ thin film. Figure 7.1b depicts the SEM image of the cut cross-section of the cell, revealing an ASL composed of porous Ni-BZCYYb with a thickness of approximately 800 μm, an AFL layer of Ni-BZCYYb with a thickness of only about 10 μm, a densely packed SDC layer electrolyte with a thickness of 15 μm, and the crushed nanofiber cathode – PBSCF with a thickness of 25 μm – coated with PrO_x nanoparticles. The ARL, serving as the thermal catalyst layer, has a thickness of 50 μm and is attached to the ASL layer, as depicted in Figure 7.1c and d.

The electrochemical performance of this type of fuel cell was evaluated at 500 °C using a CH_4 flow with a partial pressure of H_2O of 3.5%. Figure 7.1e illustrates the *I–V* and *I–P* curves of a single cell. Based on Figure 7.1e, the peak power density of such a single cell is $0.37\,W\,cm^{-2}$ at 500 °C when fueled with dry CH_4 (96.5% CH_4 + 3.5% H_2O). Notably, the power density of this cell surpasses the reported highest value, $0.142\,W\,cm^{-2}$, documented for a protonic ceramic fuel cell that utilized steam-rich CH_4 instead of nearly dry CH_4 [14]. Household-available natural gas has a composition close to the nearly dry CH_4 used in this test [5]. This cell can sustain operation for at least 380 hours with an output power of $0.28\,W\,cm^{-2}$ at 0.75 V under continuous operation without interruption. This suggests that the cell, particularly the nanocomposite anode, can stably function in both thermal catalytic conversion of natural gas and electrooxidation. Characterizations of the anode

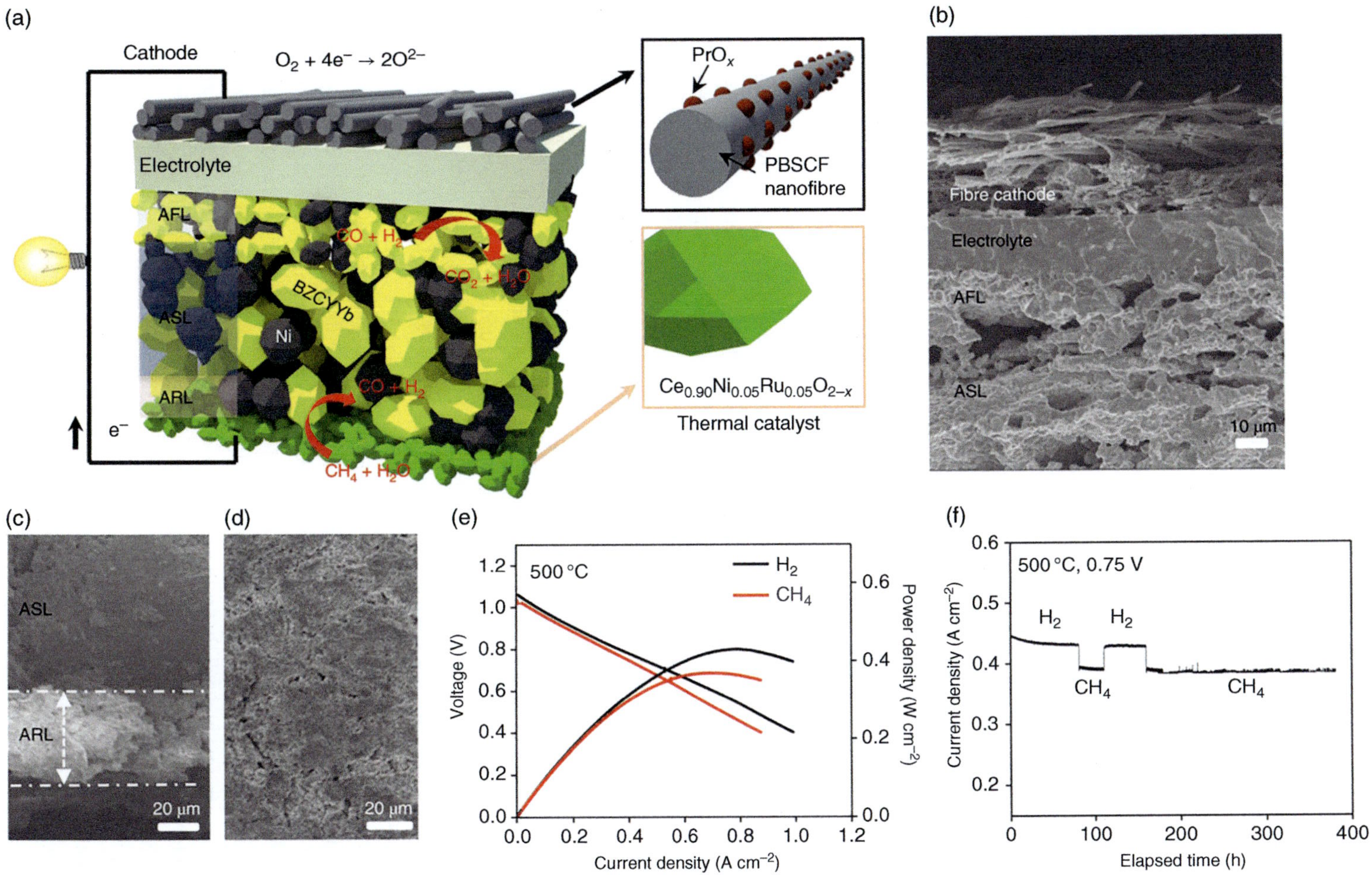

Figure 7.1 Structure and performance of an intermediate-temperature fuel cell. (a) Schematics of a single cell (the yellow and gray grains represent the $BaZr_{0.1}Ce_{0.7}Y_{0.1}Yb_{0.1}O_{3-\delta}$ [BZCYYb] and the Ni phase, respectively). (b) A scanning electron micrograph of a single cell, including a fiber cathode, a dense electrolyte, an AFL, and an ASL. (c) A cross-sectional view of the ASL and the ARL. (d) A top–down view of the ARL. (e) *I–V* and *I–P* curves of a single cell when operated at 500 °C using H_2 (black lines) or methane (red lines) (with ~3.5% H_2O) as the fuel and ambient air as the oxidant. (f) The current density of a single cell at a constant cell voltage of 0.75 V when methane (with ~3.5% H_2O) or H_2 was used as the fuel and ambient air was used as the oxidant. *Source:* Reproduced with permission from [5] / Springer Nature.

materials indicate the lack of coke formation after 380 hours of operation. The durability of this cell stands in stark contrast to the complete deactivation observed in a state-of-the-art Ni-based anode after only 10 minutes of operation, as observed in a comparative study of the two types of fuel cells using nearly dry CH_4 [5].

Another significant advantage of this fuel cell is the low cost of the fuel. It operates using nearly dry CH_4 consisting of 96.5% CH_4 and 3.5% H_2O. In contrast, a protonic ceramic fuel cell must be fueled with steam-rich methane composed of 28.6% CH_4 and 71.4% H_2O for steam reforming of CH_4. However, the intermediate-temperature fuel cell demonstrated in Figure 7.1 operates efficiently using nearly dry CH_4. This is possible due to the generation of H_2O through the coupling between O^{2-} at the anode, migrating from the cathode through the electrolyte, and H^+ generated through the oxidation of H_2 on the anode. The use of nearly dry CH_4 in Figure 7.1 significantly reduces the cost of the fuel cell, as a water management system is typically costly. This advantage also stems from the well-designed thermal catalysts, which are highly active at 500 °C and resistant to coke formation during steam reforming. The catalytic mechanism of $Ce_{0.90}Ni_{0.05}Ru_{0.05}O_2$ at 500 °C was described in the section on steam reforming of CH_4.

References

1 Linan, A., Williams, F. A. 1993. *Fundamental Aspects of Combustion*. New York, Oxford.

2 Pfefferle, L. D., Pfefferle, W. C. 1987. "Catalysis in combustion." *Catal. Rev. Sci. Eng.* 29, 219–267.

3 Lunsford, J. H. 1989. "The role of surface generated gas-phase radicals in catalysis." *Langmuir* 5, 12–16.

4 Zarur, A. J., Ying, J. Y. 2000. "Reverse microemulsion synthesis of nanostructured complex oxides for catalytic combustion." *Nature* 403, 65–67.

5 Chen, Y., deGlee, B., Tang, Y., Wang, Z., Zhao, B., Wei, Y., Zhang, L., Yoo, S., Pei, K., Kim, J. H., Ding, Y., Hu, P., Tao, F., Liu, M. 2018. "A robust fuel cell operated on nearly dry methane at 500 °C enabled by synergistic thermal catalysis and electrocatalysis." *Nat. Energy* 3, 1042–1050.

6 Aliotta, C., Liotta, L. F., Deganello, F., La Parola, V., Martorana, A. 2016. "Direct methane oxidation on $La_{1-x}Sr_xCr_{1-y}Fe_yO_{3-\delta}$ perovskite-type oxides as potential anode for intermediate temperature solid oxide fuel cells." *Appl. Catal. B* 180, 424–433.

7 Kan, H., Lee, H. 2010. "Sn-doped Ni/YSZ anode catalysts with enhanced carbon deposition resistance for an intermediate temperature SOFC." *Appl. Catal. B* 97, 108–114

8 Luo, X. J., Fong, K. F. 2016. "Development of 2D dynamic model for hydrogen-fed and methane-fed solid oxide fuel cells." *J. Power Source* 328, 91–104.

9 Wang, F., Zhou, Q., He, T., Li, G., Ding, H. 2010. "Novel $SrCo_{1-y}Nb_yO_{3-\delta}$ cathodes for intermediate-temperature solid oxide fuel cells." *J. Power Source* 195, 3772–3778.

10 McAnulty, M. J., Poosarla, V. G., Kim, K. Y., Jasso-Chávez, R., Logan, B. E., Wood, T. K. 2017. "Electricity from methane by reversing methanogenesis." *Nat. Commun.* 8, 15419.

11 Lo Faro, M., Antonucci, V., Antonucci, P. L., Aricò, A. S. 2012. "Fuel flexibility: A key challenge for SOFC technology." *Fuel* 102, 554–559.

12 Sengodan, S., Choi, S., Jun, A., Shin, T. H., Ju, Y. W., Jeong, H. Y., Shin, J., Irvine, J. T. S., Kim, G. 2015. “Layered oxygen-deficient double perovskite as an efficient and stable anode for direct hydrocarbon solid oxide fuel cells.” *Nat. Mater.* 14, 205–209.

13 Zhu, W., Xia, C., Fan, J., Peng, R., Meng, G. 2006. “Ceria coated Ni as anodes for direct utilization of methane in low-temperature solid oxide fuel cells.” *J. Power Source* 160, 897–902.

14 Duan, C., Tong, J., Shang, M., Nikodemski, S., Sanders, M., Ricote, S., Almansoori, A., O’Hayre, R. 2015. “Readily processed protonic ceramic fuel cells with high performance at low temperatures.” *Science* 349, 1321–1326.

8

Brief of Thermodynamics of Transformation of Methane at Low Temperature

8.1 Feasibility of Methane Conversion at Low Temperature through Oxidation

The large-scale conversion of methane into value-added chemicals at a low temperature is a critical strategic pursuit in the transition from petroleum-based to methane-based resources for manufacturing industries. However, the chemical transformation of CH_4 poses significant challenges in both catalysis science and chemical industries. For instance, the efficient conversion of methane into methanol has been identified as 1 of the 10 challenges in catalysis. This challenge arises from the unique chemical properties of CH_4 – it is the smallest organic molecule with the strongest C–H bond. The bond dissociation energy of the C–H bond in methane is approximately 46 kJ mol^{-1} higher than that of methanol, leading to low selectivity in methanol production from CH_4. This is primarily due to the easier oxidation of the product methanol compared to the reactant CH_4 itself. The lower proton affinity, reduced electrophilicity, higher ionization potential, and lower deprotonation capability of CH_4 compared to methanol contribute to the low selectivity for product formation during oxidation at low temperatures.

Apart from selectivity challenges, the activity in terms of the capability to be transformed into methanol or other partially oxidized products is another hurdle. The oxidation states of carbon and hydrogen in CH_4 are the lowest −4 of C and highest +1 of H, respectively. Hydrogen typically exhibits a +1 oxidation state in most compounds, such as H_2O and CH_3OH. Thus, the primary reactions involving the transformation of CH_4 must be oxidation aside from nonoxidative processes like $CH_4 \rightarrow C + 2H_2$, $2CH_4 \rightarrow CH_3CH_3 + H_2$, and $6CH_4 \rightarrow C_6H_6 + 9H_2$. Unfortunately, these nonoxidative processes are thermodynamically unfavorable at temperatures of ≤250 °C at all. In contrast, the oxidation of CH_4 to form organic oxygenates such as CH_3OH or CH_3COOH is thermodynamically favorable at temperatures below 250 °C, making it the predominant approach for transforming CH_4 to value-added chemicals at a relatively low temperature (<250 °C) or even room temperature.

In principle, a variety of oxidants can be used for the oxidation of CH_4, including O_3, H_2O_2, $S_2O_8{}^{2-}$, $IO_4{}^{-}$, H_2SO_4, SO_3, Cu^{2+}, Ce^{4+}, Ce^{3+}, Co^{3+}, Cl_2, SbF_5, and many others. Many of these oxidants have high oxidizing potential in contrast to O_2, making them capable of oxidizing CH_4 more effectively than O_2 from a thermodynamic standpoint. However, from

Low-temperature Activation and Catalytic Transformation of Methane to Non-CO_2 Products,
First Edition. Franklin Tao.

an economic perspective, O_2 is the preferred choice, with air being a nearly cost-free oxidant. However, despite its economic advantage, only a small portion of reported catalysts can catalyze the direct oxidation of CH_4 by O_2 under a mild condition (<250 °C). The majority of these reactions in this area have utilized non-O_2 oxidants such as H_2O_2, which can be synthesized from H_2 and O_2.

8.2 Why Should We Pursue a Low-temperature CH_4 Transformation Route?

From a thermodynamic perspective, transforming CH_4 into chemicals at room or low temperatures (≤250 °C) is indeed feasible. The question then arises: why opt for low-temperature routes instead of utilizing existing high-temperature processes? A key characteristic of high-temperature routes for CH_4 transformation is the substantial energy input they require. For instance, consider the reforming of CH_4 with steam, a widely adopted process in chemical industries for syngas and hydrogen generation. The reaction, $CH_4 + H_2O \rightarrow CO + 3H_2$, typically occurs on nickel nanoparticles supported on inert oxides at temperatures ranging from 800 to 1000 °C, with a feed gas pressure of 30–60 atm. Due to the thermodynamic constraints of this reaction, CH_4 reforming with steam is highly endothermic, necessitating a heat input of 60 kcal mol^{-1} [1]. In other words, the utilization of 1 mol of CH_4 requires at least 60 kcal of heat. In contrast, a reaction transforming CH_4 into a value-added chemical such as CH_3OH or acetic acid at room temperature or near-room temperature requires significantly less energy input.

Furthermore, most sources of CH_4 are situated in remote areas with shale rocks and exist as methane hydrates in the ocean, being primary examples. Shale rocks are typically found in mountains or rural regions, while methane hydrates are located offshore, far from existing industrial facilities for methane transformation. Building high-temperature reactors and related costly facilities near these inconvenient source locations incurs significant costs. Therefore, opting for expensive high-temperature reactors is not economically viable. Alternatively, transporting CH_4 from rural regions to existing facilities incurs high transportation cost. This presents a challenge for countries seeking to utilize domestic methane to produce essential intermediate compounds for chemical industries. Hence, it is imperative to develop methods for transforming CH_4 into value-added, easily transportable liquid chemicals under mild conditions in terms of near-room temperature or at least at a relatively low temperature (≤250 °C), enabling local operation on a large scale.

Crude oil still serves as the primary source of crucial intermediate compounds in chemical industries. To replace crude oil with CH_4, an array of new chemical reactions must be developed. Figure 1.1 outlines a portion of the intermediate compounds and value-added chemicals used by chemical industries. Most of these intermediate compounds exampled in this chart are currently produced from crude oil under high-temperature reaction conditions. While most reported studies focus on the production of methanol, ethanol, acetic acid, and formate, the development of new catalytic routes for transforming CH_4 into important intermediate compounds of chemical and manufacturing industries other than methanol, ethanol, acetic acid, and formate is a crucial aspect of the low-temperature methane activation and catalytic transformation.

8.3 Significance of Catalyst Design for Compensating Slow Kinetics of Methane Conversion at Low Temperature

The oxidative transformation using an appropriate oxidant is thermodynamically viable at temperatures below 250 °C and even at room temperature. However, activating the C–H bond in methane presents an exceptionally challenging task due to its high bond strength, low electrophilicity, high ionization potential, and low deprotonation propensity compared to other hydrocarbons. The key to achieving low-temperature catalytic transformation lies in designing a catalyst that initiates a pathway with a low activation barrier for methane transformation at temperatures below 250 °C. Furthermore, such a catalyst is expected to remain inert in potential oxidations of the desired product. This feature is particularly crucial for achieving a low-temperature methane catalytic transformation with high selectivity for a desired oxidation product.

Reference

1 Steinberg, M. 1999. "Fossil fuel decarbonization technology for mitigating global warming." *Int. J. Hydrogen Energy* 24, 771–777.

9

Activation of CH_4 by Free-standing Cations (M^+ or M_a^{n+}) of Single Atom or Cluster at Room Temperature and Its Significant Indication for CH_4 Low-Temperature Activation

In this book, the term "low-temperature transformation of CH_4" refers to a type of reaction performed at temperatures ranging from room temperature to approximately 250 °C. This reaction can take either a noncatalytic or catalytic process. In general, terms such as C–H activation and CH_4 activation have been used interchangeably, denoting a noncatalytic reaction that transforms CH_4 into value-added chemicals. Low-temperature catalytic transformation of CH_4 is typically oxidative catalysis of CH_4. Oxidants, including H_2SO_4, SO_3, O_3, N_2O, NO, O_2, and H_2O_2, have been utilized in the literature. Numerous studies on the catalytic oxidation of CH_4 at temperatures up to 250 °C have been reported. These activations or catalytic transformations of CH_4 can be performed in a gas phase, liquid phase, or at solid–liquid interfaces as reported in the literature. Chapters 9–21 will cover the low-temperature activation and catalytic transformation of CH_4 into value-added chemicals, with a focus on the fundamental understanding of the activation and catalytic mechanisms at a molecular level.

Metals can take various forms in the activation or catalysis of CH_4, ranging from supported metal nanoparticles and more intricate bimetallic nanoparticles to supported singly dispersed bimetallic sites, single atoms of metal anchored on support, metals coordinating with ligands, metal-oxide ions (MO^+), and even naked metal atoms (M^+). Undoubtedly, the naked, free-standing, single-atom (M^+) or dual-atom cation (M_2^+) serve as the simplest model for investigating how a metal can activate an organic molecule, with our focus here specifically on CH_4. It is worth emphasizing that the naked M^+ or single-atom cation provides the most straightforward model. Fundamental studies of the reaction between M^+ and CH_4 offer a clear chemistry picture, providing significant insight for a deep understanding of catalytic reactions performed on a metal catalyst. Low-pressure gas-phase studies involving a single reactant, CH_4, in a clean environment provide an opportunity to explore reaction thermodynamics and kinetics in an unperturbed "model" setting.

In this chapter, the activations of CH_4 to form MCH_2^+ or ethylene by the simplest model catalysts, including naked single-atom cations (M^+), dual atom clusters (M_2), or double atomic cations (M_2^+), at room temperature in CH_4 at a low pressure were discussed.

Low-temperature Activation and Catalytic Transformation of Methane to Non-CO_2 Products, First Edition. Franklin Tao.

9.1 Activity in Dehydrogenation of CH_4 and Reaction with Other Hydrocarbons on Free-standing Cation of Single-atom M^+ of the First-row (3d) Transition Metals and Its Indication for CH_4 Low-Temperature Activation

As early as 1981, it was reported that the gas-phase reaction of Cr^+ with CH_4 resulted in the formation of $CrCH_2{}^+$ and H_2, with additional byproducts $CrCH_3{}^+$ and CrH^+ [1]. This reaction, depicted schematically in Figure 9.1, is exothermic [1]. Notably, thermodynamic calculations reveal that the reaction between ground-state Cr^+ and CH_4 is inherently endothermic. Therefore, the observed exothermic reaction is attributed to the high energy state of Cr^+ in terms of its excited state. This finding aligns with previous reports indicating the coexistence of ground and excited states of Cr^+ cations generated from the electronic impact of $Cr(CO)_6$ observed by an ion cyclotron resonance spectrometer.

For these first-row transition metals, it is reported that only Sc^+, Ti^+, V^+, and Cr^+ can dehydrogenate CH_4 to form $MCH_2{}^+$ and H_2. In addition, in general, M^+ of the first-row transition metals exhibit a much lower activity in contrast to the third-row metal in the same group [2]. The difference in activity is rationalized in Section 10.3. The activity is influenced by both electron configuration and spin state of M^+. Specifically, if Cr^+ is switched from a ground state to an electronically excited state, the thermodynamics of this activation changes from the endothermic thermodynamics associated with ground-state Cr^+ to an exothermic one. In the second row of transition metal cations, M^+, only Zr^+ has been identified to be active for CH_4 dehydrogenation [3].

In contrast to the second-row transition metal ions, many third-row transition metal ions exhibit activity to activate C–H of CH_4, and several can even undergo sequential reactions to form a hydrocarbon chain bonded to the metal atom, $MC_nH_{2n}{}^+$ [4–6]. $M^+ + nCH_4 \rightarrow MC_nH_{2n}{}^+ + nH_2$. The formation of hydrocarbon chains (C_nH_{2n}) on M^{3+} of the third row of transition metals suggests that a single-atom metal center may be sufficient for Fischer–Tropsch synthesis (FTS) [2], challenging the conventional notion that supported metal or metal carbide nanoparticles are the active phase in FTS, as commonly reported in the literature.

Some M^+ cations of the first-row transition metals such as Cr^+ is active for both β-hydrogen migration to lose H_2 to form a dehydrogenation product such as $M{=}C_2H_4$ and β-alkyl migration to lose a smaller alkane to form a small alkane molecule such as CH_4 although the M^+ of the third-row metals such as W^+ does not conduct a β-alkyl migration reaction. Thus, there is no product of C–C cleavage formed on W^+. This difference is discussed in Section 9.3.

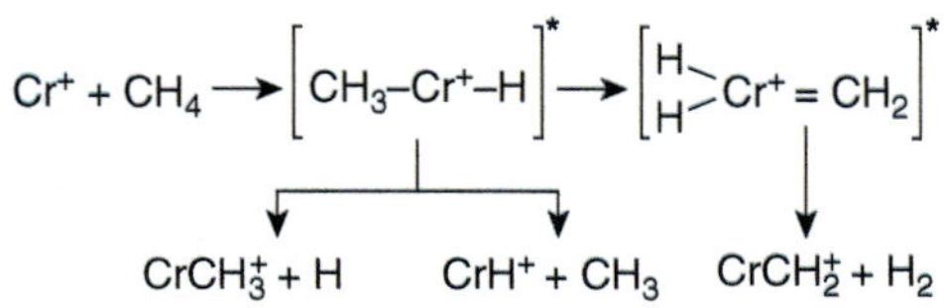

Figure 9.1 Schematics showing the reaction between Cr^+ and CH_4. *Source:* Reproduced with permission from [1] / American Chemical Society.

9.2 Activity in Dehydrogenation of CH_4 on Free-standing Cation of Single-atom M^+ of the Third-row (5d) Transition Metals and Its Indication for CH_4 Low-Temperature Activation

In general, M^+ of the third-row transition metal is much more active in the dehydrogenation of CH_4 than the first-row and second-row metal in the same group. The deep reasons are discussed in Section 10.3. For instance, Ta^+ can dehydrogenate CH_4 to form $TaCH_2^+$ and H_2 [7]. $TaCH_2^+$ can even further couple with another CH_4 to form $TaC_2H_4^+$ till $TaC_4H_8^+$. As shown in Figure 9.2, being one of the products, $TaC_2H_4^+$ was clearly observed by mass spectroscopy [7]. These observed $TaC_2H_4^+$ were formed by the reaction of Ta^+ in the CH_4 gas with a pressure of only 3.3×10^{-6} Torr. An isotope-labeled reaction between Ta^+ and a mixture of CH_4 and CD_4 (bottom panel of Figure 9.2) found that $TaCHD^+$ was formed. It suggests the dissociation of $TaCH_2^+$ and $TaCD_2^+$ must happen, forming CH_2^+ and CD_2^+ radicals. Then, CH_2^+ and CD_2^+ scatter to form CHD^+.

W^+ exhibits high activity in activating CH_4. W^+ can couple with CH_4 to form WCH_2^+ [2]. A subsequent coupling between WCH_2^+ and CH_4 can even form $WC_8H_{16}^+$. The ligand $C_8H_{16}^+$ of $WC_8H_{16}^+$ is much longer than the C_2H_4 formed in $TaC_2H_4^+$ [2]. In terms of Re^+, the reaction between Re^+ and CH_4 to form $ReCH_2^+$ is endothermic [2]. It makes sense that $ReCH_2^+$ can readily dissociate to Re^+ and CH_4 with the aid of H_2 since this reaction ($ReCH_2^+ + H_2 \rightarrow Re^+ + CH_4$), the reverse reaction of CH_4 activation ($ReCH_2^+ + H_2 \rightarrow Re^+ + CH_4$) is exothermic. Simply, the Re^+ is nearly inactive for reaction with CH_4. Interestingly, ReO^+ exhibits a reaction rate similar to W^+ [2]. In other words, ReO^+ is much more active than Re^+. Similar to ReO^+, a great number of MO^+ (M = transition metal) clusters are active for oxidative of CH_4 to form CH_3OH and M^+. Chapter 11 will discuss the oxidation of CH_4 by MO^+ cluster at room temperature in low-pressure CH_4. ReO^+ can be produced through electron impact on the vapor of heated ReO_3 [2].

Regarding Os^+, there was no report of the activity of Os^+ for CH_4 activation, although OsO_n^+ (n =0–4) is active for oxidating CH_4 [4]. Ir^+ is active for reacting with CH_4, forming $IrCH_2^+$, $IrC_2H_4^+$, and $IrC_3H_6^+$ [2]. Different from the activation of CH_4 by Ir^+, Ir^+ can activate C_2H_6 to form $IrC_6H_8^+$ and $IrC_6H_{10}^+$. Pt^+ can couple with CH_4 to form $PtCH_2^+$, which subsequently reacts with CH_4 till the formation of $PtC_5H_{10}^+$ [2]. Different from the reactive nature of Ir^+ and Pt^+ to CH_4, Au^+ is not active for the dehydrogenation of CH_4 to form $AuCH_2^+$ and H [8]. Similar to Au^+, Hg^+ is not active for the dehydrogenation of CH_4 to form $HgCH_2^+$ and H_2.

These studies show in general, the first two row transition metal ions, M^+, are not active. In addition, some third-row transition metal ions, including La^+, Hf^+, Re^+, and Au^+, are not active. However, these third-row transition metal ions, including Ta^+, W^+, Re^+, Ir^+, and Pt^+, are active for dehydrogenation of CH_4 to form $MC_nH_{2n}^+$ species at low pressure of CH_4 (10^{-6}–10^{-5} Torr CH_4) [2].

The activity of M^+ in the dehydrogenation of CH_4 to MCH_2^+ also depends on the valence state [2]. Compared to M^+, the valence d orbitals of M^{2+} contract substantially. This contraction makes the size of valence orbitals $(n\text{-}1)$d of M^{2+} similar to the size of open-shell orbitals of the CH_2 fragment [2]. This makes a more effective overlap between the $(n-1)$d of M^{2+} and orbitals of CH_2. In addition, the increase of charge (from +1 of M^+ to +2 of

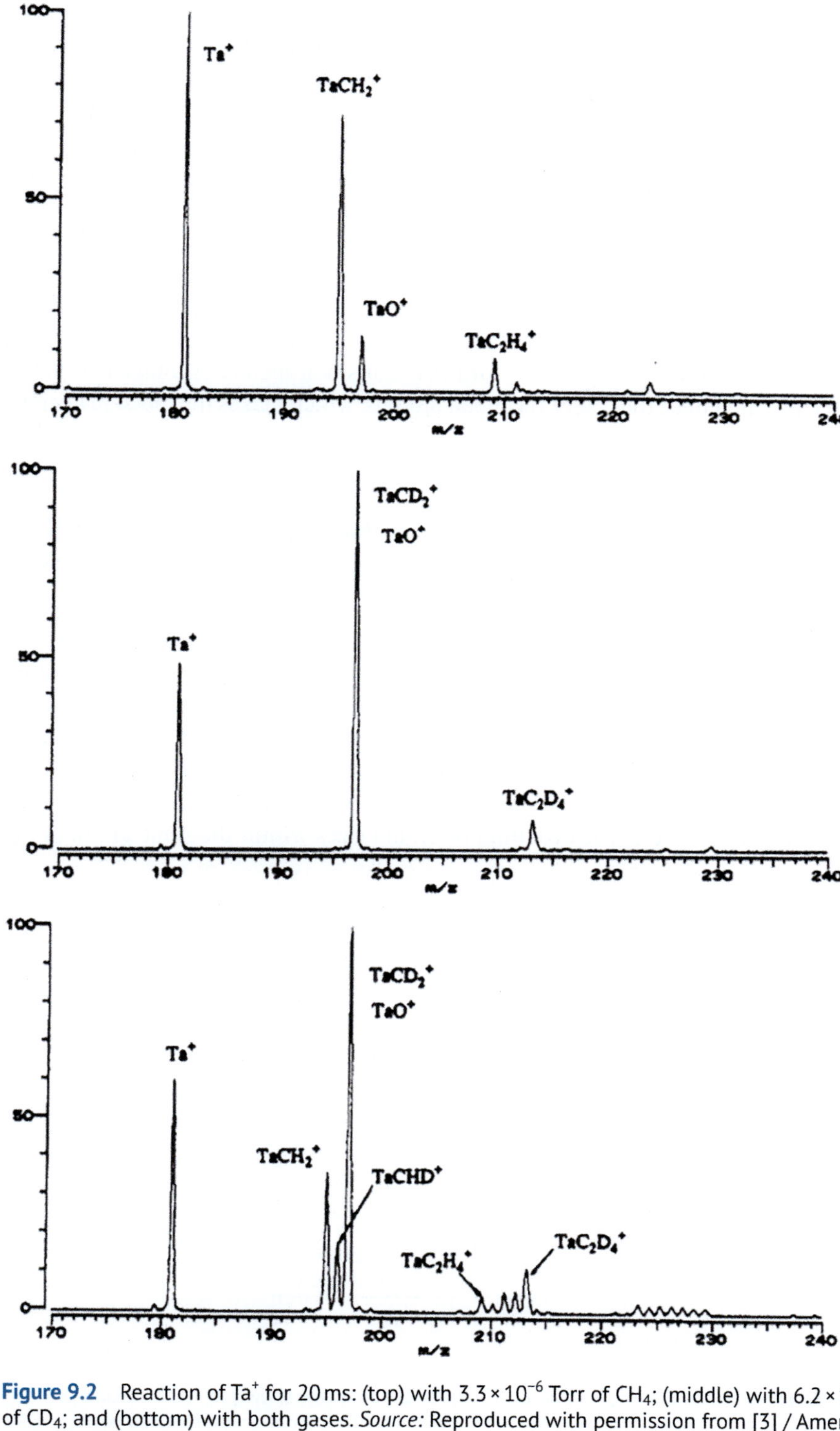

Figure 9.2 Reaction of Ta^+ for 20 ms: (top) with 3.3×10^{-6} Torr of CH_4; (middle) with 6.2×10^{-6} Torr of CD_4; and (bottom) with both gases. *Source:* Reproduced with permission from [3] / American Chemical Society.

M^{2+}) has a larger stabilization effect on (n – 1)d than ns. Thus, the σ bond of M=C of $[M{=}CH_2]^{2+}$ has the least s character of the metal. Since the σ bond of M=C mainly has the character of the d orbital of the metal in the case of a high charge, the third-row (6d) transition metal can form two strong bonds in MCH_2^{2+} [2].

Different from the M^+ of the first-row transition metals, the bonding of $M{=}CH_2^+$ with one or more CH_2 to form $MC_2H_4^+$ or $MC_nH_2^{n+}$ can be rationalized with a few factors. One factor is the exchange energy loss. The exchange energy loss is the loss of exchange energy upon bonding between M^+ and CH_2 or between $MC_2H_4^+$ and CH_2. For instance, bonding between Re^+ and CH_2 causes Re^+ to lose exchange energy of 58 kcal mol^{-1}. The bonding between $ReCH_2^+$ and CH_2 loses 36 kcal mol^{-1} [2]. Based on the reference [2], since the first CH_2 pays enough loss of exchange energy, the second CH_2 can strongly bonds to MCH_2^+ to form $MC_2H_4^+$ and even $MC_nH_{2n}^+$ for the third-row transition metals. Another reason is that the first CH_2 can reduce the size of the metal orbitals, which makes the metal ready for bonding a second ligand [2].

9.3 Factors Leading to the Difference between High Activity of 5d Transition Metal Ion to CH_4 Dehydrogenation and Nearly Inertness of 3d or 4d Metal Ion

As represented in Figure 9.3, the literature reports that the M–C bond strengths of 5d transition metals are stronger than those of 3d and 4d transition metals [9]. Thus, the activities in the dehydrogenation of CH_4 by M^+ of 5d transition metals are stronger than 3d and 4d metals. This notable difference can be explained by considering the size of the valence orbitals and the stability of M^+ [2]. The contraction effect of lanthanide series elements and relativistic effects contribute to the evolution of orbital sizes within the same group [10].

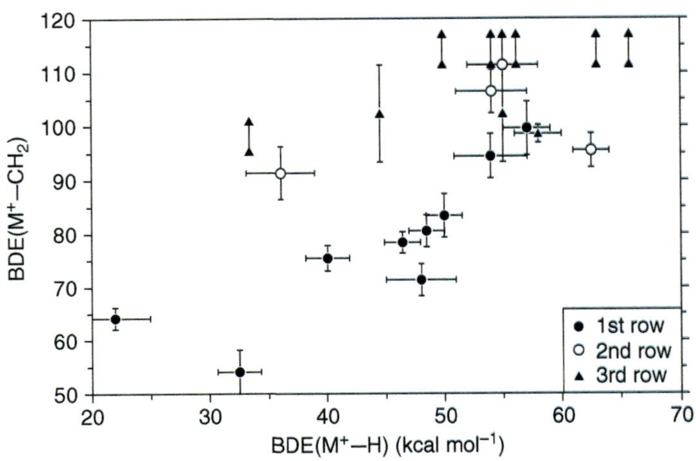

Figure 9.3 Bond strengths $D(M^+{=}CH_2)$ and $D(M^+{-}H)$ for the first row shown in •, the second row shown in (O), and the third-row transition metals shown in triangle. *Source:* Reproduced with permission from [2] / American Chemical Society.

These favorable changes in orbital size give a better bonding overlap between M^+ and CH_4 and decrease the loss of exchange energy in bonding [9]. Thus, 5d transition metals exhibit higher activation in the dehydrogenation of CH_4 at room temperature. However, the size of d orbitals increases from the first-row transition metal to the third row ($r_{3d} < r_{4d} < r_{5d}$), the size of s orbitals does not follow this trend. According to the literature, the sizes of 4s orbitals of first-row transition metals (also called 3d transition metals) are very similar to those of 6s orbitals [11]. The sizes of 6s orbitals are smaller than the 5s orbitals of second-row transition metals ($r_{4s} \approx r_{6s} < r_{5s}$), called orbital contracts. Consequently, the evolutions of sizes of 3d, 4d, and 5d orbitals, as well as 4s, 5s, and 6s orbitals, result in valence orbitals 5d and 6s of the third-row metals having similar sizes ($r_{5d} \approx r_{6s}$). This leads to a more effective 6s–5d hybridization, making them more active toward carbon atoms of CH_4 and forming a strong $M{-}CH_2$ bond for third-row transition metals (also called 5d transition metals).

On the contrary, the activity of M^+ in dehydrogenating CH_4 to form $MCH_2{}^+$ depends on orbital size since the lengths of σ and π bonds of M=C in $MCH_2{}^+$ are partially determined by the size of valence orbitals, *n*s and (*n* – 1)d [2]. For the first and second rows of transition metals, their 4s and 5s orbitals are larger than their 3d and 4d orbitals, respectively. The σ bond of the M=C is mainly contributed from the *n*s orbital, while the π bond is substantially contributed from (*n* – 1)d orbital [12]. A large difference in size between valence orbitals *n*s and (*n* – 1)d makes the σ and π bonds of M=C compromise to reach the same size for the σ and π bonds [2]. More compromise makes the M=C bond weaker. For the first-row (3d) and second-row (4d) transition metals, the M=C is weaker as the difference in orbital size between *n*s and (*n* – 1)d is larger. Compared to the first-row and second-row metals, the similar sizes of 6s and 5d orbitals of the third-row (6d) transition metals do not request an obvious compromise, and thus the M=C is strong. Thus, the bond strengths of M=C in terms of bond dissociation energy of the third-row metals are larger than those of the first-row and second-row transition metals (Figure 9.3). The strong M=C bond suggests that M^+ cations of the third-row (6d) transition metals are more active in the dehydrogenation of CH_4. The orbital size effect can readily explain the lack of activity of an M^+ of the first-row transition metal such as Cr^+, the low activity of Mo^+ in reacting with CH_4, and the high activity of W^+ for the VIB group metals.

The correlation of the orbital size effect with the activity in the dehydrogenation of CH_4 can be interpreted from the angle of kinetics. This understanding results from the high activity of M^+ of the third-row transition metals in the dehydrogenation of CH_4, but the lack of activity in the de-methanation of ethane through C–C cleavage. In the activation of C_2H_6, there are competing reaction channels, including β-hydrogen migration to lose H_2 and β-alkyl migration to lose a smaller alkane [2, 13]. The involved orbitals of β-alkyl to be migrated are more directional, while the participating orbital of an H atom to be migrated are not. As the valence orbitals of the late transition metals, particularly the third-row metals to be used in bonding with H or alkyl have a larger d character [11], a coupling between the orbitals of M^+ having a larger d character with the least directional orbitals of H atom is favorable in contrast to the quite directional orbital of alkyl. In other words, the larger d character of M^+ of the third-row transition metal makes it preferentially perform β-hydrogen migration to lose H_2 through dehydrogenation. But for M^+ of the first-row metal, the valence orbital involved in the reaction is less d character, thus being less directional. Thus, the coupling of less directional valence orbitals of these M^+ of 3d or

3d transition metal elements can couple with the directional orbitals of alkyl of a hydrocarbon such as ethane to perform β-alkyl migration other than β-hydrogen migration, forming products of both C–C and C–H cleavage of C_2H_6 [2, 13–15].

9.4 Activity in CH_4 Dehydrogenation or C_2H_4 Formation on Free-standing Cluster $[M_a]^0$ or Cluster Cation $[M_a]^{n+}$ and Its Indication to CH_4 Low-Temperature Activation

The preceding sections have discussed the activity of various single-atom cations in their reactions with CH_4 molecules. These comparisons bear significant relevance to catalysis on single-atom sites anchored on a support or coordinated with ligands. In the realm of heterogeneous catalysis, real catalysts often involve metal or oxide nanoparticles supported on nonmetallic supports, primarily inert oxides, and transition metal oxides. In such catalysts, clusters consisting of more than one metal atom typically engage with a reactant molecule. Hence, on most supported catalysts, an active site comprises a couple, a few, or even several atoms of metal or even a couple of metal elements supported on an oxide or other compounds such as a carbide, nitride, or even sulfide. A model catalyst for such a supported polyatomic site is a free-standing polyatomic cation, also called a cluster cation. Investigation of the activity of a cluster cation is significant for designing new catalysts for a chosen reaction. For instance, it was found that single-atom cation, Au^+, is not active in activating C–H of CH_4 or bonding with CH_4 [2, 16]. However, Au^{2+} can bond with several CH_4 molecules and dehydrogenate to form ethylene under multiple-collision conditions [17].

The literature has systematically investigated the reaction between CH_4 and unsupported multiatom clusters, including alloy clusters. For instance, strong size-dependent activity of unsupported, neutral Pt_n clusters has been reported [18]. Platinum dimers (Pt_2) through platinum pentamers (Pt_5) are highlighted as the most active clusters in activating CH_4, with Pt_{6-24} clusters exhibiting lower activity under similar conditions. The decrease in activity, along with an increase in Pt_n cluster size, is likely attributed to the higher coverage of H atoms remaining on the Pt_n surface after CH_4 activation, paralleling the trend observed for H_2 adsorption on Pt_n. The size dependence is expected to originate from the unsaturated coordination, where a Pt atom with low coordination exhibits higher adsorption energy to a reactant molecule [18]. This size-dependent activity in reacting with CH_4 on metal clusters mirrors the size-dependent catalytic activity, and even catalytic selectivity, observed for reactions performed on metal nanoparticle catalysts [19].

Similar size dependence has been found for multiatom cations and even multiatom anions, such as Pt_n^+ or Pt_n^-, where multiatom cations tend to be more active than their anionic counterparts, likely due to the favorable electrophilic attack of Pt_n^+ to the C–H bond of CH_4 [20–23]. In addition to pure metal multiatom clusters or cluster cations, Pt-based alloy cluster cations, $Pt_mAu_n^+$ $(m + n \leq 4)$ $(m + n \leq 4)$ such as $PtAu^+$, have demonstrated activity in C–H activation of CH_4 and N–H activation of NH_3, as reported in the literature [24, 25].

Beyond Pt-based clusters, Rh-based clusters, including Rh_2^+ and $RhPt^+$, exhibit high activity for C–H activation of CH_4 and N–H activation of NH_3. Notably, Au_2^+ has been

Figure 9.4 Schematics representing the proposed catalytic cycle of this CH_4 dehydrogenation coupling on Au_2^+. *Source:* Reproduced with permission from [17] / John Wiley & Sons.

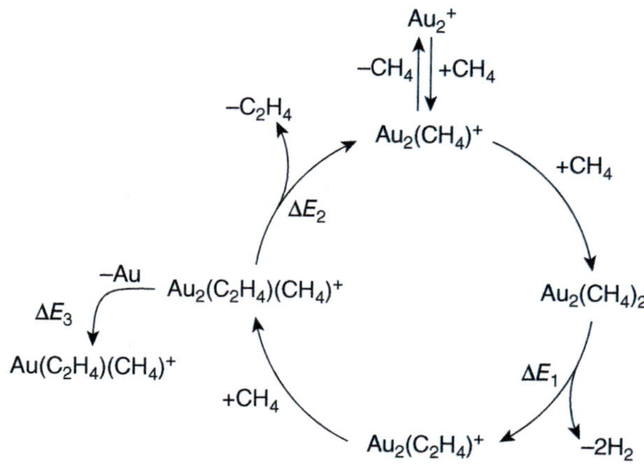

found to be active for the dehydrogenation coupling of CH_4 to form ethylene, exemplified by the reaction $Au_2^+ + 2CH_4 \rightarrow Au_2^+ + C_2H_4 + 2H_2$ [17]. This reaction sparks interest in the possibility of directly producing crucial intermediate compounds such as C_2H_4, one of the most demanded intermediate compounds of chemical industries, through the catalysis of CH_4 at room temperature or relatively low temperatures on polyatomic cluster cations. Figure 9.4 schematically presents the proposed catalytic cycle for this CH_4 dehydrogenation coupling on Au_2^+.

Computational studies [17] aided the interpretation of the CH_4 dehydrogenation to form C_2H_4 on Au_2^+. As shown in Figure 9.5, the chemisorption of a CH_4 on a naked Au_2^+ is exothermic in terms of releasing heat of 0.88 eV (see value of y-axis in Figure 9.5). This is accomplished by overlapping sp^3 orbitals of CH_4 with the d orbitals of Au_2^+. Computational studies found that the configuration with the lowest energy is that one of the Au atoms of Au_2^+ directly bonds to a CH_4 molecule. Subsequently, a second CH_4 molecule is adsorbed to the $Au_2CH_4^+$. This adsorption results in a further decrease of energy by 0.57 eV (see B in Figure 9.5). In the optimized structure (B), C and H of each CH_4 bridges on an Au atom. Structure S in Figure 9.5 is the $Au_2C_2H_4^+$ formed by losing two H_2 molecules of structure B, $Au_2(CH_4)_2^+$. The transformation of B to S experiences nearly ten elementary steps, as detailed in Figure 9.5. The formation and desorption of the first H_2 molecule is done from I to K through J in Figure 9.5 since K is $Au_2(CH_3)_2^+$. The release of this H_2 molecule is a thermodynamically favorable step. The release of the second H_2 molecule is the step from R to S, as shown in Figure 9.5. It is also a downhill step.

Based on the energy profile in Figure 9.5, the highest activation barrier in the transformation of $Au_2(CH_4)_2^+$ to $Au_2C_2H_4^+$ is the barrier from intermediate K to M [17]. In this step, the H atom of a CH_4 bonds to another Au atom, making a CH_4 bridge the two Au atoms of a Au_2^+. This is a key step toward binding CH_3 on the same Au atom before the C–C coupling of the two CH_3 can be conducted. Notably, the coupling in the step of M to O clearly confirms the necessity and significance of a biatom or polyatom site in a chain-propagating reaction, such as a C–C or C–N coupling on a catalyst surface.

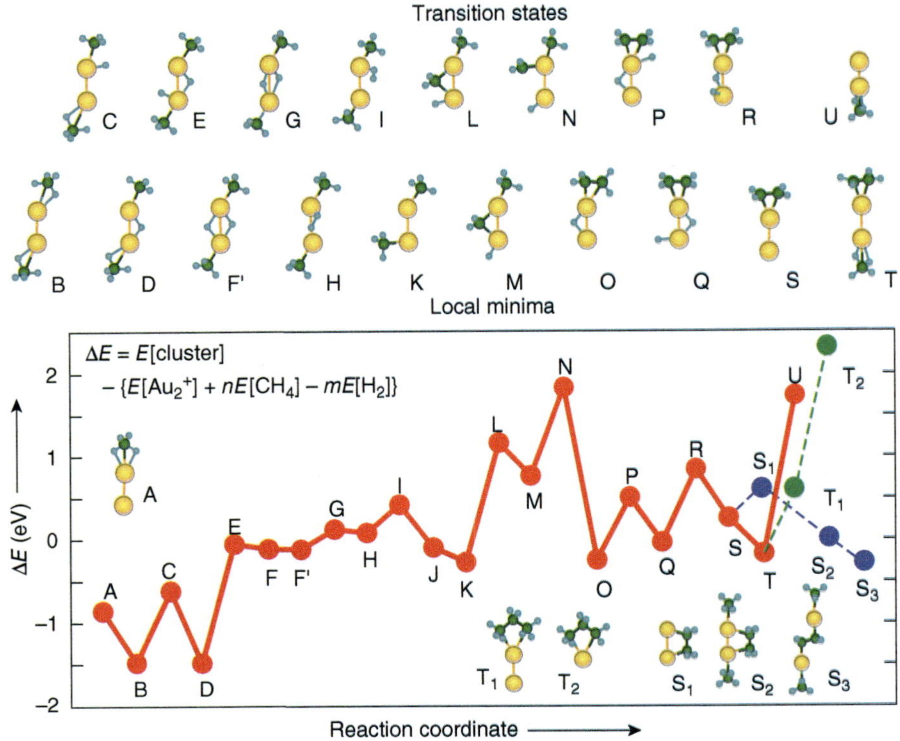

Figure 9.5 Calculated reaction pathways and the corresponding energy profiles. Atomic configurations corresponding to transition state barriers and local potential energy minima are shown in the top two rows (Au yellow, C green, H light blue). The pathway leading to the formation and desorption of ethylene, ending with $Au_2(CH_4)^+$ (denoted as U), is connected by a solid line. Alternative pathways are denoted by dashed lines: T→T2 (dashed green line) describes the dissociation of the gold cluster; S→S3 (dashed blue line), describes the formation of $Au_2(C_4H_{12})^+$. All the species identified in the mass spectra, and occurring in the reaction mechanism cycle (Figure 9.4), are found on the calculated reaction path. For configuration U, the curly bracket should include {... $E[C_2H_4]$}, and for T_2 it should include {... E[Au]}. *Source:* Reproduced with permission from [17]/ John Wiley & Sons..

References

1 Halle, L. F., Armentrout, P. B., Beauchamp, J. L. 1981. “Formation of chromium carbene ions by reaction of electronically excited chromium ions with methane in the gas phase.” *J. Am. Chem. Soc.* 103, 962–963.

2 Irikura, K. K., Beauchamp, J. L. 1991. “Electronic structure considerations for methane activation by third-row transition-metal ions.” *J. Phys. Chem.* 95, 8344–8351.

3 Ranasinghe, Y. A., MacMahon, T. J., Freiser, B. S. 1991. “Formation of thermodynamically stable dications in the gas phase by thermal ion-molecule reactions: Tantalum(2+) and zirconium(2+) with small alkanes.” *J. Phys. Chem.* 95, 7721–7726.

4 Irikura, K. K., Beauchamp, J. L. 1989. “Osmium tetroxide and its fragment ions in the gas phase: Reactivity with hydrocarbons and small molecules.” *J. Am. Chem. Soc.* 111, 75–85.

5 Armentrout, P. B., Beauchamp, J. L. 1989. "The chemistry of atomic transition-metal ions: Insight into fundamental aspects of organometallic chemistry." *Acc. Chem. Res.* 22, 315–321.

6 Irikura, K. K., Beauchamp, J. L. 1991. "Methane oligomerization in the gas phase by third-row transition-metal ions." *J. Am. Chem. Soc.* 113, 2769–2770.

7 Buckner, S. W., MacMahon, T. J., Byrd, G. D., Freiser, B. S. 1989. "Gas-phase reactions of niobium(1+) and tantalum(1+) with alkanes and alkenes. Carbon-hydrogen bond activation and ligand-coupling mechanisms." *Inorg. Chem.* 28, 3511–3518.

8 Chowdhury, A. K., Wilkins, C. L. 1987. "Reactions of atomic gold ions with aliphatic and aromatic hydrocarbons and alkyl halides." *J. Am. Chem. Soc.* 109, 5336–5343.

9 Simoes, J. A. M., Beauchamp, J. L. 1990. "Transition metal-hydrogen and metal-carbon bond strengths: The keys to catalysis." *Chem. Rev.* 90, 629–688.

10 Pyykko, P. 1988. "Relativistic effects in structural chemistry." *Chem. Rev.* 88, 563–594.

11 Ohanessian, G., Goddard III, W. A. 1990. "Valence-bond concepts in transition metals: Metal hydride diatomic cations." *Acc. Chem. Res.* 23, 386–392.

12 Ohanessian, G., Brusich, M. J., Goddard III, W. A. 1990. "Theoretical study of transition-metal hydrides. 5. Hafnium to mercury (HfH+ through HgH+), barium and lanthanum (BaH+ and LaH+)." *J. Am. Chem. Soc.* 112, 7179–7189.

13 Tolbert, M. A., Mandich, M. L., Halle, L. F., Beauchamp, J. L. 1986. "Activation of alkanes by ruthenium, rhodium, and palladium ions in the gas phase: Striking differences in reactivity of first- and second-row metal ions." *J. Am. Chem. Soc.* 106, 5675–5683.

14 Byrd, G. D., Freiser, B. S. 1982. "Gas-phase reactions of rhodium(+) ion with alkanes." *J. Am. Chem. Soc.* 104, 5944–5950.

15 Schilling, J. B., Beauchamp, J. L. 1988. "What is wrong with gas-phase chromium? A comparison of the unreactive chromium(1+) cation with the alkane-activating molybdenum cation." *Organometallics* 7, 194–199.

16 Li, F. X., Armentrout, P. B. 2006. "Activation of methane by gold cations: Guided ion beam and theoretical studies." *J. Chem. Phys.* 125, 133114.

17 Lang, S. M., Bernhardt, T. M., Barnett, R. N., Landman, U. 2010. "Methane activation and catalytic ethylene formation on free Au_2^+." *Angew. Chem. Int. Ed.* 49, 980–983.

18 Trevor, D. J., Cox, D. M., Kaldor, A. 1990. "Methane activation on unsupported platinum clusters." *J. Am. Chem. Soc.* 112, 3742–3749.

19 Cao, S., Tao, F., Tang, Y., Li, Y., Yu, J. 2016. "Size- and shape-dependent catalytic performances of oxidation and reduction reactions on nanocatalysts." *Chem. Soc. Rev.* 45, 4747–4765.

20 Achatz, U., Berg, C., Joos, S., Fox, B. S., Beyer, M. K., Niedner-Schatteburg, G., Bondybey, V. E. 2000. "Methane activation by platinum cluster ions in the gas phase: Effects of cluster charge on the Pt_4 tetramer." *Chem. Phys. Lett.* 320, 53.

21 Kummerlçwe, G., Balteana, I., Sun, Z., Balaj, O. P., Bondybey, V. E., Beyer, M. K. 2006. "Activation of methane and methane-d4 by ionic platinum clusters." *Int. J. Mass Spectrom.* 254, 183–188.

22 Adlhart, C., Uggerud, E. 2006. "Reactions of platinum clusters $Pt_n^{\pm}$, n = 1–21, with CH_4: to react or not to react." *Chem. Commun.* 2581–2581.

23 Adlhart, C., Uggerud, C. 2007. "Mechanisms for the dehydrogenation of alkanes on platinum: Insights gained from the reactivity of gaseous cluster cations, Pt_n + n =1–21." *Chem. Eur. J.* 13, 6883–6890.

24 Koszinowski, K., Schröder, D., Schwarz, H. 2003. "Probing cooperative effects in bimetallic clusters: Indications of C–N coupling of CH_4 and NH_3 mediated by the cluster ion $PtAu^+$ in the gas phase." *J. Am. Chem. Soc.* 125, 3676–3677.

25 Koszinowski, K., Schröder, D., Schwarz, H. 2004. "C–N coupling of methane and ammonia by bimetallic platinum–gold cluster cations." *Organometallics* 23, 1132–1139.

10

Oxidization of CH_4 by Free-standing MO^+ Clusters at Room Temperature in Low-pressure CH_4

10.1 Brief

One of the most potent oxidizing agents is cation. The early exploration of methane oxidation with cations anchored on oxide was likely inspired by the oxidation of CH_4 molecules by free-standing MO^+ clusters prepared in gas phase. Since the primary objective of this oxidation is to produce methanol, oxygen atoms are incorporated as part of the oxidizing agent. Thus, MO^+ can serve as an effective oxidizing agent, with M representing a metal, spanning a wide range of possibilities. Well-studied examples include FeO^+, CoO^+, and OsO^+. Reflecting on the early studies of oxidizing methane with these cations to produce methanol, it is crucial to acknowledge that these early investigations laid the foundation for the latest developments in the activation or catalytic oxidation of methane with M–O species anchored in micropores. Subsequent oxidation of the metal cation to form MO^+ is a prerequisite before the reuse of the M^{n+} cation. To facilitate the separation of metal atoms from products and reactants, approaches such as immobilizing them into micropores or onto the surface of support have been employed.

Oxidizing CH_4 with free-standing MO^+ represents an appropriate approach to gaining insights into reactivity trends, the role of ligand atoms, the formation of intermediates, and the efficiency and selectivity of ideal products at a molecular level.

10.2 Preparation of MO^+ Clusters

MO^+ cations, also known as clusters, are typically prepared using mass spectrometry techniques such as electron impact ionization, chemical ionization, ion/molecular reactions of bare M^+ cations with a stable oxidant and splitting methods. The electron impact ionization method can generate MO^+ from volatile precursors like anhydrous copper (II) for CuO_n^+ and osmium (VIII) tetraoxide [1, 2]. In the chemical ionization method, organometallic compounds and oxidants are mixed. For example, FeO^+ can be produced from the mixture of $Fe(CO)_5$ and N_2O [3] or $Fe(CO)_5$ and O_2 [4]. This chemical ionization method is commonly used for studying reactivities of MO^+ since MO^+ can be readily

Low-temperature Activation and Catalytic Transformation of Methane to Non-CO_2 Products,
First Edition. Franklin Tao.

produced from metals and simple oxidants such as O_2, O_3, or N_2O. Thermoionization in flames and photoionization are also methods for preparing MO^+ [5–9]. However, this preparation method may not allow for a simultaneous or immediate study of their reactivity with CH_4.

10.3 Experimental and Computational Approaches for Studying Reaction between MO^+ Cluster and CH_4

Experimental studies on the oxidation of MO^+ were typically conducted using ion cyclotron resonance (ICR) mass spectrometry. MO^+ cations are trapped in the magnetic field of the ion cyclotron setup, and the reactions are usually performed at room temperature. Checking whether a product can be detected serves as an indicator of the thermodynamic feasibility of the reaction at room temperature. For instance, reactivity in oxidizing CH_4 by MO^+ was investigated by monitoring the time dependence of the oxidant, MO^+ or/and the product ions, such as Fe^+ signals. The obtained time dependence of the ion signals of MO^+ or/and M^+ is analyzed by assuming a pseudo-first-order kinetics, generating a rate constant k_R. Similarly, the capture rate k_C is measured, and a reaction efficiency, defined as $\varphi = k_R/k_C$ is measured. φ serves as a measure of the reactivity of MO^+ and is essential for understanding thermodynamic or kinetic restrictions. By using isotope-labeled reactants [10, 11], the kinetics isotope effects on k_R and product branching ratios can be studied, providing insight into reaction mechanisms. Ligand-exchange reactions [12–14] and collision-induced dissociation experiments [15, 16] allow for identifying ionic products of the reactions. Tandem mass spectrometry can be used to characterize possible reaction intermediates, as reported in the literature [17]. Neutralization-reionization mass spectrometry (NRMS) [18–20] was employed to examine transition-metal-mediated oxidation reactions. The guided ion beam (GIB) technique was used to study thermochemistry [21–23] and reactivity [24, 25]. With GIB techniques, the bond dissociation energy (BDE) of Fe–O of FeO^+ was measured by finding the experimental threshold for the formation of M^+ by colliding MO^+ with xenon. GIB was also used to measure the thermodynamics of ions and neutral molecules [26, 27]. MO^+ in the gas phase was examined with the flame-ionization method [5–7, 28–35]. These complementary experimental techniques/methods allow for the development of a potential-energy surface for the oxidation of methane by MO^+ [36].

In addition to experimental techniques/methods, a computational study is another major approach to studying the oxidation of CH_4 with MO^+. Given the relatively small size of both reactants, CH_4 and MO^+, computational studies provided reliable predictions of transition states and intermediates [37–40]. Computational approaches can offer information on thermodynamics and kinetics, including activation barriers and vibrational signatures of intermediates or products, allowing for validation against experimental measurements. Particularly, the integration of experimental and theoretical approaches has led to a reliable prediction of the reaction pathway [41].

10.4 Chemical Properties of MO⁺ and Their Indications for Activity in Oxidizing CH_4

Bond dissociation energies (BDE) of M–O bond in MO^+ of 3d transition metals were experimentally measured. Figure 10.1 presents the BDE of M–O of MO^+ as a function of the atomic number of all 3d transition metals [36]. ScO^+, TiO^+, and VO^+ have higher BDE than CrO^+, MnO^+, FeO^+, CoO^+, NiO^+, CuO^+, and ZnO^+ [42], consistent with the higher oxophilicities of Sc, Ti, and V. These values are, in fact, lower than the M–O of the neutral metal oxide. In general, the BDEs of M–O of MO^+ increase for heavier elements in the same group [43, 44]. Considering MO^+ as a molecule, their molecular orbitals can be assigned. This information is crucial for studying thermodynamics and predicting symmetries and even multiplicities of ground states of MO^+. Figure 10.1 lists the occupancy of electrons at the ground state of MO^+ [41]. Based on the information in Figure 10.1, 1σ, 2σ, and 1π are bonding valence orbitals of ScO^+ that are fully occupied at low-energy orbital. This occupancy gives rise to a low-spin ground state ($^1\Sigma^+$) and makes the BDE as high as 164 kcal mol^{-1} (Table 10.1) [36]. Different from ScO^+, FeO^+ is at a high-spin ground state ($^6\Sigma^+$), as shown in Figure 10.1. Its 1δ, 2π, and 3σ orbitals are singly occupied. As these orbitals have high energies in contrast to 1σ, 2σ, and 1π of ScO^+, electrons in 3σ can be readily excited. Thus, the BDE of FeO^+, 81 kcal mol^{-1}, is much lower than ScO^+. Unlike FeO^+, the high-energy orbitals of ZnO^+ are fully occupied. Compared to the singly occupied 1δ, 2π, and 3σ in FeO^+, these orbitals in ZnO^+ are fully occupied. The puling repulsion makes an electron in 3σ of ZnO^+ be excited readily in contrast to FeO^+. Thus, the BDE of ZnO^+, 38 kcal mol^{-1}, is lower than FeO^+.

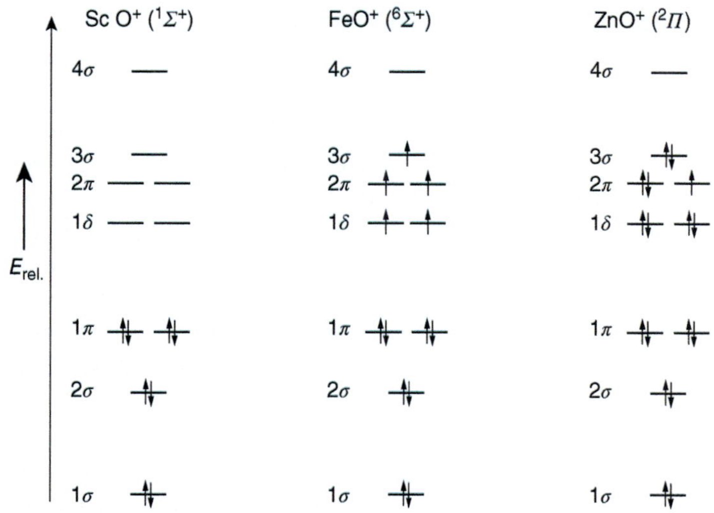

Figure 10.1 Schematic molecular orbital diagram for the representative first-row tradition metal oxide cations ScO^+. FeO^+ and ZnO^+ (only the valence orbitals are depicted for a discussion). *Source:* Reproduced with permission from [36] / John Wiley & Sons.

Table 10.1 Experimental bond dissociation energies (BDEs) in kcal mol^{-1} with error limits, F of diatomic transition metal oxide cations.

MO^+	BDE	F	MO^+	BDE	F	MO^+	BDE	F
ScO^+	164 [a]	±2	YO^-	170 [c]	±5	LaO^+	206 [f]	±4
TiO^+	159 [a]	±2	ZrO^-	201 [c]	±8	HfO^-	173 [c]	±5
VO^+	134 [a]	±2	NbO^-	207 [d]	±53	TaO^+	188 [c]	±15
CrO^+	86 [b]	±3	MoO^+	113 [c]	±19	WO^+	126 [c]	±10
MnO^-	68	±3	TcO^+	–	–	ReO^+	–	–
FeO^+	81 [b]	±2	RuO^+	88 [i]	±2	OsO^+	100 [g]	±12
CoO^+	77 [b]	±2	RhO^+	70 [i]	±2	IrO^+	59 [c]	[e]
NiO^-	63 [b]	±2	PdO^+	34 [i]	±3	PtO^+	59 [c]	±7
CuO^+	37 [b]	±3	AgO^+	28 [i]	±2	AuO^+	[h]	–
ZnO^+	38 [b]	±3	CdO^+	–	–	HgO^+	–	–

Note: References [a–h] can be found in the reference [36].
Source: Reproduced with permission from [36] / John Wiley & Sons.

From a catalysis point of view, the BDE of M–O of MO^+ should be low enough so that M–O can be broken to form $M–H^+$ or $M–CH_3^+$. On the contrary, BDE (M^+–O) must be high enough to allow for re-oxidizing M^+ back to MO^+ by an oxidant to regenerate the site, MO^+.

10.5 Fundamental Understanding of the Evolution of the Activity of MO^+ in Oxidizing CH_4 and Its Indication for Catalytic Oxidation of CH_4

To gain a fundamental understanding of how an M–O cation oxidizes methane, the reaction mechanisms of several M–O clusters with CH_4 were explored. $M–O^+$ clusters were generated using ICR mass spectrometry, as described in the literature [45, 46]. $M–O^+$ was produced in a low-vacuum environment from precursors prepared through electron impact ionization of volatile metal oxides or ion/molecule reactions of bare M^+ cations with suitable oxidants or sputtering. These clusters were introduced into a glass chamber filled with CH_4 at pressures ranging from 0.01 to 0.5 Torr. For example, the reaction between FeO^+ and CH_4 formed CH_3OH and Fe^+. This reaction is typically exothermic, and products such as methanol can be easily identified using a quadrupole mass filter.

In terms of reaction environment, these gas-phase reaction conditions of MO^+ with CH_4 at low pressure are markedly different from the oxidation of methane through a catalyst buried in a liquid solvent. In the latter scenario, methane oxidation must occur on catalytic sites located on the surface or in micropores and the catalyst particles are typically in a solvent. The gas phase CH_4 cannot directly access the catalyst surface. Typically, CH_4 has a quite low solubility in aqueous solutions. As the solubility of methane in a solvent is proportional to the pressure of methane gas, the CH_4 gas phase above the liquid containing catalyst particles is kept at high pressure, by which CH_4 molecules can access to the catalytic sites at the interface between the aqueous solution and catalyst particles.

MnO^+ [47], CoO^+ [48], FeO^+ [49], and NiO^+ [36] were investigated for the oxidation of methane to produce methanol. Among these metal-oxygen clusters, MnO^+ [47], CoO^+ [48], FeO^+ [49], and NiO^+ [36] displayed activity in oxidizing methane, while ScO^+, TiO^+, and VO^+ did not. Density functional theory (DFT) B3LYP calculations were employed to study the kinetics and thermodynamics of the oxidation of CH_4 by these first-row metal-oxygen ($M{-}O^+$) clusters. Both high-spin and low-spin potential energy surfaces were simulated in these computational studies. A spin inversion near a crossing region of potential energy surfaces was identified for FeO^+, CoO^+, and NiO^+, and this spin inversion played a significant role in the activation barrier of two transition states of CH_4 oxidation. Reactions between early transition metal-oxygen clusters (ScO^+, TiO^+, and VO^+) and CH_4 were found to be energetically unfavorable, rendering them inactive. These computational findings are consistent with experimental studies of the oxidation of CH_4 by MO^+. Experimentally, early MO^+ clusters, including ScO^+, TiO^+, and VO^+, exhibited low activity in oxidizing alkanes or alkenes.

FeO^+ demonstrated considerable oxidation activity toward CH_4, resulting in a high methanol yield of up to 41% [49]. MnO^+ exhibited high activity in oxidizing CH_4, although its selectivity for methanol was relatively low [49]. CoO^+ showed activity in methane oxidation, but its selectivity for methanol was low. Despite CoO^+ exhibiting low reactivity toward methane, it achieved a 100% branching ratio to methanol. In comparison to CoO^+ and MnO^+, NiO^+ displayed high activity and selectivity for methanol in the oxidation of CH_4 [36]. This trend in reactivity is consistent with the behavior of transition metal ions in the oxidation of organic molecules, highlighting the higher oxidizing power of an MO^+ with a higher atomic number.

Valence bond calculations indicated that the bond strength of MO^+ depends on how the d-block orbitals of these transition metal atoms are filled [36]. Early transition metal ions (M^+) form a triple bond with O, while late transition metal ions establish a weak reactive double bond with oxygen. Computational studies revealed that bond length (r_{M-O}) increases and BDE decreases with an increase in the number of d-electrons of the 3d transition metal, as summarized in Table 10.2 [51]. Molecular orbital perspectives showed that four pairs of electrons of ScO^+ occupied the low-lying 1σ, 2σ, 1π, and 1π orbitals, resembling the electron configuration of inert N_2. Additionally, the calculated BDEs of TiO^+ and VO^+ were 155.1 and 137.2 kcal mol^{-1}, respectively (Figure 10.1). In contrast to early 3d transition metals, the BDE of late transition metal-oxygen clusters was notably lower. Calculated dissociation energy values for Fe–O (FeO^+ at $^6\Sigma_+$ state), Co–O (CoO^+ at $^5\Delta$ state), and Cu–O (CuO^+ at $^3\Pi$ state) were 75.2, 73.3, and 69.3 kcal mol^{-1}, respectively. The computational results of these bond dissociation energies were consistent with experimental measurements [36], where FeO^+, CoO^+, and NiO^+ demonstrated the ability to oxidize CH_4 to CH_3OH, while ScO^+, TiO^+, and VO^+ were inactive. Thus, BDE serves as a general index for evaluating the activity of $M{-}O^+$.

Thermodynamic calculations were performed for the oxidation of CH_4 by MO^+: $MO^+ + CH_4 \rightarrow M^+ + CH_3OH$ (M: 3d metal). The overall energy changes for this reaction were found to be 73.5 kcal mol^{-1} for Sc, 72.4 kcal mol^{-1} for Ti, 54.5 kcal mol^{-1} for V, −1.3 kcal mol^{-1} for Cr, −26.2 kcal mol^{-1} for Mn, −12.6 kcal mol^{-1} for Fe, −25.6 kcal mol^{-1} for Co, −26.5 kcal mol^{-1} for Ni, and −50.0 kcal mol^{-1} for Cu. These calculations were consistent with the observed activity of FeO^+, CoO^+, and NiO^+ and the inertness of ScO^+, TiO^+, and VO^+ [36, 51].

Table 10.2 Representation of the reaction efficiencies in terms of activity and product branching in terms of product selectivity for oxidation of CH_4 with $[Mn-O]^+$, $[Fe-O]^+$, and $[Co-O]^+$.

$[M-O]^+$	$[Mn-O]^+$	$[Fe-O]^+$	$[Co-O]^+$
Reaction efficiency	40	20	0.5
$[MOH]^+ + CH_3$	100	57	0
$[MCH_2]^+ + H_2O$	0	2	0
$M^+ + CH_3OH$	<1	41	100

Source: Reproduced with permission from [54] / American Chemical Society.

Computational studies proposed a pathway involving the breaking of a C–H bond of CH_4 to form methyl and the migration of methyl to bond with O atom of M-O to form C–O, leading to the production of methanol. Figure 10.2 illustrates the proposed pathway [51]. Activation barriers of TS1 is defined as a dissociation limit (ΔE_2). For CrO^+, MnO^+, FeO^+, and NiO^+ at low-spin states, they were found to be negative [51], aligning with experimental observations. Therefore, ΔE_2 is considered another index for the reactivity of oxidizing CH_4.

Corresponding to the proposed pathway (Figure 10.2), the potential energy diagrams were calculated as represented in Figure 10.3 [51]. According to computational studies in Figure 10.3, the first step is to form a reactant complex of MO^+ and CH_4 through the formation of an M–C bond. The binding energies of CH_4 to ScO^+, TiO^+, and VO^+ were determined to be 13.5, 15.4, and 16.8 kcal mol^{-1}, respectively. The formed metal-hydroxy intermediate CH_3–M^+–OH is quite stable, with energies 10–20 kcal mol^{-1} lower than the dissociation limits (Figure 10.3).

Based on Figure 10.3, the second step involves the formation of CH_3OH through a product complex, where O of CH_3OH bonds to the M atom. The product complex is formed through the second transition state, as schematically shown in Figure 10.2. Notably, the energies of the complex of MO^+ and CH_3OH in the low-spin state and high-spin state differ

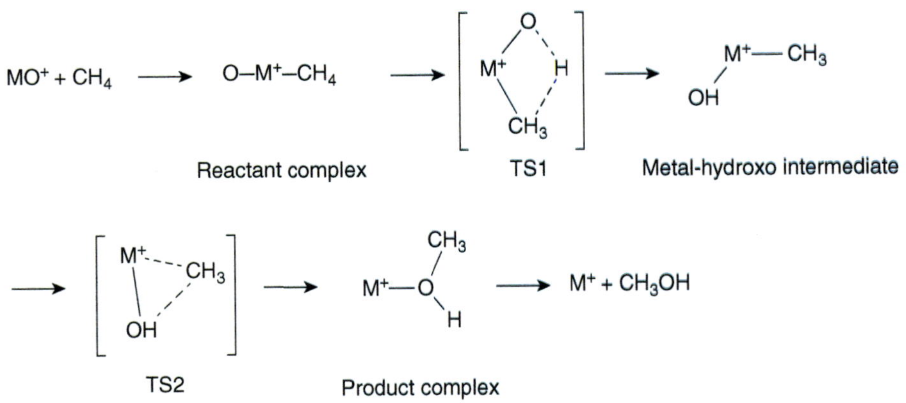

Figure 10.2 Schematic showing the reaction pathway proposed in the literature. *Source:* Reproduced with permission from [51] / American Chemical Society.

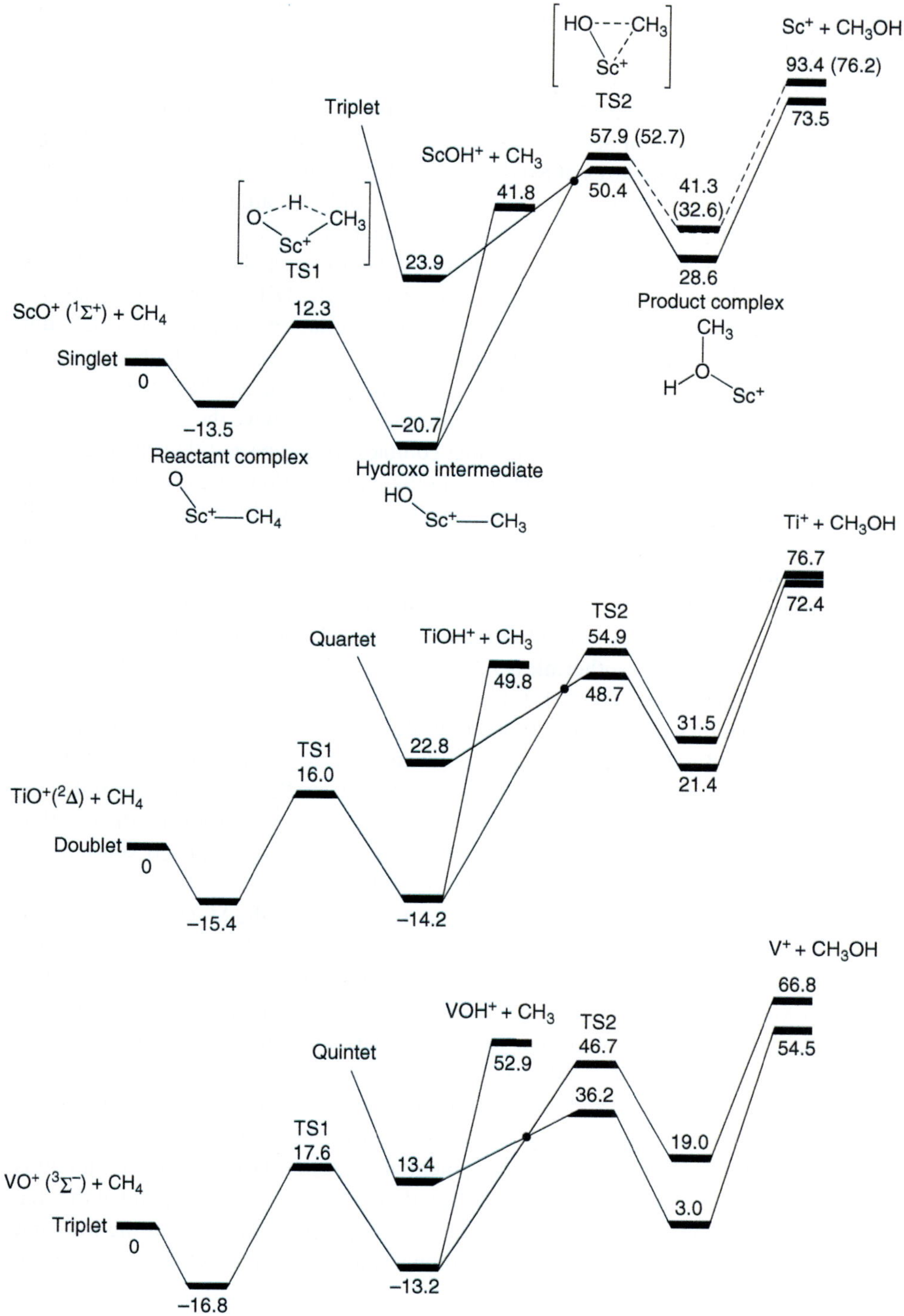

Figure 10.3 Potential energy diagrams (including zero-point energy) along the reaction pathway, $MO^+ + CH_4 \rightarrow M^+ + CH_3OH$, in the high-spin and the low-spin states. Relative energies are in kcal mol^{-1}. The dotted lines in the region of TS2, Sc^+ (CH_3OH), and Sc^+ indicate the use of the spin-unrestricted method. Values in parentheses are energies in the open-shell singlet. *Source:* Reproduced with permission from [51] / American Chemical Society.

significantly, as illustrated in Figure 10.3. The complexes of ScO^+, TiO^+, and VO^+ are quite unstable, with energies higher than the dissociation limit by 3–30 kcal mol^{-1}. Following the formation of the product complex in Figure 10.3, the subsequent step is the breaking of the M–O bond to form CH_3OH and M^+. The activation barriers for this dissociation step are quite high: 73.5 kcal mol^{-1} for ScO^+, 72.4 kcal mol^{-1} for TiO^+, and 54.5 kcal mol^{-1} for VO^+. Evidently, the product complexes of ScO^+, TiO^+, and VO^+ cannot overcome the high barrier at room temperature, explaining why ScO^+, TiO^+, and VO^+ are inactive for oxidizing CH_4 to form CH_3OH. Thus, their inertness is well explained.

Cr and Mn are transition metals in the middle of the fourth period. Experimental studies indicate that CrO^+ are nearly inactive in oxidizing CH_4 to CH_3OH, although they can oxidize CH_3CH_3 to CH_3CH_2OH [52]. The methane hydroxylation pathways of CrO^+ and MnO^+ were proposed by computational studies (Figure 10.4) [51]. Both high-spin sextet and low-spin quartet were considered in these computational studies. The potential energy surface of the high-spin sextet in the hydroxo intermediate lies above the low-spin quartet, but they reverse in the formation of a product complex. It is expected that the high-spin energy surface and low-spin one should cross in the pathway. Compared to the high-spin sextet of CrO^+, the energy pathway of its low-spin quartet is quite different. In the first half of the pathway, the energy diagram of the quartet state is lower than the sextet state of CrO^+. A notable change is that the ordering of energy between the sextet and quartet states reverses in the second half of the pathway, in contrast to the first half. The reaction performed with the quartet of CrO^+ is endothermic by 34.3 kcal mol^{-1}, while the reaction with the sextet of CrO^+ is nearly exothermic (−1.3 kcal mol^{-1}). Thus, computational studies suggest that CrO^+ could hydroxylate CH_4 [51].

Experimental studies have demonstrated that MnO^+ in a quintet ground state can oxidize CH_4 to form $MnOH^+$ and CH_3 (main products) as well as Mn^+ and CH_3OH (main product) [47]. This finding aligns with computational studies [51]. As illustrated in Figure 10.4, the activation barrier of TS1 is −6.8 kcal mol^{-1} scaled from the dissociation limit, and the activation barrier of TS2 is −16.5 kcal mol^{-1}. These low barriers substantiate the observed high activity in the oxidation of CH_4 [51].

Both experimental and theoretical studies were performed on the $M–O^+$ clusters of late 3d transition metals, including FeO^+, CoO^+, and NiO^+. The insights derived from FeO+, CoO^+, and NiO^+ complement the findings from the investigations of ScO^+, TiO^+, VO^+, CrO^+, and MnO^+. The activities of FeO^+, CoO^+, and NiO^+ are notably high. Figure 10.5 presents the energy diagrams of the reaction between CH_4 and FeO^+, CoO^+, and NiO^+ [51]. Remarkably, the energy profiles of the same $M–O^+$ (FeO^+, CoO^+, or NiO^+) at the high-spin state and low-spin state differ, and they cross at the early steps of the pathway. Thus, a low-energy reaction pathway for oxidizing CH_4 to CH_3OH by FeO^+, CoO^+, and NiO^+ is available by crossing the high-spin and low-spin potential energy surfaces. The crossing points are marked with solid dots in Figure 10.5. In fact, there is another crossing point at the step near the product. Thus, the inversion between the high-spin pathway and the low-spin pathway occurs twice based on these computational studies [51]. This is the unique feature of FeO^+, CoO^+, and NiO^+.

In Figure 10.5, the activation barrier for TS1 on FeO^+ at the sextet state is 8.3 kcal mol^{-1} higher than the dissociation limit. Alternatively, it can be avoided by switching to the quartet state of FeO^+, which gives an activation barrier of −0.7 kcal mol^{-1}. The switch

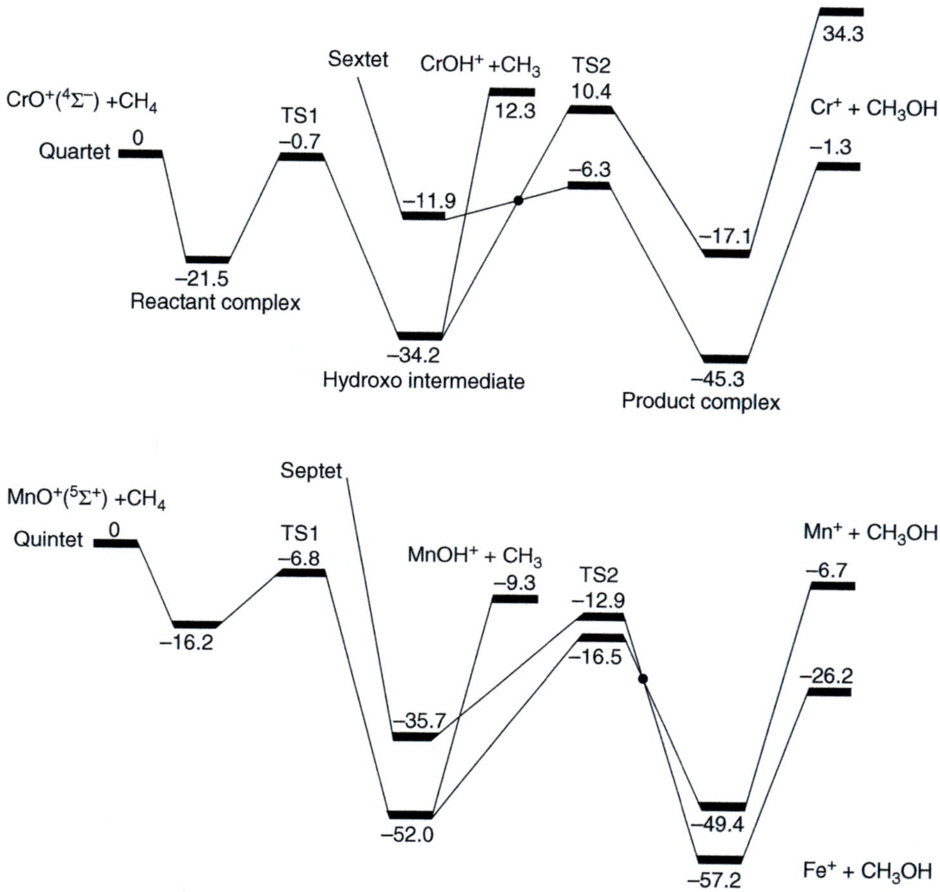

Figure 10.4 Potential energy diagrams (including zero-point energy) along the reaction pathway, $MO^+ + CH_4 \rightarrow M^+ + CH_3OH$, in the high-spin and the low-spin states. Relative energies are in kcal mol^{-1}. MO^+: CrO^+ or MnO^+. *Source:* Reproduced with permission from [51] / American Chemical Society.

between high-spin and low-spin energy profiles allows the reaction to take a pathway with a low activation barrier. Thus, FeO^+ exhibits high activity for the oxidation of CH_4. Similar computational studies were conducted for the oxidation of CH_4 by CoO^+ and NiO^+ (Figure 10.5). Similar features, in terms of two reversals between the high-spin pathway and the low-spin pathway, were found in the middle and lower panels in Figure 10.5. The feature of crossing two optional pathways makes the oxidation of CH_4 on FeO^+, CoO^+, and NiO^+ have low activation barriers in the formation of TS1 and TS2, making the oxidation kinetically favorable. Notably, such a crossing feature could not be found in the oxidation of CH_4 by MO^+ of early 3d transition metals.

The evolution of the catalytic activity of the MO^+ of the 3d transition metal can be interpreted through an analysis of the interaction between molecular orbitals of CH_4 and MO^+ involved in the transition states [50–51]. The formation of TS1 can be considered as the overlapping of molecular orbitals. As shown at the right side of Figure 10.6a–c, a CH_4

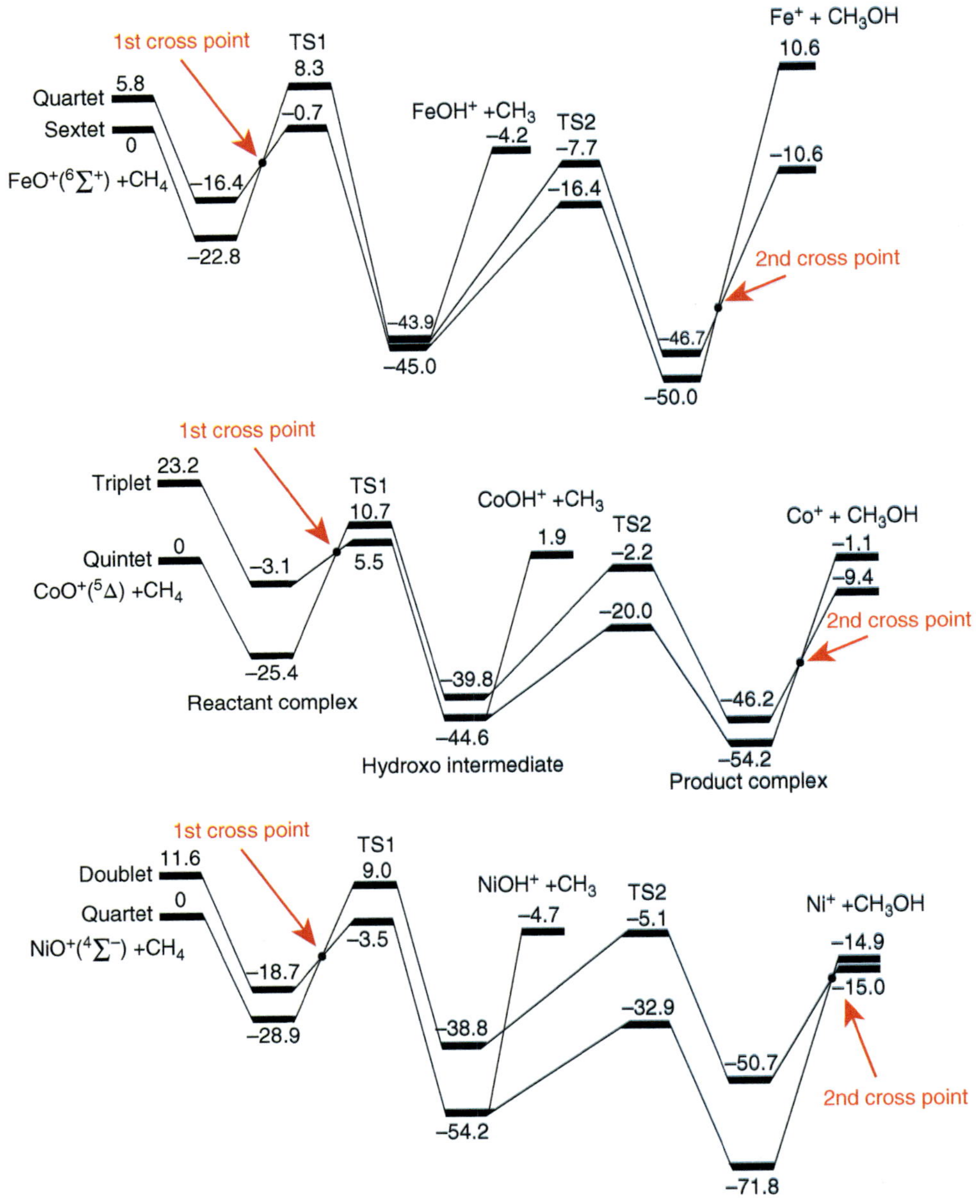

Figure 10.5 Potential energy diagrams (including zero-point energy) along the reaction pathway, $MO^+ + CH_4 \rightarrow M^+ + CH_3OH$, in the high-spin and the low-spin states. Relative energies are in $kcal\,mol^{-1}$. MO^+: FeO^+, CoO^+, or NiO^+. *Source:* Reproduced with permission from [51] / American Chemical Society.

fragment consists of three C–H bonding orbitals originating at the 3-d degenerate HOMO of methane with T_d symmetry [51]. In terms of MO^+, it consists of low-lying ligand orbitals of 2σ and 1π and high-lying 3σ, 1δ, and 2π. These orbitals of MO^+ are marked at the left side of Figure 10.6a-c. The interaction of the fragment orbitals of MO^+ and CH_4 constitutes the molecular orbital of TS1 which are located at the center of Figures 10.6a–c. Obviously, the

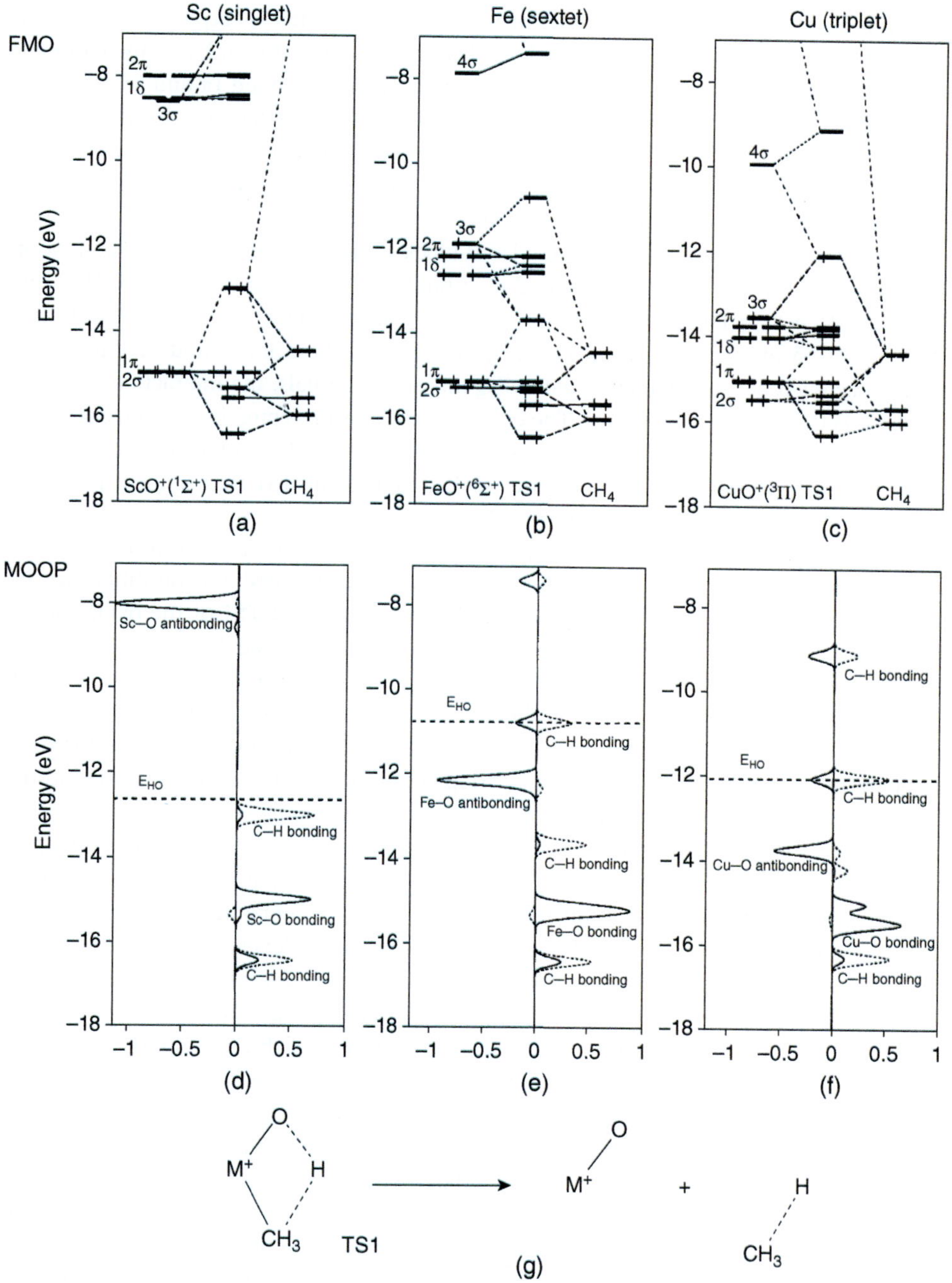

Figure 10.6 Fragment molecular orbital (FMO) and molecular orbital overlap population (MOOP) analyses of TS1 for the singlet state of ScO^+ (a, d), the sextet state of FeO^+ (b, e), and the triplet state of CuO^+ (c, f). Solid and dotted curves in MOOP indicate the contribution to the M–O bond and the C–H bond being dissociated, respectively. *Source:* Reproduced with permission from [51] / American Chemical Society.

orbital diagrams of these three TS1 states of ScO^+, FeO^+, and CuO^+ are quite different. For example, the very high-lying orbitals, such as 1δ and 2π of ScO^+, do not have any net interaction with C–H bonding orbitals in Figure 10.6a. Only the low-lying 1π and 2σ occupied orbitals of ScO^+ interact with the C–H bonding orbitals of CH_4. However, this interaction forms two orbitals filled with four electrons. This interaction is not effective since it leads to repulsion between the two molecular fragments from CH_4 and ScO^+. This ineffective interaction makes CH_4 activation on ScO^+ ineffective.

Different from ScO^+, the partially occupied d-block orbitals, including 3σ, 1δ, and 2π of FeO^+ and CuO^+, largely interact with bonding orbitals of CH_4 due to the partially filled 3σ, 1δ, and 2π of FeO^+ in Figure 10.6b and c and their relatively low energy in contrast to the high-lying 3σ, 1δ, and 2π orbitals of ScO^+ in Figure 10.6a. As these 3σ, 1δ, and 2π of FeO^+ are partially occupied before interacting with CH_4 molecular orbitals, the interaction between orbitals of FeO^+ and CH_4 is effective. The effective orbital overlapping lowers the energy of the whole molecular system [51]. For instance, the interaction of 3σ orbitals of FeO^+ at the quartet state and HOMO of CH_4 in Figure 10.6b stabilizes the TS1 state since it is a two-orbital-two-electron type of interaction [51, 53].

In addition to the orbital interactions between CH_4 and MO^+, molecular orbital population (MOOP) analyses were conducted in the literature to understand the difference in activity of different MO^+ in oxidizing CH_4 [51]. The structure of TS1 of the oxidation of CH_4 to CH_3OH by MO^+ can be found on the left side of Figure 10.6g. Figure 10.6d–f illustrates the overlap population of molecular orbitals of CH_4 and MO^+ to the four-centered TS1. Contributions of the C–H bond and the M–O bond are represented with dotted and solid curves, respectively. The highest occupied orbital of the TS1 is marked with a horizontal dotted line. In each of Figure 10.6d–f, overlap populations of bonding and antibonding orbitals of TS1 are placed to the right and left, respectively.

Figure 10.6 suggests the significance of the number of d-electrons of the 3d transition metal in the formation of TS1, which is a key step in the oxidation of CH_4 by MO^+. Weakening a C–H bond is crucial for forming new bonds, H····O and C····M in TS1. The interaction of orbitals of CH_4 and MO^+ could alter the energy levels of the C–H bonding orbitals, making these C–H bonding orbitals in TS1 have a higher or lower energy level. This is a consequence of the orbital interaction in terms of population. As shown in Figure 10.6d, the C–H bonding energy level is below the E_{HO} of the TS1 of CH_4 and ScO^+. In this TS1, the C–H bonding orbital is still occupied, indicating that the C–H bond is not weakened. In other words, the interaction between orbitals of CH_4 and ScO^+ is ineffective. This is consistent with the Fragment molecular orbital analysis in Figure 10.6a. In contrast to TS1 of ScO^+ and CH_4, the C–H bonding orbital in TS1 of FeO^+ and CH_4 was pushed up. Thus, as shown in Figure 10.6e, the C–H bonding level at about −11 eV in TS1 of CH_4 and FeO^+ is only half occupied, suggesting a substantial weakening of the C–H bond. Such a weakening of the C–H bonding orbital is favorable for the formation of the H–O bond, leading to the production of CH_3OH. Similarly, due to the interaction of orbitals of CH_4 and CuO^+, a C–H bonding in TS1 of CH_4 and CuO^+ was pushed up to a quite higher energy level at −9 eV and thus it is unoccupied (Figure 10.6f). Thus, in this TS1, the C–H bond is significantly weakened.

Overall, the MOOP analysis reasonably rationalizes the trend in the activity of M–O in the oxidation of CH_4 to form CH_3OH [51]. The reaction, $MO + H{-}CH_3 \rightarrow MO{-}H + CH_3\cdot$ (M = Fe, Cu, or other metals), is very effective in homolytic C–H bond activation since it occurs at room temperature. This reaction is thermodynamically favorable since the hydrogen affinities of these metal oxide cations are larger than the C–H bond strength of CH_4 [51]. These studies indicate that a doped cation and its bonded oxygen atom, such as $M–O^+$, on the surface of a metal oxide can be a good candidate for the activation of C–H of $H–CH_3$. These indications were proved by later findings discussed in Chapters 14–20.

10.6 Fundamental Understanding of Product Selectivity for CH_3OH in Oxidation of CH_4

Selectivity for producing methanol is another important issue interested in fundamental studies. Experimental reports show a significant difference in selectivity for producing methanol among CoO^+, FeO^+, and MnO^+ (Table 10.2). Computational studies well rationalized this difference. Among the three MO^+ clusters (Table 10.2), MnO^+ exhibits the highest activity but its branching ratio to product methanol is nearly zero. CoO^+ exhibits the lowest activity in oxidizing CH_4 but 100% selectivity for producing methanol. FeO^+ exhibits relatively high activity for oxidizing CH_4 and reasonable selectivity for producing methanol.

As shown in Figure 10.7, there are two optional pathways from the reactants MO^+ and CH_4 to form M^+ and CH_3OH [54]. One is through the methoxy intermediate, and the other is through the hydroxy intermediate. Two transition states are involved in each of the pathways. Before forming the first transition state ($TS1_M$ or $TS1_H$), the reactants form a reactant complex, $O–M^+–CH_4$. Experimental studies found that the hydroxy intermediate $HO–Fe^+–CH_3$ is the main intermediate in the oxidation of CH_4 by FeO^+ [49, 55]. A similar

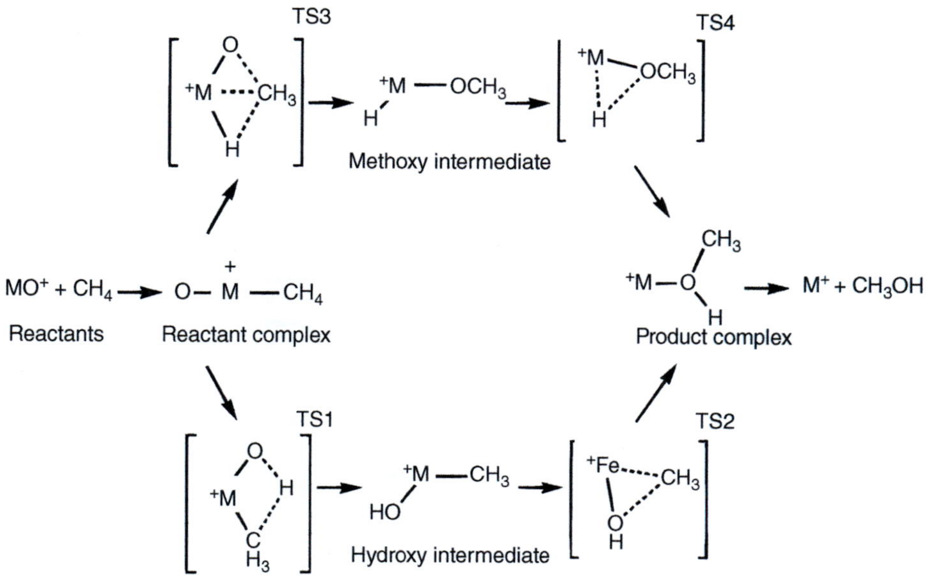

Figure 10.7 Representation of potential reaction pathways from MO^+ and CH_4 to the formation of CH_3OH and M^+. A modification was made by replacing FeO^+ with MO^+. *Source:* Reproduced with permission from Yoshizawa et al. [54] / American Chemical Society.

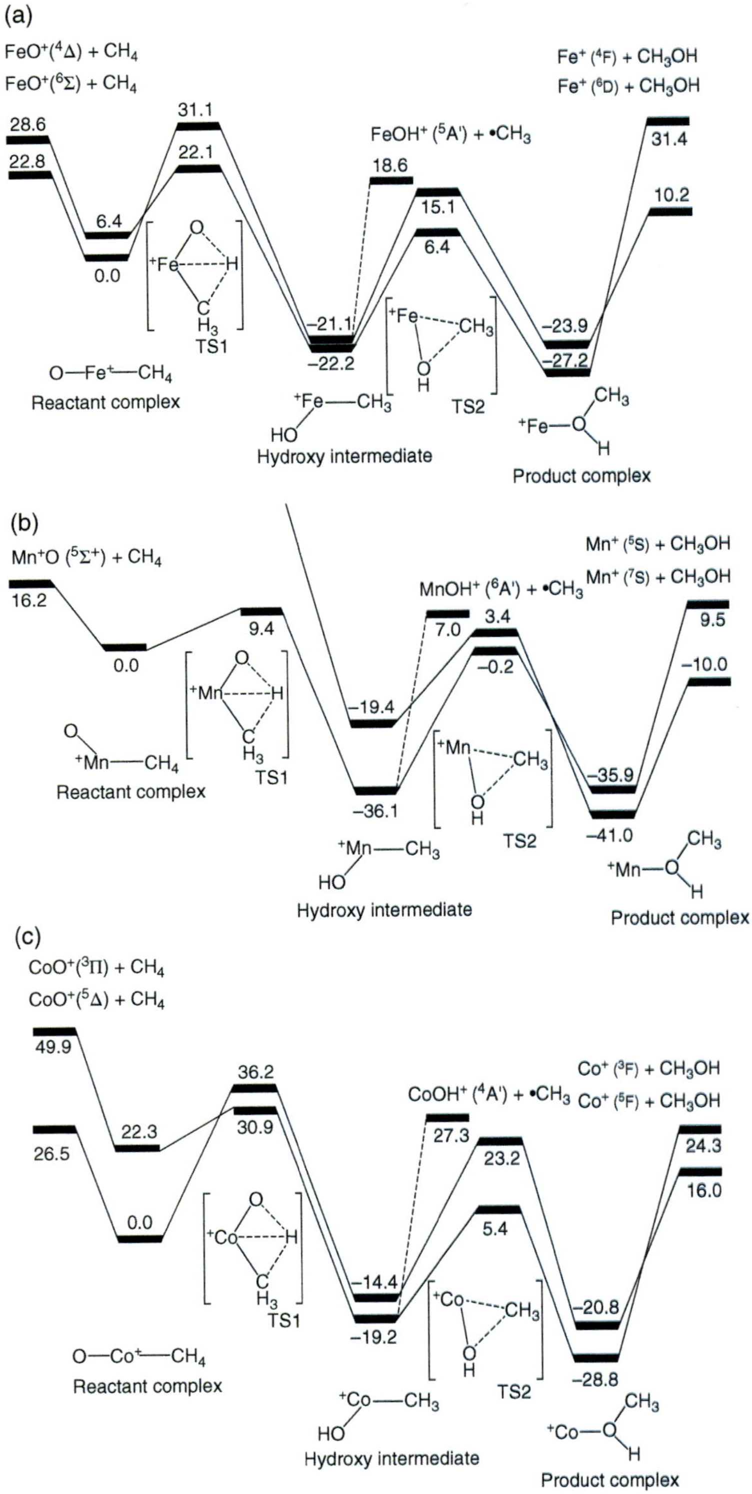

Figure 10.8 Potential energy diagrams along the reaction pathway, $MO^+ + CH_4 \rightarrow M^+ + CH_3OH$, in the sextet and quartet states. Relative energies are in kcal mol^{-1}. (a) For FeO^+. (b) For MnO^+. (c) For CoO^+. *Source:* Reproduced with permission from [54] / American Chemical Society.

Table 10.3 Representation of activation barriers of TS1 and TS2 along the reaction pathway of Figure 10.7 on MnO^+, FeO^+, and CoO^+.

MO^+ (spin state)	Barrier for TS1 (kcal mol^{-1})	Barrier for TS2 (kcal mol^{-1})
MnO^+ (quintet)	9.4	35.9
FeO^+ (Sextet-quartet)	22.1	
FeO^+ (quartet)		28.6
CoO^+ (quintet-triplet)	30.9	
CoO^+ (triplet)		24.6

Source: Reproduced with permission from [54] / American Chemical Society.

hydroxy intermediate, $HO–Co^+–CH_3$, was also proposed in the oxidation of CoO^+ to CH_4 in literature [48]. Ab initio studies suggest the pathway through the hydroxy intermediate, $HO–Fe^+–CH_3$ is more energetically favorable over the one through a methoxy intermediate [54], consistent with experimental findings [49, 55].

As the methoxy intermediate has a high energy in the pathway that involves methoxy (Figure 10.7), this particular route in Figure 10.7 was not considered in the fundamental studies of the selectivity for producing CH_3OH. Therefore, only the pathway forming a hydroxy intermediate in Figure 10.8 was considered. Figure 10.8 presents the reaction pathway and energy profile for the oxidation of CH_4 to form CH_3OH and M^+ by MO^+ through the pathway of hydroxy intermediate. Table 10.3 lists the activation barriers of TS1 and TS2 in the pathways of Figure 10.8 for FeO^+, MnO^+, and CoO^+ [54]. These barriers, obtained from the calculated potential energy diagrams in Figure 10.8, can rationalize the differences in selectivity for producing methanol in terms of the branching ratio of the three oxidants. The decrease of reactivity from MoO^+, FeO^+ to CoO^+ is consistent with the increase of activation barrier for forming TS1 from 8.4 kcal mol^{-1} for MnO^+, 22.1 kcal mol^{-1} for FeO^+, to 30.9 kcal mol^{-1} for CoO^+. The high branching ratio for methanol on CoO^+, but quite low on MnO^+, can be well explained by the activation barrier for forming TS2, which is 35.9 kcal mol^{-1} for MnO^+, 28.6 kcal mol^{-1} for FeO^+, and 24.6 kcal mol^{-1} for CoO^+ (Table 10.3) [54]. Clearly, these computational studies effectively rationalized the differences in the branching ratio of methanol and the selectivity for methanol.

References

1 Sulzle, D., Schwarz, H., Moock, K. H., Terlouw, J. K. 1991. "On the existence of novel nitrides and oxides of copper; CuN, CuO_2 and CuNO." *Int. J. Mass Spectrom. Ion Processes* 108, 269–272.

2 Irikura, K. K., Beauchamp, J. L. 1989. "Osmium tetroxide and its fragment ions in the gas phase: Reactivity with hydrocarbons and small molecules." *J. Am. Chem. Soc.* 111, 75–85.

3 Schroder, D., Fiedler, A., Hrusak, J., Schwarz, H. 1992. "Experimental and theoretical studies toward a characterization of conceivable intermediates involved in the gas-phase oxidation of methane by bare FeO^+. Generation of four distinguishable [Fe, C, H4, O]+ isomers." *J. Am. Chem. Soc.* 114, 1215–1222.

4 Lane, K., Sallans, L., Squires, R. R. 1984. "Reactions of transition-metal carbonyl anions with dioxygen in the gas phase." *J. Am. Chem. Soc.* 106, 2719–2721.

5 Jensen, D. E., Jones, G. A. 1973. "Iron compounds in flames. Relative stabilities of Fe, FeO, FeOH and $Fe(OH)_2$." *J. Chem. Soc. Faraday Trans. 1.* 69, 1448–1454.

6 Ham, N. S., McAlister, T. 1983. "Flame ionization of iron and lead." *Aust. J. Chem.* 36, 1299–1304.

7 Goodings, J. M., Hassanali, C. S., Patterson, P. M., Chow, C. 1994. "A new flame-ion mass spectrometer: Chemi-ionizatino of lanthanum observed in hydrogen—oxygen—argon flames." *Int. J. Mass Spectrom. Ion Processes* 132, 83–96.

8 Dyke, J. M. 1987. "Properties of gas-phase ions. Information to be obtained from photoelectron spectroscopy of unstable molecules." *J. Chem. Soc., Faraday Trans. 2.* 83, 69–87.

9 Green, J. C. 1994. "Variable photon energy photoelectron spectroscopy of transition metal molecules." *Acc. Chem. Res.* 27, 131–137.

10 Thornton, E. K., Thornton, E. R. 1970. "Isotope effects in chemical reactions." *ACS Monogr. Ser.* 167, 274.

11 Thibblin, A., Ahlberg, P. 1989. "Reaction branching and extreme kinetic isotope effects in the study of reaction mechanisms." *Chem. Soc. Rev.* 18, 209–224.

12 Buckner, S. W., Freiser, B. S. 1988. "Reactivity, photochemistry and thermochemistry of simple metal—Ligand ions in the gas phase." *Polyhrdron* 7, 1583–1603.

13 Eller, K., Schwarz, H. 1991. "Organometallic chemistry in the gas phase." *Chem. Rev.* 91, 1121–1177.

14 Hertwig, R. H., Koch, W., Schröder, D., Schwarz, H., Hrušák, J., Schwerdtfeger, P. 1996. "A comparative computational study of cationic coinage metal–ethylene complexes (C_2H_4) M+ (M = Cu, Ag, and Au)." *J. Phys. Chem.* 100, 12253–12260.

15 Burnier, R. C., Cody, R. B., Freiser, B. S. 1982. "Collision-induced dissociation of proton-bound alcohol dimers by Fourier-transform mass spectrometry." *J. Am. Chem. Soc.* 104, 7436–7441.

16 Cody, R. B., Freiser, B. S. 1982. "Collision-induced dissociation in a Fourier-transform mass spectrometer." *Int. J. Mass Spectrom. Ion Processes* 41. 199–204.

17 Busch, K. L., Glish, G. L., McLuckey, S. A. 1988.*"Mass Spectrometry: Techniques and Applications of Tandem Mass Spectrometry."* Weinheim, VCH.

18 McLafferty, F. W. 1990. "Studies of unusual simple molecules by neutralization-reionization mass spectrometry." *Science* 247, 925–929.

19 McLafferty, F. W. 1992. "Neutralization-reionization mass spectrometry." *Int. J. Mass Spectrum. Ion Processes* 118-119, 221–235.

20 Goldberg, N., Schwarz, H. 1994. "Neutralization-reionization mass spectrometry: a powerful "laboratory" to generate and probe elusive neutral molecules." *Acc. Chem. Res.* 27, 347–352.

21 Lias, S. G., Liebman, J. F., Levin, R. D. 1984. "Evaluated gas phase basicities and proton affinities of molecules; heats of formation of protonated molecules." *J. Phys. Chem. Ref: Data* 13, 695–808.

22 Knyazev, V. D., Dubinsky, I. A., Slagle, I. R., Gutman, D. 1994. "Unimolecular decomposition of t-C4H9 radical." *J. Phys. Chem.* 98, 5279–5289

23 Berkowitz, J., Ellison, G. B., Gutman, D. 1994. "Three methods to measure RH bond energies." *J. Phys. Chem.* 98, 2744–2765.

24 Ervin, K. M., Armentrout, P. B. 1985. "Translational energy dependence of Ar++XY→ArX++Y (XY=H2,D2,HD) from thermal to 30 eVc.m." *J. Chem. Phys.* 83, 166–189.

25 Schultz, R. H., Armentrout, P. B. 1991. "Reactions of N+4 with rare gases from thermal to 10 eV center-of-mass energy: collision-induced dissociation, charge transfer and ligand exchange." *Int. J. Mass Spectrum. Ion Processes* 107, 29–48.

26 Armentrout, P. B. 1990. "Electronic state-specific transition metal ion chemistry." *Annu. Rev. Phys. Chem.* 41, 313–344.

27 Armentrout, P. B. 1991. "Chemistry of excited electronic states." *Science* 251, 175–179.

28 Schultz, R. H., Crellin, K. C., Armentrout, P. B. 1991. "Sequential bond energies of iron carbonyl Fe(CO)x+ (x = 1−5): Systematic effects on collision-induced dissociation measurements." *J. Am. Chem. Soc.* 113, 8590–8601.

29 Ham, N. S., McAlister, T. 1988. "Graphite furnace mass spectrometry of cobalt." *Spectrochim. Acta Part B: Atomic Spectrosc.* 43, 789–797.

30 Patterson, P. M., Goodings, J. M. 1996. "Flame-ion chemistry of the lanthanide metals Ce, Pr and Nd." *Int. J. Mass Spectrom. Ion Processes.* 152, 43–60.

31 Hassanali, C. S., Goodings, J. M. 1990. "Ion chemistry of phosphorus in hydrocarbon flames: Part II. Positive ion formation." *Int. J. Mass Spectrom. Ion Processes.* 99, 191–205.

32 van Doren, J. M., Barlow, S. E., Depuy, C. H., Bierbaum, V. M. 1987. "The tandem flowing afterglow-shift-drift." *Int. J. Mass Spectrom. Ion Processes.* 81, 85–100.

33 Clemmer, D. E., Honma, K., Koyano, I. 1993. "Kinetics of excited-state titanium(a5F) depletion by nitric oxide, oxygen, nitrous oxide, and nitrogen." *J. Phys. Chem.* 97, 11480–11488.

34 Cooks, R. G., Kaiser, Jr., R. E. 1990. "Quadrupole ion trap mass spectrometry." *Acc. Chem. Res.* 23, 213–219.

35 Matsuo, Y., Maeda, H., Takami, M. 1993. "Gas-phase ion—Molecule reaction in an RF ion trap. Reactivity of the 5d transition series of metal ions (Lu+, Hf+, Ta+, and W+) with O2." *Chem. Phys. Lett.* 201, 341–344.

36 Schröder, D., Schwarz, H. 1995. "C–H and C–C bond activation by bare transition-metal oxide cations in the gas phase." *Angew. Chem. Int. Ed.* 34, 1973–1995.

37 Musaev, D. G., Koga, N., Morokuma, K. 1993. "Ab initio molecular orbital study of the electronic and geometric structure of methylenerhodium(1+) and the reaction mechanism: RhCH2+ + H2 .fwdarw. Rh+ + CH4." *J. Phys. Chem.* 97, 4064–4075.

38 Holthausen, M. C., Heiuemann, C., Cornehl, M. -C., Koch, W., Schwarz, H. 1995. "The performance of density-functional/Hartree–Fock hybrid methods: Cationic transition-metal methyl complexes MCH+3 (M=Sc–Cu,La,Hf–Au)." *J. Chem. Phys.* 102, 4931–4941.

39 Koga N., Morokuma, K. 1991. "Ab initio molecular orbital studies of catalytic elementary reactions and catalytic cycles of transition-metal complexes." *Chem. Rev.* 91, 823–842.

40 Sargent, A. L., Hall, M. B., Guest, M. F. 1992. "Theoretical studies of inorganic and organometallic reaction mechanisms. 3. The origin of the difference in the barrier for the kinetic and thermodynamic products for the oxidative addition of dihydrogen to a square-planar iridium complex." *J. Am. Chem. Soc.* 114, 517–522.

41 Fiedler, A., Schroder, D., Shaik, S., Schwarz, H. 1994. "Electronic structures and gas-phase reactivities of cationic late-transition-metal oxides." *J. Am. Chem. Soc.* 116, 10734–10741.

42 Fisher, E. R., Elkind, J. L., Clemmer, D. E., Georgiadis, R., Loh, S. K., Aristov, N., Sunderlin, L. S., Armentrout, P. B. 1990. "Reactions of fourthperiod metal ions (Ca^+–Zn^+) with O_2: Metaloxide ion bond energies." *J. Chem. Phys.* 93, 2676–2691.

43 Siegbahn, P. E. M. 1993. "A comparison of the bonding in the second-row transition-metal oxides and carbenes." *Chem. Phys. Lett.* 201, 15–23.

44 Siegbahn, P. E. M. 1993. "Binding in second-row transition metal dioxides, trioxides, tetraoxides, peroxides, and superoxides." *J. Phys. Chem.* 97, 9096–9102.

45 Grosshans, P. B., Marshall, A. G. 1991. "General theory of excitation in ion cyclotron resonance mass spectrometry." *Anal. Chem.* 63, 2057–2061.

46 Marshal, A. G., Grosshans, P. B. 1991. "Fourier transformation ion cyclotron resonance mass spectrometry: The teenage years." *Anal. Chem.* 63, 215A–229A.

47 Ryan, M. F., Fiedler, A., Schroeder, D., Schwarz, H. 1995. "Radical-like behavior of manganese oxide cation in its gas-phase reactions with dihydrogen and alkanes." *J. Am. Chem. Soc.* 117, 2033–2040

48 Chen, Y. M., Clemmer, D. E., Armentrout, P. B. 1994. "Conversion of CH_4 to CH_3OH: Reactions of CoO^+ with CH_4 and D_2, Co+ with CH_3OD and D_2O, and Co+(CH_3OD) with Xe." *J. Am. Chem. Soc.* 116, 7815–7826.

49 Schröder, D., Schwarz, H. 1990. "FeO^+ activates methane." *Angew. Chem. Int. Ed.* 29, 1433–1434.

50 Schröder, D., Schwarz, H. 1990. "$Fe^{\oplus}$-catalyzed gas-phase oxidation of ethane by N_2O." *Angew. Chem. Int. Ed.* 29, 1431–1432.

51 Shiota, Y., Yoshizawa, K. 2000. "Methane-to-methanol conversion by first-row transition-metal oxide ions: ScO^+, TiO^+, VO^+, CrO^+, MnO^+, FeO^+, CoO^+, NiO^+, and CuO^+." *J. Am. Chem. Soc.* 122, 12317–12326.

52 Kang, H., Beauchamp, J. L. 1986. "Gas-phase studies of alkane oxidation by transition-metal oxides. Selective oxidation by CrO^+." *J. Am. Chem. Soc.* 108, 7502–7509.

53 Yoshizawa, K., Shiota, Y., Yamabe, T. 1997. "Reaction paths for the conversion of methane to methanol catalyzed by FeO^+." *Chem. Eur. J.* 3, 1160–1169.

54 Yoshizawa, K., Shiota, Y., Yamabe, T. 1998. "Methane–methanol conversion by MnO^+, FeO^+, and CoO^+: A theoretical study of catalytic selectivity." *J. Am. Chem. Soc.* 120, 564–572.

55 Schwarz, H. 1991. "Activation of methane." *Angew. Chem., Int. Ed.* 30, 820–821.

11

Catalytic Oxidation of Methane through Free-standing M^+ in Gas Phase at Low Temperature

Oxidation of CH_4 with MO^+ has been extensively investigated, as discussed in Chapter 10, demonstrating the ability of MO^+ to oxidize CH_4. The oxidation potential trend of first-row transition metal-based MO^+ clusters has been thoroughly studied. Certain free-standing clusters, such as PtO^+, exhibiting catalytic activity in CH_4 oxidation. Decades ago, a Pt-based molecular catalyst was reported for the oxidation of CH_4. Notably, compared to a Pt-based molecular catalyst, PtO^+ clusters in the gas phase have also demonstrated activity in catalyzing the oxidation of CH_4.

Other than the direct oxidation of CH_4 by MO^+ cluster, CH_4 can be oxidized by O_2 through the catalysis of Pt^+. Third-row transition metal cations, denoted as M^+, exhibit activity in the dehydrogenation of CH_4 to form a carbene species, $M{=}CH_2$. It is reported that oxygen atoms from an oxidant can transfer to $M{=}CH_2$, resulting in the formation of oxygenated products [1]. Interestingly, scientists successfully designed a catalytic cycle involving naked transition metal cations, CH_4, and a source of oxygen atoms [1]. Figure 11.1 illustrates the proposed reaction pathway for the catalytic oxidation of CH_4 to CH_3OH or HCHO and H_2O through the naked transition metal cation Pt^+ and the source of oxygen atoms from O_2. The initial step involves the dehydrogenation of CH_4 by a third-row transition metal cation such as Pt^+, forming a platinum carbene species $[PtCH_2]^+$. Subsequently, the platinum carbene undergoes oxidation by molecular O_2 to form PtO^+, Pt^+, and organic oxygenates.

Notably, N_2O is not effective in oxidizing $[PtCH_2]^+$ in contrast to O_2, despite its common use in activating metal atoms anchored in zeolite for oxidation of CH_4 as to be discussed in Chapters 17 and 19. It is well known that the resonance structure of N_2O allows for a single, double, or triple bond in the N–O bond. Although the average bond order of N–O in N_2O is similar to O–O in O_2, the N–O bond in N_2O is anticipated to be stronger than the O–O bond in O_2. In addition, the steric hindrance of N_2O, owing to its larger size compared with O_2, complicates the oxidation of $Pt{=}CH_2$ as opposed to O_2. Compared to O_2, the larger steric effect of N_2O in coupling with $Pt{=}CH_2$, is supported by the observation that N_2O can oxidize the small Pt^+ at room temperature but is inactive toward the much larger $Pt{=}CH_2$.

Given that Pt^+ is a gas-phase catalyst, it undergoes regeneration to actively participate in subsequent catalytic cycles. PtO^+ serves as an intermediate, facilitating the delivery of oxygen atoms for the formation of oxygenated compounds. As demonstrated in the preceding

Low-temperature Activation and Catalytic Transformation of Methane to Non-CO_2 Products,
First Edition. Franklin Tao.

chapter, PtO^+ exhibits activity in the oxidation of CH_4, leading to the formation of CH_3OH. The proposed reaction pathway, depicted in Figure 11.1, was further validated by experiments using an isotope labeled CH_2D_2. The credibility of the proposed pathway is supported by the measured temporal evolution of species containing Pt, as illustrated in Figure 11.2 [1]. For instance, at the onset of catalytic oxidation, the concentration of $[PtCH_2]^+$ steadily increases within the initial 20 seconds. Subsequently, it experiences a gradual decrease and stabilizes at a specific concentration at the 120th second, suggesting

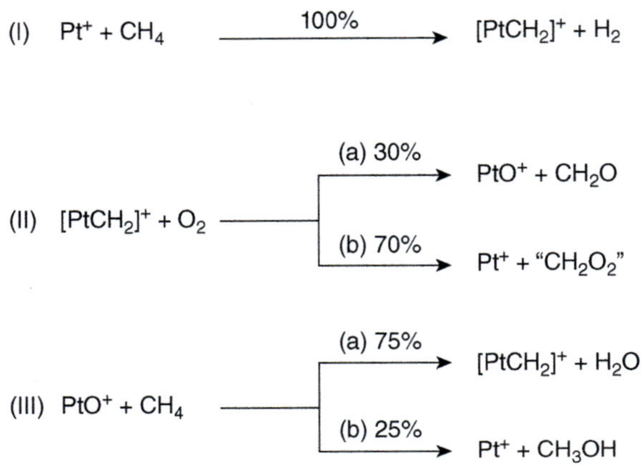

Figure 11.1 Formation and reactions of $[PtCH_2]^+$ and $[PtO]^+$, as well as the formation of the methane oxidation products CH_3OH, CH_2O. and "CH_2O_2". *Source:* Reproduced with permission from Wesendrup et al. [1] / John Wiley & Sons..

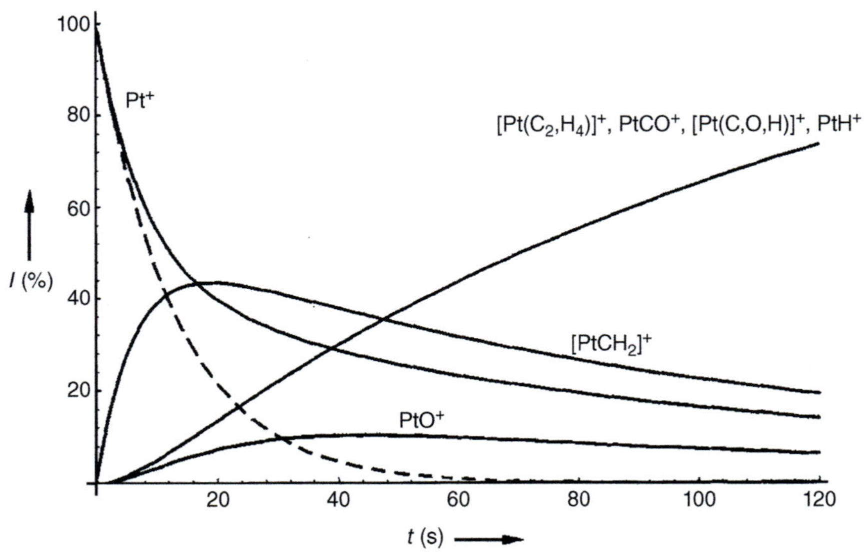

Figure 11.2 Temporal evolution of the species involved in the Pt^+-catalyzed methane oxidation (CH_4:O_2 = 1:20). The dashed curve shows the decrease of the Pt^+ signal in the presence of methane only. All curves are based on a numerical solution of the appropriate differential equations on use of the k, values given in the text. For the sake of clarity, the experimentally measured data are omitted: these data agree with the computed ion abundancies within experimental error (±5%). Ion losses at long storage time (>80 s) are taken into account. *Source:* Reproduced with permission from Wesendrup et al. [1] / John Wiley & Sons.

that the formed $[PtCH_2]^+$ functions as an intermediate in the pathway of oxidizing CH_4 with O_2 to form CH_3OH catalyzed by Pt^+. Similarly, the role of PtO^+ as another intermediate is evidenced by the initial increase in the concentration of PtO^+, followed by a slight decrease, and then maintained a relatively constant concentration. It is reported the turnover number for Pt^+ catalyzing oxidation of CH_4 to CH_3OH with O_2 is approximately 6 under the condition where the pressures of CH_4 and O_2 were 4×10^{-9} and 8×10^{-8} mbar, respectively [1]. Under this condition, the selectivities for CH_3OH, HCHO, and HCOOH are 10%, 25%, and 65%, respectively.

Reference

1. Wesendrup, R., Schröder, D., Schwarz, H. 1994. "Catalytic Pt^+-mediated oxidation of methane by molecular oxygen in the gas phase." *Angew. Chem. Int. Ed.* 24, 1174–1176.

12

Activation and Catalytic Oxidation of CH_4 through M_1O_n Clusters Anchored on Open Support at Low Temperature

12.1 Context

Chapter 10 demonstrated the ability of MO^+ clusters to oxidize CH_4 into oxygenates, highlighting that MO^+ clusters of certain metal elements exhibit high activity for CH_4 oxidation [1]. However, the MO^+ clusters discussed in Chapter 10 are free-standing and generally only stable under high-vacuum conditions. Developing a durable catalytic process to oxidize CH_4 in the gas phase using an oxidant like O_2 with MO^+ as a catalyst remains challenging. Although these clusters readily activate the C–H bond of CH_4 at room temperature, their practical application is limited. This chapter and the following ones will explore catalytic oxidation of CH_4 through reactions performed on M-O sites, either on open surfaces of transition metal oxides such as Co_3O_4 or anchored within the micropores of zeolites. Specifically, this chapter focuses on two cases: the catalytic oxidation of CH_4 on Ni–O sites located on the surface of $NiCo_2O_4$ nanoparticles at 200–300°C in a CH_4 and O_2 flow, and the activation of the C–H bond of CH_4 by Ir-O sites on an $IrO_2(110)$ thin film at 150 K under low-pressure CH_4 exposure.

12.2 Cations Doped on Open Surface of Transition Metal Oxide

In the realm of heterogeneous catalysis, the surface of an oxide, particularly a reducible oxide, can be modified through the introduction of cations [1, 2]. A guest cation can bond to a few oxygen atoms of the topmost atomic layer through "siting" on them, termed the cap-on mode. In addition, the guest cation may be inserted into the M–O bond of the topmost atomic layer.

Beyond these anchoring modes, the guest cations can replace a small portion of existing cations (M^{n+}) of the topmost atomic layer and the subsurface region of oxide support, which is called a substitution mode. The substitution mode can be energetically favorable, especially when the size of the guest cations is similar to that of the native cations of the support oxide.

An illustrative example is the formation of single-atom Ni sites on the surface of the Co_3O_4 lattice, on which Ni cations substitute for Co cations, resulting in singly dispersed Ni sites on the surface of Co_3O_4, $NiCo_2O_4$ [3]. It is considered a catalyst containing a M_1O_n cluster on its

Low-temperature Activation and Catalytic Transformation of Methane to Non-CO_2 Products, First Edition. Franklin Tao.

support. It is closely related to the M_1O_n cluster formed in micropores of zeolites to be discussed in the next several chapters, although here the M_1O_n cluster is anchored on an open surface of support instead of a confined, internal surface of a micropore. This type of single-atom catalyst can perform oxidative catalysis, such as the complete oxidation of CH_4. Notably, CO_2, a product of the complete oxidation of CH_4, is not a welcomed product in terms of mitigation of the greenhouse effect and global warming. Here, the reason for choosing complete oxidation of CH_4 on $NiCo_2O_4$ is to discuss the activity of an M_1O_n cluster anchored on an open support at a relatively low temperature [3]. In addition, from the point of view of mitigating global warming for oxidizing CH_4 to CO_2 at a relatively low temperature is valuable for meeting the goal of limiting the environmental temperature rise to less than 1.5°C since the effect of CH_4 on global warming is 26 times of CO_2.

An instance of oxidative catalysis on a single-atom site anchored on an oxide is the catalytic oxidation of CH_4 on $NiCo_2O_4$ [3]. $NiCo_2O_4$, $Ni_{0.75}Co_{2.25}O_4$, and $Ni_{1.25}Co_{1.75}O_4$ were prepared through co-precipitation followed by calcination in O_2. None of these catalysts formed detectable NiO phases, as indicated by the absence of the strongest diffraction peak of NiO at 43.5° [3]. It suggests that Ni cations occupy the lattice of Co cations or insert into the Co–O bond. However, the insertion of Ni between Co–O must induce an obvious lattice distortion, given the high concentrations of Ni in $Ni_{0.75}Co_{2.25}O_4$ (25%), $NiCo_2O_4$ (33%), and $Ni_{1.25}Co_{1.75}O_4$ (42%). Thus, the absence of the NiO phase in the XRD patterns of $NiCo_2O_4$, $Ni_{0.75}Co_{2.25}O_4$, and $Ni_{1.25}Co_{1.75}O_4$ implies that Ni atoms substituted for Co atoms in the Co_3O_4 lattice [3]. The size of catalyst particles is in the range of 3–6 nm. To study the amount of Co cations in the surface region of the catalysts, ambient pressure X-ray photoelectron spectroscopy (AP–XPS) was employed. AP–XPS studies revealed that the atomic fraction of Ni cations remains constant at temperatures ≤350°C, indicating the robust anchoring of Ni cations.

$NiCo_2O_4$ exhibits high activity for complete oxidation of CH_4 at 250°C, as illustrated in Figure 12.1a. The pivotal role of Ni cations in catalyzing the complete oxidation of CH_4 is evident in Figure 12.1a, where Co_3O_4, lacking Ni cations, is nearly inert under the same catalytic condition as $NiCo_2O_4$. The mechanism by which Ni cations catalyze the complete oxidation of CH_4 at a relatively low temperature was elucidated through density functional theory (DFT) studies. TEM studies of the morphology of $NiCo_2O_4$ nanoparticles suggest a preferentially exposed (110) surface [3]. The optimized surface structure of $NiCo_2O_4$(110)-B is depicted in Figure 12.1b, where oxygen, cobalt, and nickel atoms are represented in red, blue, and yellow, respectively. Clearly, Ni atoms are singly dispersed on the surface of the Co_3O_4 lattice. The first coordination shell of a Ni atom consists of four oxygen atoms, and the second shell consists of six Co atoms (Figure 12.1b).

The significant contrast in activity for oxidizing CH_4 between singly dispersed Ni atom sites on $NiCo_2O_4$ and NiO(100) and NiO(111) in Figures 12.1c and 12.1e underscores the uniqueness of the electronic state of the singly dispersed Ni atoms in Co_3O_4. On the $NiCo_2O_4$(110)-B in Figure 12.1b, the coordination number of oxygen atoms to a Ni atom on $NiCo_2O_4$ is four, similar to the coordination environment of a Ni atom on NiO(110) of a planar NiO sheet in Figure 12.1d, previously reported as a catalyst highly active for complete oxidation of CH_4. The high activity in oxidizing CH_4 on $NiCo_2O_4$(110)-B and NiO(110) which have a similar coordination shell of four O atoms is supported by the lack of activity for this reaction on either a Ni atom coordinating with five O atoms on NiO(100) of a cubic NiO nanoparticle or a Ni atom coordinating with six O atoms on NiO(111) of a tetrahedral NiO nanoparticle [4].

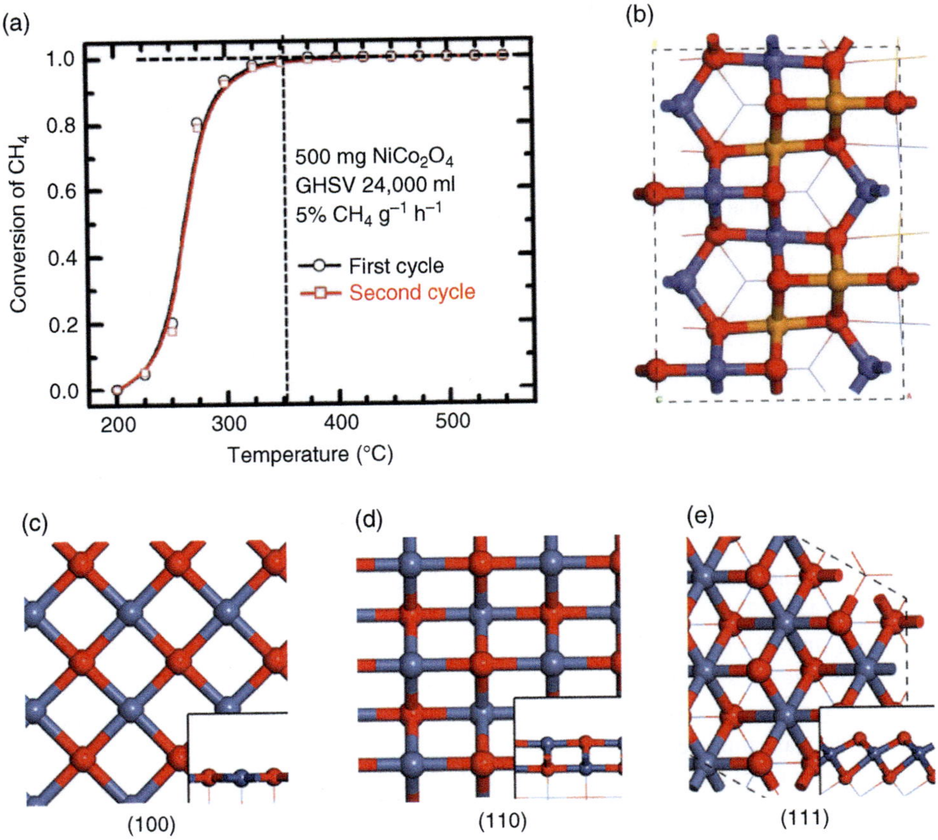

Figure 12.1 Catalytic performance of $NiCo_2O_4$ for complete oxidation of CH_4 and optimized surface structures of $NiCo_2O_4$(110)-B, NiO(100), NiO(110), and NiO(111). (a) Catalytic performances of $NiCo_2O_4$ in the temperature range of 200–550°C. (a) CH_4 can be completely oxidized at a temperature up to 550°C (black line), and the catalytic performance remains the same after heating to 550°C and then cooling to 200°C(red line). (b–e) Optimized surface structures of $NiCo_2O_4$(110)-B, NiO(100), NiO(110), and NiO(111). *Sources:* (a and b) Reproduced with permission from [3] / Springer Nature. (c–e) Reproduced with permission from [4] / American Chemical Society.

These differences clearly emphasize the crucial role of an appropriate coordination environment of cations in activating CH_4 and subsequent steps in the oxidation of CH_4.

The role of Ni^{3+} in $NiCo_2O_4$ was thoroughly investigated, and four potential surface structures of a catalyst nanoparticle including (100), (110)-A, (110)-B, and (111) are depicted in Figure 12.2. In all these structures, Ni atoms are situated in the positions of Co^{3+}, implying that both Ni^{3+} and Co^{3+} share the same coordination environment with four oxygen atoms [3]. The initial step of CH_4 activation, involving the cleavage of the first C–H bond, was examined. Table 12.1 shows that the activation barrier for cleaving the first C–H bond on Ni^{3+} of (110)-B (0.52 eV) is lower than that on Co^{3+} (0.90 eV). Consequently, Ni^{3+} is a more favorable site kinetically for cleaving the C–H bond of CH_4 than Co^{3+} with the same coordination environment. Consistently, computational studies reveal that the adsorption energy of CH_3 on Ni^{3+} is higher than that on Co^{3+} by 0.21 eV. Thus, the presence of Ni^{3+} with an appropriate coordination (four oxygen atoms) is crucial for activating the C–H bond of CH_4.

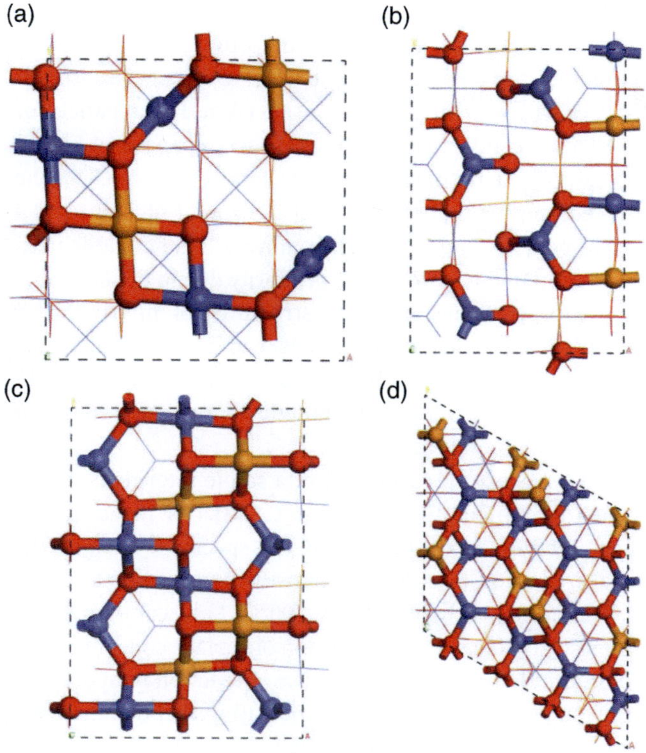

Figure 12.2 Top views of the most exposed surfaces of $NiCo_2O_4$ (crystal type I). (a) $NiCo_2O_4$(100), (b) $NiCo_2O_4$(110)-A, (c) $NiCo_2O_4$(110)-B, and (d) $NiCo_2O_4$(111). The carbon, hydrogen, oxygen, cobalt, and nickel atoms are shown in gray, white, red, blue, and yellow, respectively. The use of colors representing different atoms is implemented through the main text in the supporting information. The atoms on the top and sub-surfaces are illustrated in the ball-and-stick model, and the rest are in the line model. This illustration is used throughout this report. *Source:* Reproduced with permission from [3] / Springer Nature.

Table 12.1 The activation barriers of the first C–H of CH_4 on three different facets (100, 110-A, and 110-B) of crystal types I and II.

Crystal type	Type I		Type II	
Site for the dissociation of the first C–H of CH_4	Co^{3+}	Ni^{3+}	Co^{3+}	Ni^{3+}
100 surface	0.91	1.22	0.83	N/A
110-A surface	1.06	1.12	0.65	N/A
110-B surface	0.90	0.52	0.80	0.73

In addition, our calculations show that the activation barriers on the surface of pure Co_3O_4(110-B) are 0.38 eV higher than the $NiCo_2O_{4-x}$(110-B) of type I crystal, which suggests that doping nickel cations in Co_3O_4 could significantly enhance the reactivity by decreasing the activation barrier of the first C–H of CH_4.
Source: Reproduced with permission from [3] / Springer Nature.

Following the cleavage of the first C–H bond on the Ni atom, CH_3 is bound to the Ni atom. Subsequently, the bound CH_3 transfers to adjacent oxygen, forming CH_3···O–Ni, as shown in the shadowed section in panel 1 of Figure 12.3. The CH_3···O–Ni species undergoes dehydrogenation to form Ni...CO, with two proposed pathways (A and B) in panel 1 of Figure 12.3 [3]. The CO bound to Ni atom can be found in the top part of panel 1 of Figure 12.3. The bound intermediate CO can couple with one adjacent O atom to form CO_2. The two pathways are presented with red and black fonts in panel 1. Panel 2 of Figure 12.3 illustrates the energy profiles for pathways A and B. Clearly, pathway B is favorable both kinetically and thermodynamically compared to pathway A, primarily due to the favorable dehydrogenation step from CH_2O to CHO. The optimized structures of intermediates CH_2O and CHO in pathways A and B are presented in panels 3 and 4 of Figure 12.3. Consequently, computational studies suggest that pathway B (depicted in black fonts in panel 1 and the black line in panel 2 of Figure 12.3) is the kinetically favorable route for oxidizing CH_4 to CO_2 at a relatively low temperature.

Both AP–XPS and *in situ* IR studies suggest the formation of a stable intermediate, the OCHO species, on the surface at low temperatures (60–250°C) [3]. As illustrated in Figure 12.4a, peak 2 of C 1s is attributed to the C 1s of the OCHO species formed on the surface. It was observed at temperatures as low as 60°C. Its intensity increased up to 100°C and started to decrease at 150°C, becoming nearly invisible at 200°C. The evolution of the intensity of this OCHO species suggests its volcano-like evolution of surface concentration, forming at low temperatures and accumulating with the increase of reaction temperature. Then, its intensity decreases as the temperature rises further, indicating that this intermediate gains enough energy to transform into the next intermediate toward forming the product to desorb from the surface. The evolution of the intensity of peak 2, representing the feature of a stable intermediate OCHO, is presented in Figure 12.4b. The formation of the stable intermediate OCHO is supported by the observation of formate species OCHO at 1561 cm^{-1} through *in situ* IR studies in Figure 12.4c.

Computational studies in Figure 12.5 support the formation of a stable intermediate OCHO at low temperatures. Figure 12.5a presents two sub-pathways (CO oxidation and OCHO dehydrogenation) of pathway B. The two sub-pathways of pathway B are also presented in panel 1 of Figure 12.3. The black line in Figure 12.5b represents the two elementary steps (CHO → CO → CO_2) of the sub-pathway of CO oxidation of Figure 12.5a. In addition, CHO can form the OCHO intermediate and then transform into the final products CO_2 and H_2O through two alternative steps (CHO → OCHO → CO_2) shown with red fonts in Figure 12.5a. The energy profiles of the two sub-pathways including CHO → CO → CO_2 (black line) and CHO → OCHO → CO_2 (red line) are presented in Figure 12.5b. Figures 12.5c and d present the optimized structures of these intermediates and transition states of the two sub-pathways in pathway B. In fact, the first step (CHO → OCHO) in the sub-pathway of OCHO dehydrogenation is thermodynamically favorable at low temperatures in contrast to the sub-pathway of CO oxdidation (black line in Figure 12.5b). But at low temperatures, the second alternative step (OCHO → CO_2) is not kinetically favorable compared with the second step of the sub-pathway of CO oxidation (CO → CO_2) (black line in Figure 12.5b) because the activation barrier from OCHO intermediate to the final products CO_2 (OCHO → CO_2) is much higher than the barrier for transforming CO to CO_2 (red line in Figure 12.5b). Thus, the formation of the OCHO intermediate is favorable at low temperatures; but it cannot transform into products at low temperatures due to the high-activation barrier, thus making OCHO observable at low temperatures. It is

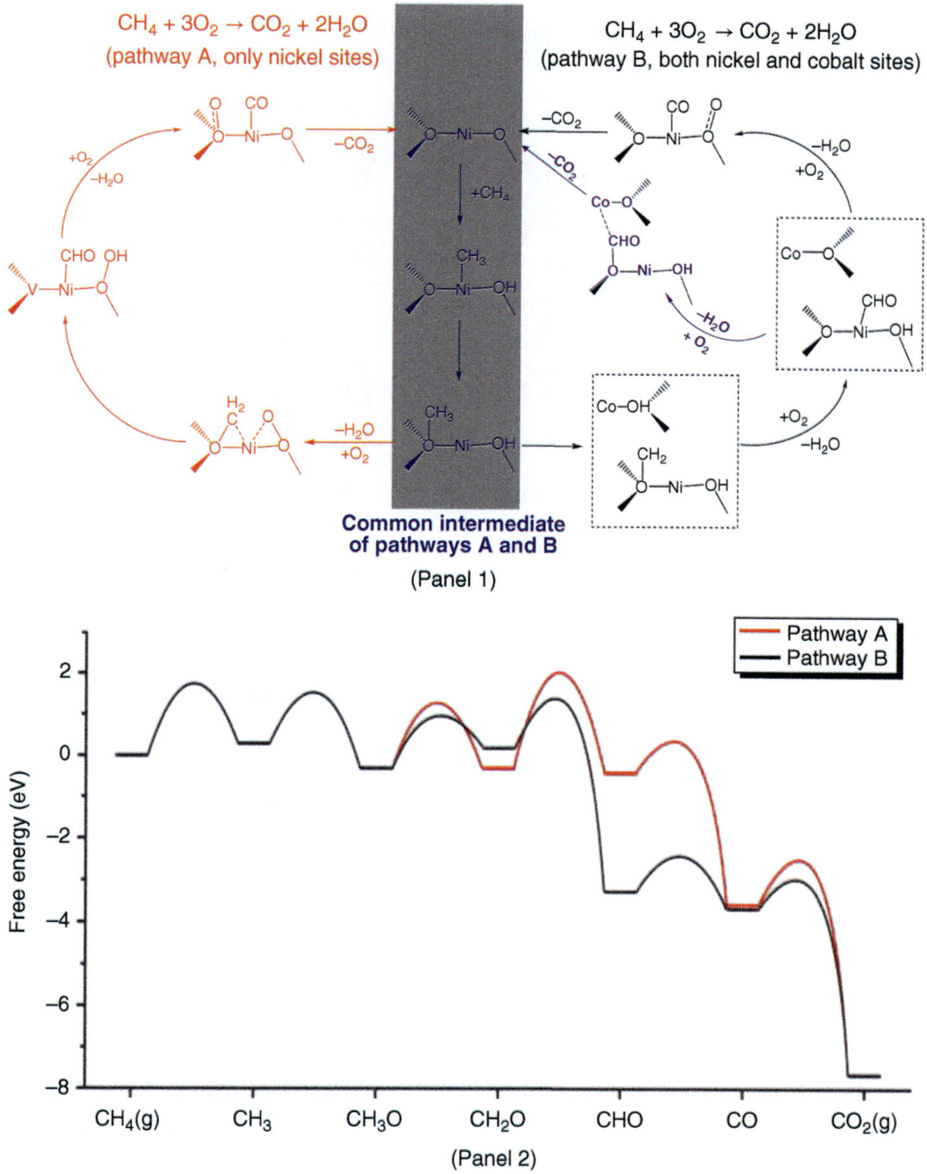

Figure 12.3 Reaction mechanism for complete oxidation of CH_4 to CO_2 on B-type (110) surface of $NiCo_2O_4$. (Panel 1) Schematics of two different reaction pathways of methane complete oxidation on $NiCo_2O_4$ (110-B) of type I crystal. The structures shown in blue in the middle of the schematic are the common intermediates for both pathways A and B; they are CH_3 formed through C–H dissociation, and CH_3O formed through a coupling of the carbon atom of CH_3 and the oxygen atom of the surface lattice. After CH_3O is formed, there are two different possible pathways A and B. Pathway A (shown in red) is the dehydrogenation by OH species of nickel site only, while pathway B (shown in black) is the dehydrogenation by oxygen species of nearby cobalt site. (Panel 2) Energy profile of pathways A (shown in red) and B (shown in black). The entropies of all the gas-phase species are obtained at 823.15 K. (Panel 3) Structures of intermediates of pathway A from reactants to products; Pathway A includes the $CH_2O + O{-}O$ (a1 for top view and a2 for side view), $CHO + HO{-}O$ (b1 for top view and b2 for side view), and $CO + O{-}O$ (c1 for top view and c2 for side view). (Panel 4) Structures of intermediates of pathway B from reactants to products; Pathway B includes $CH_2O + OH$ (d1 for top view and d2 for side view), $CHO + O$ (e1 for top view and e2 for side view), and $CO + O{-}O$ (f1 for top view and f2 for side view). It is noted that the elementary steps for the transformation of intermediate CHO to products CO_2 and H_2O presented here is the sub-pathway of CO oxidation. *Source:* Reproduced with permission from [3] / Springer Nature.

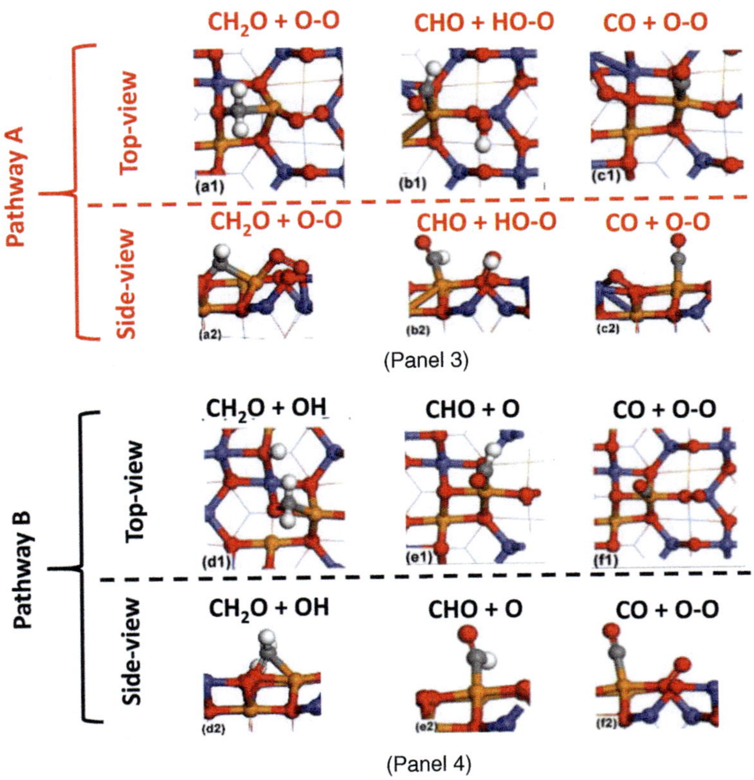

Figure 12.3 (Continued)

called a stable intermediate in the reference [3]. At a relatively high temperature, OCHO can cross the activation barrier of 1.64 eV to form product molecules. This is why the peak 2 in Figure 12.4a representing OCHO became weak at 150°C and disappeared at 200°C along the increase of catalysis temperature from 25 to 400°C.

Other than the Ni cations doped on Co_3O_4, several examples capable of activating CH_4 have been reported in the literature. Ni, Co, and Fe cations on $Ce_2O_3(001)$, $CeO_2(111)$, or ceria nanoparticles can activate C–H bonds of CH_4 at room temperature or above [5–11].

12.3 Cations on the Surface of Iridium Oxide Thin Film

In contrast to the doped Ni, Co, or Fe cations on Co_3O_4 or CeO_2, it is reported that cations of coordinately unsaturated Ir (Ir_{cus}) and their adjacent oxygen anions on $IrO_2(110)$ surface (Figure 12.6) are active for activating C–H through forming a η^2–σ complex at a temperature as low as 150 K [12]. The $IrO_2(110)$ thin film model catalyst was prepared by exposing a clean Ir(100) single crystal to 5 Torr O_2 at 502°C. The thickness used for activation C–H of CH_4 is 46 monolayers in terms of 3.5 nm [12]. The temperature-programmed reaction spectroscopy (TPRS) studies suggest that C–H dissociation of CH_4 on $IrO_2(110)$ is performed through a precursor-mediated mechanism [12]. In addition, the analysis of low-temperature CH_4 TPRS data gives an estimated binding energy of 38 kJ mol^{-1}. Thus, the activation barrier for dissociating C–H of CH_4 is about 28.5 kJ mol^{-1} [12]. This value is quite close to the

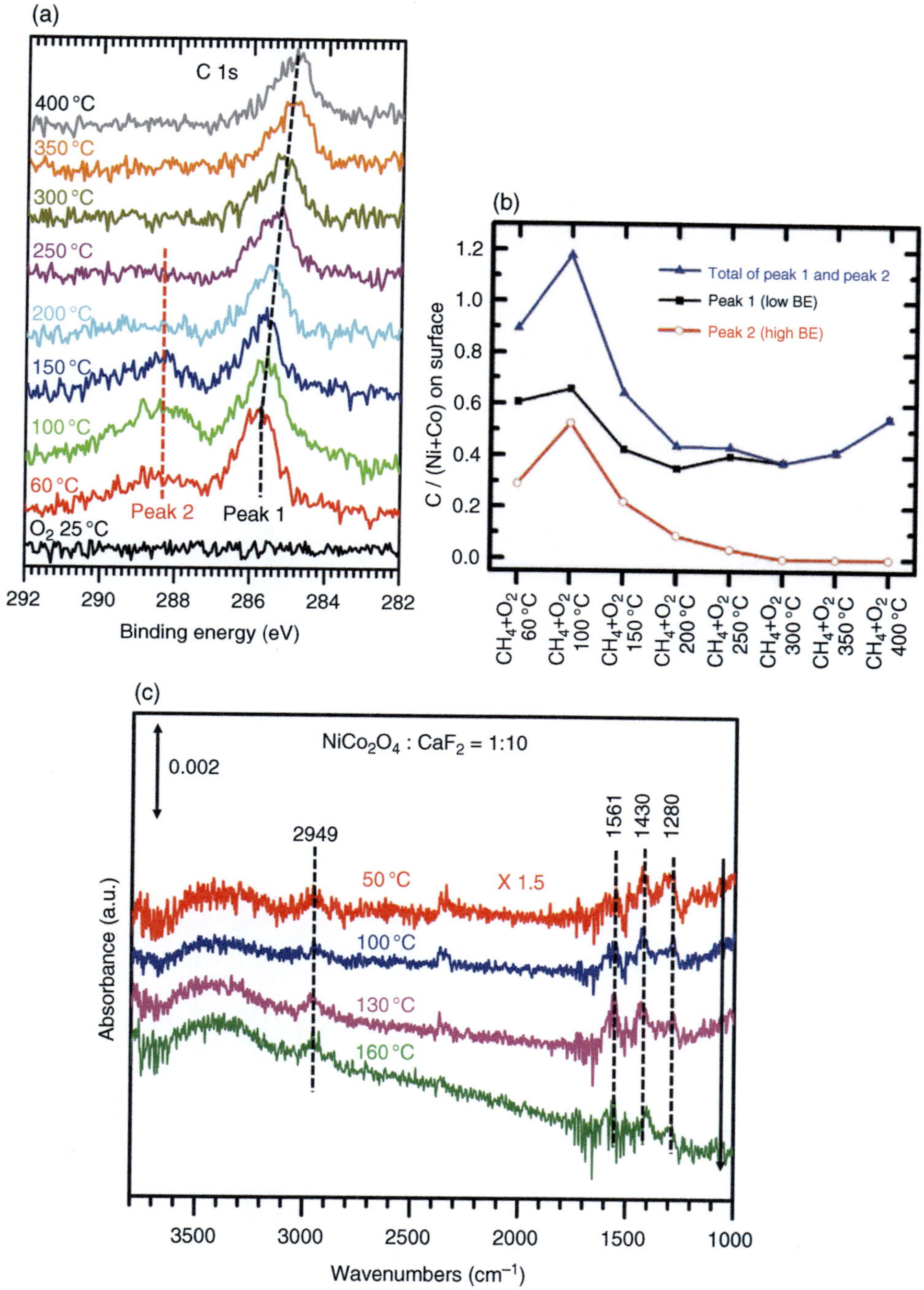

Figure 12.4 Formation of intermediate at low temperature and transformation to products at relatively high temperature. (a) Evolution of C 1s XPS peaks of intermediate (peak 2) and spectator (peak 1) formed on catalyst surface in 60–400°C. The relative intensity of peak 2 increases along the increase of temperature and reaches a maximum at 100–150°C, then starts to decrease and then completely disappears at 200–250°C. The Intensity of the spectator peak remains, although its peak position downshifts slightly due to a possible change in the molecular structure of the spectator. (b) The plot of the C/(Ni+Co) ratio in 60–400°C shows the evolution of the surface concentration of carbon-containing species during catalysis at different temperatures based on in situ studies of $NiCo_2O_4$ surface by AP–XPS. (c) *In situ* infrared studies of the surface species on $NiCo_2O_4$ catalyst in the mixture with CaF_2 under the same catalytic condition as the measurement of catalytic performance in a fixed-bed flow reactor. *Source:* Reproduced with permission from [3] / Springer Nature.

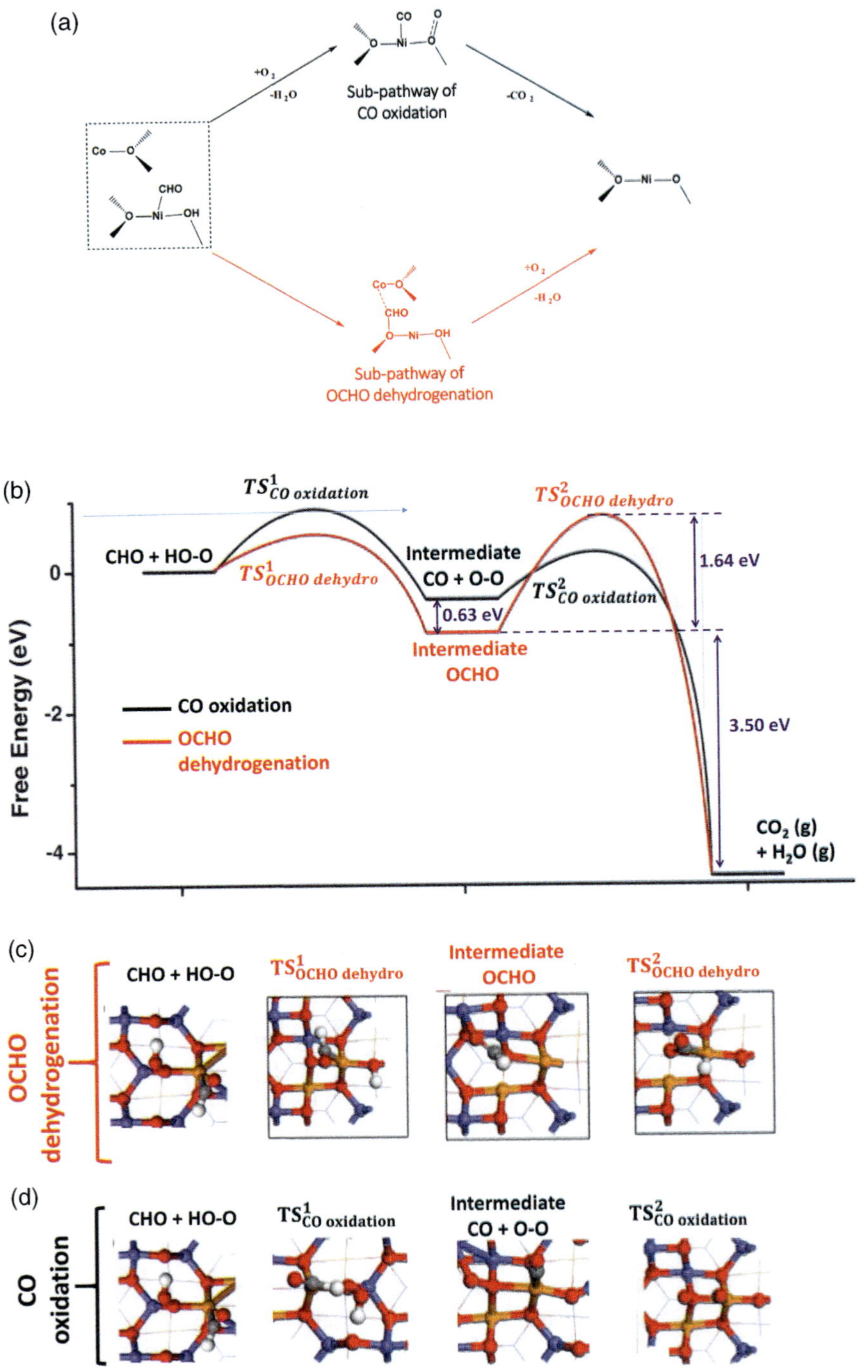

Figure 12.5 DFT calculations of sub-pathways for transforming CHO to product molecules. Each sub-pathway includes two elemental steps with two transition states and one intermediate. (a) Two different sub-pathways for the transformation of CHO to products, including the sub-pathway of CO oxidation (shown in black), which is the last part of pathway B in Figure 12.3, and the sub-pathway of OCHO dehydrogenation (shown in red). (b) Calculated energy profiles and optimized intermediates (CO + O–O versus OCHO) and transition states for the transformation of CHO to product CO_2 on $NiCo_2O_4$ (110-B) of type I crystal after CHO is formed in pathway B. The elemental steps involved in this transformation (from CHO to products) are called a sub-pathway. For this transformation, the sub-pathway of CO oxidation is the last part of pathway B in (a). Other than this sub-pathway, the sub-pathway of OCHO dehydrogenation (red line) was proposed and calculated. (c) Optimized structures of transition states and intermediate of the sub-pathway of OCHO dehydrogenation. (d) Optimized structure of transition states and intermediate of the sub-pathway of CO oxidation. *Source:* Reproduced with permission from [3] / Springer Nature.

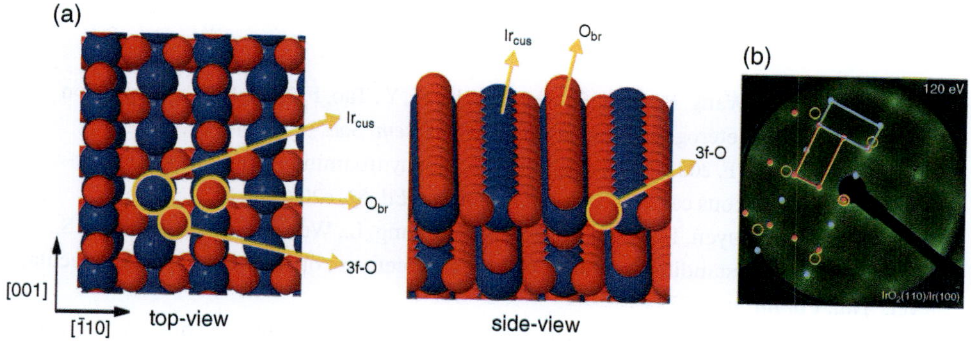

Figure 12.6 $IrO_2(110)$ surface. (a) Model representations of the stoichiometric $IrO_2(110)$ surface with Ir_{cus}, O_{br}, and 3f-O atoms labeled. The Ir and O atoms are represented as blue and red spheres, respectively. (b) LEED pattern obtained from an ~3.5-nm $IrO_2(110)$ film grown on Ir(100) by oxidizing in 5 torr of O_2 at 775 K. The orange circles mark the LEED spot positions from the Ir(100) substrate, and the light blue and pink spots represent reciprocal space points computed for two orientations of the rectangular $IrO_2(110)$ unit cell with dimensions of 1.16 Å × 2.34 Å, where *a* is the Ir(100) lattice constant. *Source:* Reproduced with permission from [12] / AAAS.

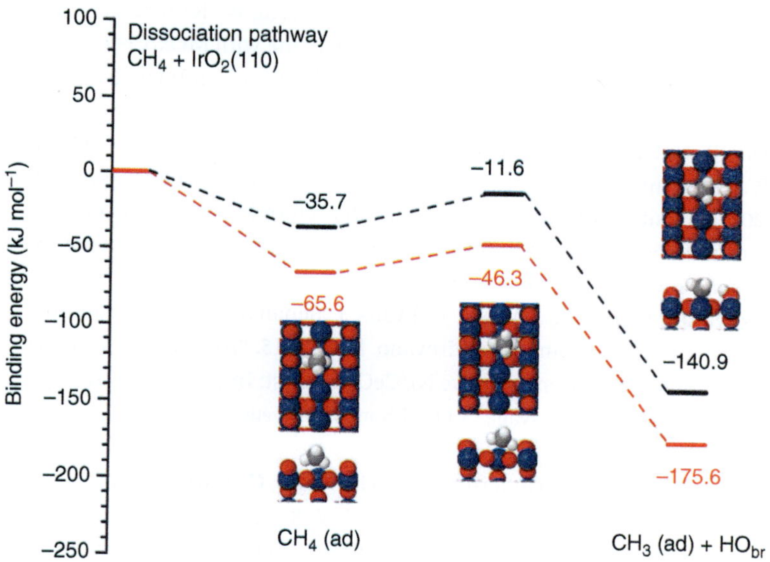

Figure 12.7 Energy diagrams for the formation and C–H bond cleavage of a CH_4 complex on $IrO_2(110)$. The results were computed with conventional DFT (black) and DFT-D_3 (red), as well as images of the initial, transition, and final states. *Source:* Reproduced with permission from [12] / AAAS.

values 24.1 or 20.8 kJ mol^{-1} suggested by DFT calculations shown in Figure 12.7 [12]. As shown in Figure 12.7, the adsorption energy of CH_4 on $IrO_2(110)$ is −35.7 kJ mol^{-1} and the dissociation of C–H of an adsorbed CH_4 needs to cross a barrier of about 24.1 kJ mol^{-1} based on DFT calculation, which is consistent with the experimentally estimated barrier, 28.5 kJ mol^{-1}. In summary, the confirmed activity of the Ir_{cus}–O pair in dissociating C–H of CH_4 at a temperature as low as 150 K suggests that IrO_2 is a promising catalyst for catalytic conversion of CH_4 at a low temperature.

References

1 Zhang, X., Sun, Z., Wang, B., Tang, Y., Nguyen, L., Li, Y., Tao, F. 2018. "C-C coupling on single-atom based heterogeneous catalyst." *J. Am. Chem. Soc.* 140, 954–962.

2 Li, Y., Tang, Y., Tao, F. 2023. "C-N coupling through hydroaminoalkylation on a single-atom Rh heterogeneous catalyst", *Angew. Chem. Int. Ed.* 62, e202214332.

3 Tao, F., Shan, J., Nguyen, L., Wang, Z., Zhang, S., Zhang, L., Wu, Z., Huang, W., Zeng, S., Hu, P. 2015. "Understanding complete oxidation of methane on spinel oxide at a molecular level." *Nat. Commun.* 6, 7798.

4 Zhang, S., Li, Y., Wang, Z., Tang, Y., Huang, X., House, S. D., Huang, H., Zhou, Y., Shen, W., Yang, J., Wang, C., Zhao, Y., Schlogl, R., Hu, P. 2021. "Coordination number-dependent complete oxidation of methane on NiO catalysts". *ACS Catal.* 11, 9837–9849.

5 Liu, Z., Grinter, D. C., Lustemberg, P. G., Nguyen-Phan, T. D., Zhou, Y., Luo, S., Waluyo, I., Crumlin, E. J., Stacchiola, D. J., Zhou, J., Carrasco, J., Busnengo, H. F., Ganduglia-Pirovano, M. V., Senanayake, S. D., Rodriguez, J. A. 2016. "Dry reforming of methane on a highly-active Ni-CeO_2 catalyst: Effects of metal-support interactions on C–H bond breaking". *Angew. Chem. Int. Ed.* 55, 7455–7459.

6 Zhang, F., Yao, S., Liu, Z., Gutierrez, R. A., Vovchok, D., Cen, J., Xu, W., Ramírez, P. J., Kim, T., Senanayake, S. D., Rodriguez, J. A. 2018. "Reaction of methane with MOx/CeO_2 (M = Fe, Ni, and Cu) catalysts: In situ studies with time-resolved X-ray diffraction." *J. Phys. Chem. C* 50, 28739–28747.

7 Liu, Z., Lustemberg, P., Gutierrez, R. A., Carey, J. J., Palomino, R. M., Vorokhta, M., Grinter, D. C., Ramírez, P. J., Matolín, V., Nolan, M., Ganduglia-Pirovano, M. V., Senanayake, S. D., Rodriguez, J. A. 2017. "In-situ investigation of methane dry reforming on M-CeO_2(111) {M= Co, Ni, Cu} surfaces: Metal-support interactions and the activation of C–H bonds at low temperature." *Angew. Chem., Int. Ed.* 56, 13041–13045.

8 Carrasco, J., Lopez-Duran, D., Liu, Z., Duchon, T., Evans, J., Senanayake, S. D., Crumlin, E. J., Matolin, V., Rodriguez, J. A., Ganduglia-Pirovano, M. V. 2015. "In situ and theoretical studies for the dissociation of water on an active Ni/CeO_2 catalyst: Importance of strong metal–support interactions for the cleavage of O–H bonds." *Angew. Chem., Int. Ed.* 254, 3917–3921.

9 Lustemberg, P., Ramirez, P. J., Liu, Z., Gutierrez, R. A., Grinter, D. G., Carrasco, J., Senanayake, S. D., Rodriguez, J. A., GandugliaPirovano, M. V. 2016. "Room temperature activation of methane and dry reforming with CO_2 on Ni-CeO_2(111) surfaces: Effect of Ce3+ sites and metal-support interactions on C–H bond cleavage." *ACS Catal.* 6, 8184–8189.

10 Zhang, F., Liu, Z., Zhang, S., Akter, N., Palomino, R. M., Vovchok, D., Orozco, I., Salazar, D., Rodriguez, J. A., Llorca, J., Lee, J., Kim, D., Xu, W., Frenkel, A., Li, Y., Kim, T., Senanayake, S. D. 2018. "In-situ elucidation of the active state of Co-CeO2 catalysts in the dry reforming of methane: The important role of the reducible oxide support and interactions with cobalt." *ACS Catal.* 8, 3550–3560.

11 Liu, Z., Huang, E., Orozco, I., Lian, W., Palomino, R. M., Ru, N., Duchon, T., Nemsak, S., Crinter, D. C., Mahapaatra, M., Liu, P., Rodriguez, J. A., Senanayake, S. D. 2020. "Water-promoted interfacial pathways in methane oxidation to methanol on a CeO_2-Cu_2O catalyst." *Science* 368, 513–517.

12 Liang, Z., Li, T., Kim, M., Asthagiri, A., Weaver, J. F. 2017. "Low-temperature activation of methane on the IrO_2(110) surface." *Science* 356, 299–303.

13

Catalytic Transformation of Methane through Organometallic Approach at Low Temperature

As mentioned in the introduction of Chapter 10, the activation or catalytic transformation of CH_4 can be performed in the gas phase, liquid phase, or at solid–liquid interfaces as reported in the literature. Chapters 10–12 discussed the activation of CH_4 by M^+ and oxidation of CH_4 by MO^+, and even catalytic oxidation of CH_4 by Pt^+. This chapter focuses on the low-temperature oxidation of CH_4 by homogeneous catalysts in liquid.

13.1 Pt-based Catalysts for Production of Methanol

Early homogeneous catalysis through K_2PtCl_4. Early homogeneous catalysis using K_2PtCl_4 was pioneered as far back as the 1970s, where it was utilized to oxidize CH_4 to CH_3OH using O_2 as the oxidant, typically performed in H_2O at temperatures below 100 °C [1]. However, this catalysis system had its drawbacks. The catalytic activity was quite low, with a turnover frequency (TOF) of $<10^{-5}$ s^{-1}, and its lifespan was short.

(bpym)$PtCl_2$. To address the instability and short lifespan of K_2PtCl_4, a different catalytic system was developed using (bpym)$PtCl_2$. In this system, Pt(II) remained stable by coordinating with N-containing neutral ligands such as bipyrimidine (bpym), preventing it from being reduced [2]. This catalyst catalyzed the oxidation of CH_4 by concentrated H_2SO_4 to form CH_3OSO_3H. The catalysis was conducted in oleum (fuming sulfuric acid) at 220 °C in CH_4 at a pressure of 34.4 bar. The reported TOF was approximately 10^{-3} S^{-1}, significantly higher than that of K_2PtCl_4. In addition, this catalyst had a turnover of 300, much greater than K_2PtCl_6 [1]. Periana et al. proposed a mechanism for this catalytic process, where the active species for C–H activation of CH_4 is (bpym)Pt^{II} (Figure 13.1) [3]. An interesting feature of this mechanism is the "self-repair" step, where the intermediate (bpym)Pt^{IV} is converted back to an active species (bpym)Pt^{II} through a reaction between (bpym)Pt^{II}–CH_3 and the formed (bpym)Pt^{IV}. The (bpym)Pt^{II}–CH_3 was pre-produced through C–H activation of CH_4 by (bpym)Pt^{II} [4]. As (bpym)Pt^{II} can be readily oxidized by H_2SO_4, the self-repair reaction proceeds at diffusion-controlled rates. An improvement to this catalytic system was achieved by Michalkiewicz et al. [5], where the product CH_3OSO_3H was extracted using a membrane-based method instead of hydrolysis, followed by subsequent hydrolysis to produce CH_3OH.

Low-temperature Activation and Catalytic Transformation of Methane to Non-CO_2 Products,
First Edition. Franklin Tao.

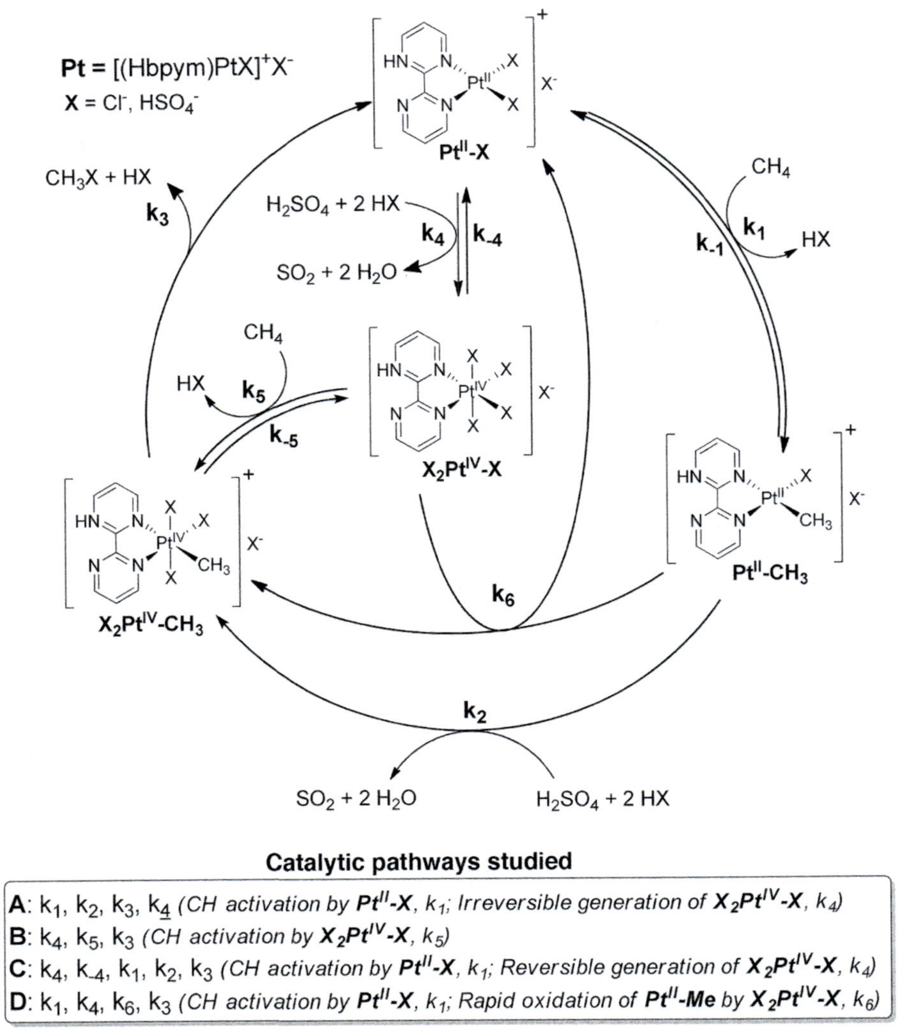

Figure 13.1 Plausible pathways for the Periana-Catalytic system that may account for the high stability. *Source:* Reproduced with permission from [3] / American Chemical Society.

Ionic-liquid-based ternary system. An ionic liquid dissolved Pt catalyst was reported by Tang et al. [6]. It is a ternary system of $PtCl_2$, PtO_2, ionic liquid, and concentrated sulfuric acid. It utilizes an environmentally friendly medium, namely, the ionic liquid. The key advantage of employing an ionic liquid lies in its ability to readily dissolve $PtCl_2$, K_2PtCl_4, PtO_2, and H_2PtCl_6. To ensure optimal performance, only methyl-based ionic liquids are suitable solvents, as indicated in the note in Table 13.1 [7]. Pyrazolium-based ionic liquids demonstrate remarkable durability under catalytic conditions. These ternary systems exhibit comparable selectivity for CH_3OSO_3H [7]. Furthermore, they demonstrate resistance to water, as the dissolved platinum salts and oxides such as $PtCl_2$, H_2PtCl_6, and PtO_2 remain insensitive to water, although (bpym)$PtCl_2$ could be sensitive to water.

Table 13.1 Catalytic oxidation of methane to methanol in 98% H_2SO_4.

Entry	Catalyst[a]	Ionic liquid[b]	T/°C	CH_3OH/mM[c]	TON
1	(bpym)$PtCl_2$		220	31	0.6
2	(bpym)$PtCl_2$	[im][Cl]	220	26	0.5
3	$PtCl_2$	[l-mim][Cl]	220	173	3.5
4	$PtCl_2$	[im][Cl]	220	100	2.0
5	$PtCl_2$	[mmim][HSO_4]	220	73	1.5
6	K_2PtCl_4	[1-mim][Cl]	220	105	2.1
7	$PtCl_4$	[1-mim][Cl]	220	89	1.8
8	PtO_2	[im][Cl]	220	52	1.0
9	PtO_2	[1-mim][HSO_4]	220	0	0
10	H_2PtCl_6	[1-mim][Cl]	220	157	3.1
11	$PtCl_2$	[pyrid][HSO_4]	220	1	~0
12	$PtCl_2$	[pyraz][HSO_4]	200	42	0.8
13	$PtCl_2$	[1-mpyraz][HSO_4]	200	79	1.6
14	$PtCl_2$	[mmpyraz][HSO_4]	200	94	1.9
15	$PtCl_2$	[triaz][HSO_4]	200	0	0

[a] Reaction conditions: 0.05 mmol Pt species + 0.3 mmol IL + 1 mL 96% H_2SO_4 + 3.4 MPa CH_4 in a 69 mL reactor at the chosen temperature for 2.5 h.
[b] Ionic liquid: im = imidazolium; 1-mim = 1-methylimidazolium; mmim = 1,3-dimethylimidazolium; pyrid = pyridinium; pyraz = pyrazolium; 1-mpyraz = 1-methylpyrazolium; mmpyraz = 1,2-dimethylpyrazolium; triaz = 1,2,4-triazolium.
[c] The methanol concentration was determined by 1H NMR from the sum of the free or protonated methanol and the ester (CH_3OSO_3H).
Source: Reproduced with permission from [7] / Royal Society of Chemistry.

13.2 Pt-based Catalysts for the Production of Acetic Acid

One route for transforming CH_4 into value-added chemicals is the oxidative carbonylation of CH_4. This reaction involves CH_4, CO, and an oxidant, generally represented as $CH_4 + [O] + CO \rightarrow CH_3COOH$. CO serves as the carbon source of the carboxylate in acetic acid. Ideally, the cost could be significantly reduced if molecular O_2 or even air proves effective. Early studies by Fujiwara et al. demonstrated that acetic acid can be synthesized through the oxidative carbonylation of methane with O_2 at 80 °C, although the TOF is as low as 10^{-5} s^{-1}, as discussed in the Rh-based and Pd-based catalysts section [4]. Apart from Pd-based and Rh-based catalysts, Bell et al. reported that $PtCl_2$ can catalyze the synthesis of acetic acid from CH_4, utilizing H_2SO_4 as the oxidant [6]. $PtCl_2$ catalyzes the carbonylation of CH_3OSO_3H to acetic acid at 180 °C under 27 bar of CH_4 with a TOF of 1.5×10^{-4} S^{-1} [6]. CO readily reduces $PtCl_2$ to Pt(0). To enhance the oxidation of the Pt(0) formed on-site back to Pt^{II}, Cu^{2+} is introduced as a cocatalyst. $PtCl_2$ used in this catalysis can be replaced with $Pt(acac)_2$, coexisting with $CuCl_2$ and O_2 in the reactor [6]. Although the active Pt(II)

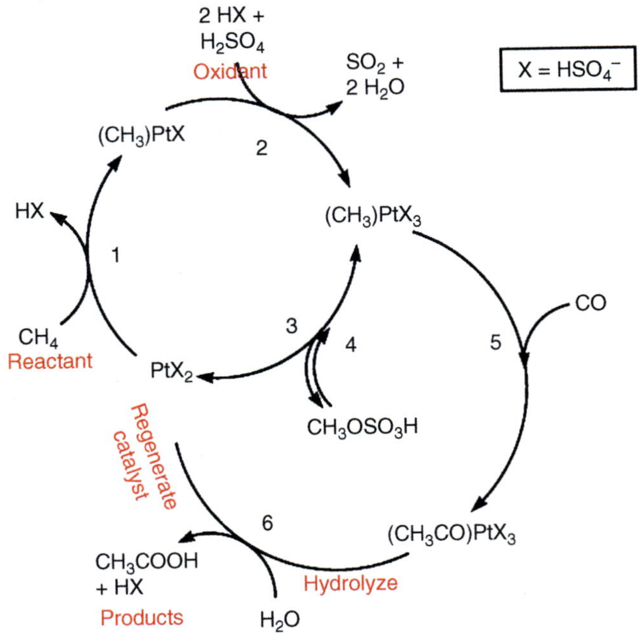

Figure 13.2 The proposed reaction mechanism for oxidative carbonylation of methane through PtX_2 at 180 °C in a solvent. *Source:* Reproduced with permission from [6] / Elsevier.

complex for the oxidative carbonylation of methane remains unidentified, a proposed reaction mechanism is illustrated in Figure 13.2. In this mechanism, the CH_3 moiety in the product, CH_3COOH, originates from CH_4 or CH_3OSO_3H. CO can be inserted into the complex $(CH_3)PtX_3$ (X = halogen) to form $(CH_3CO)PtX_3$. $(CH_3CO)PtX_3$ can be hydrolyzed to produce CH_3COOH and PtX_2. Subsequently, PtX_2 can be employed in the next catalytic cycle.

13.3 Pd-based Catalyst for Production of Methanol

Literature reported that Pd(II) salts are active for transforming CH_4 to CH_3OH in sulfuric acid solvent [2, 8–10]. The Pd(0) species formed from Pd(II) needs to be oxidized back to Pd(II). However, the reoxidation of Pd(0) to Pd(II) is too slow. Compared to Pd(II) salts, the $Pd^{II}(O_2CC_2H_5)_2$ exhibits a higher activity.

$Pd^{II}(O_2CC_2H_5)_2$. It is reported that $Pd^{II}(O_2CC_2H_5)_2$ catalyzes the oxidation of CH_4 by H_2O_2 in an inert solvent, peroxytrifluoroacetic acid, forming methyl trifluoroacetate, CF_3COOCH_3 [11]. Methanol, the ideal product of such a reaction, is unfortunately prone to further oxidation to formaldehyde, formic acid, or acetic acid. To prevent methanol from being oxidized, an inert chemical, peroxytrifluoroacetic acid, is introduced to react with methanol, forming insert CF_3COOCH_3. CF_3COOCH_3 is inert for being oxidized under the catalytic condition. Luckily, the formed CF_3COOCH_3 can be readily hydrolyzed back to methanol. The CF_3COO-group is less susceptible to electrophilic attack compared to the OH group in CH_3OH. Thus, the CF_3COOCH_3 is easily hydrolyzed to produce methanol. Here, the inert acid, peroxytrifluoroacetic acid, CF_3OOOH serves as the solvent for the oxidative transformation of CH_4 and converts the produced methanol into the inert ester, CF_3COOCH_3. The peroxytrifluoroacetic acid is generated through the oxidation of trifluoroacetic anhydride. This CH_4 oxidation is catalyzed by the $Pd^{II}(O_2CC_2H_5)_2$. In nature, this is a reaction of oxidation to CH_4 with H_2O_2 to form CH_3OH and H_2O, $CH_4 + H_2O_2 \rightarrow CH_3OH + H_2O$. It is performed at 90°C and catalyzed by Pd(II). Excess trifluoroacetic anhydride is used to remove water, a product of the reaction, as water could otherwise hydrolyze CF_3COOCH_3 to form methanol during catalysis since the formed methanol can be readily further oxidized by H_2O_2 to form byproducts under the oxidizing condition. This catalytic process involves oxidation of CH_4 to CH_3OH, formation of an inert ester, and hydrolysis of the ester to form CH_3OH. The oxidative catalysis of CH_4 by H_2O_2 occurs via an electrophilic pathway.

Other catalysts. The catalytic oxidation of CH_4 to methanol was also observed when replacing $Pd^{II}(O_2CC_2H_5)_2$ with $Pb^{II}(O_2CC_2H_5)_2$, $Fe^{II}(O_2CC_2H_5)_2$, or $Co^{II}(O_2CC_2H_5)_2$ [12]. It is anticipated that a parallel oxidation reaction occurs, as the $Pd^{II}(O_2CR)_2$ catalyst itself acts as an oxidant. The catalyst $Pd^{II}(O_2CR)_2$ can be reduced to Pd(0) by CH_4. When an oxidizing agent converts the formed Pd(0) back to Pd(II), the catalytic activity and lifespan increases [13]. Moreover, studies have demonstrated that Pd(II) can catalyze the oxidation of methane with H_2SO4 containing 30% SO_3 at 160°C to form CH_3OSO_3H [14].

13.4 Pd-based Catalyst for the Production of Acetic Acid

One of the current methods for synthesizing acetic acid from CH_4 involves a three-step process [15]. In the first step, syngas, a mixture of CO and H_2, is produced at high temperatures through the coal gasification or steam reforming of CH_4. Then, methanol is synthesized from CO and H_2 at a relatively high temperature (300°C or so). The third step involves the carbonylation of methanol at a relatively high temperature (250–300°C). While the first two steps employ heterogeneous catalysis, the third step utilizes homogeneous catalysis at around 200–300 °C. Thus, a direct synthesis of acetic acid from CH_4 at low or relatively low temperatures (≤150 °C) holds the potential to produce acetic acid with zero impact on the environment and low-energy consumption.

$PdSO_4$ catalyst with sulfuric acid. Periana et al. reported a new method for producing acetic acid directly from CH_4 oxidation with H_2SO_4, catalyzed by $PdSO_4$ [10]. This approach represents a significant advancement compared to the three-step process typically required for the transformation of CH_4 into acetic acid. In this innovative approach, H_2SO_4 serves as the oxidant and does not take CO as a reactant. The reaction equation is thus simplified to $2CH_4 + 4H_2SO_4 \rightarrow CH_3COOH + 4SO_2 + 6H_2O$. Isotope experiments have confirmed that the carbon atoms in both the carbonyl and methyl groups of acetic acid originate from CH_4. An additional ^{13}C-labeled methanol was added to a reactor containing H_2SO_4, $PdSO_4$, and CH_4 before catalysis, and the ^{13}C was observed in the carboxylate group of the produced $CH_3{}^{13}COOH$, confirming the carbon source of COOH in CH_3COOH produced in the above reaction is CH_4. The readily catalytic oxidation of CH_3OH to CO and then CO_2 by sulfuric acid suggests that the carbonylation agent of CH_3 is CO since CO is an intermediate product in the oxidation of CH_3OH to CO_2. Pd(II) is active for catalyzing the C–H activation of CH_4 through electrophilic attack of Pd(II) to C–H forming Pd–CH_3. The Pd–CH_3 can be readily transformed to CH_3OH. CH_3OH can be oxidized to CO, the intermediate product CO can couple with Pd–CH_3 to form CH_3COOH [10].

Based on the proposed mechanism in literature [10], Pd(II)-CH_3 was formed from PdX_2 through electrophilic activation of CH_4. Then, carbonylation of the –CH_3 was conducted by a self-produced CO, which is formed through the oxidation of an intermediate compound, methanol. The methanol is produced from the oxidation of the activation of CH_3 in Pd–CH_3 by sulfuric acid. Here, the CO origins at another CH_4 molecules in the overall equation, $2CH_4 + 4H_2SO_4 \rightarrow CH_3COOH + 4SO_2 + 6H_2O$. The formed CO can inserted into Pd(II)–CH_3 to form X–Pd–(CO)(CH_3) and then give product CH_3COX as shown at the right-top corner in Figure 13.3 [10]. Upon hydrolysis of CH_3COX, acetic acid is formed.

The role of the CO molecule in the proposed mechanism of acetic acid synthesis in Figure 13.3 was confirmed by another isotope experiment in ^{13}CO and CH_4 were added to the headspace of a reactor before catalysis and then $CH_3{}^{13}COOH$ was clearly observed after catalysis [16]. Although CO is a necessary reactant in the oxidative carbonylation of CH_4 to acetic acid in other synthetic route of acetic acid, a high CO pressure can decrease the conversion of CH_4 and yield of acetic acid [10]. The consequence of high CO pressure in this catalysis was supported by the observation of Pd-black in a contrast experiment where neither CH_4 nor O_2 was added to the reactor, but 1 atm CO was introduced to the headspace of a reactor containing $PdSO_4$ and H_2SO_4 before heating to 180 °C [10].

Role of molecular O_2 in $PdSO_4$ catalyzed synthesis of acetic acid. The above section clearly shows that acetic acid can be synthesized through oxidation of CH_4 with 96 wt% sulfuric acid at 180°C catalyzed by $PdSO_4$ [10]. This transformation does not include molecular O_2. The oxidant is H_2SO_4 or/and SO_3 [10]. A later report shows that O_2 is significant for promoting the yield of acetic acid and the life of the Pd(II) catalyst [16]. As shown in Table 13.2 [16], the yield of acetic acid increases along with the increase of O_2 pressure while acetic acid was produced with molecular O_2. Another significant feature is the

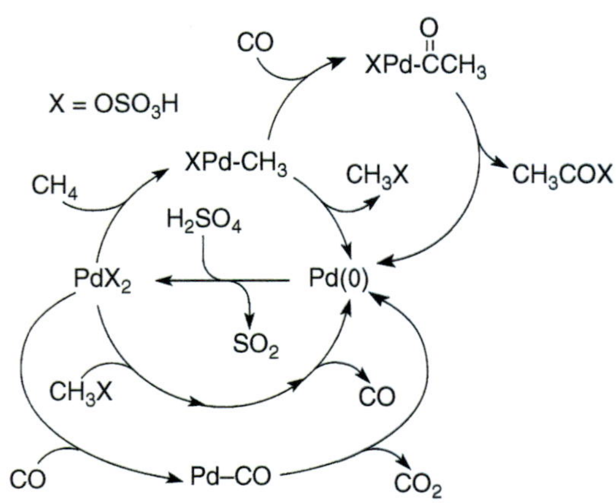

Figure 13.3 Proposed tandem catalysis scheme consistent with observations. *Source:* [10] / AAAS.

Table 13.2 Effect of O_2 pressure at 400 psi CH_4.

	O_2 pressure (psi)			
	0	**30**	**100**	**150**
CH_3COOH (mM)	78.5	118	191	284
CH_3OSO_3H (mM)	3.2	5.1	5.4	6.8
CH_3SO_3H (mM)	9.2	8.2	27.0	55.5
HO_3SCH_2COOH (mM)	8.3	10.1	12.2	16.3
$CH_2(SO_3H)_2$ (mM)	44.6	26.3	25.4	36.8
S_{CH_3COOH} (%)	51	57	50	44
S_{CO_x} (%)	27	31	41	47
Pd retained (%)	3.8	3.3	16	40

[a] Reaction conditions: 3 mL 96% H_2SO_4; 0.0121 g (20 mM) $PdSO_4$; 400 psi CH_4; X psi O_2; 180 °C; 4 h.
Source: Reproduced with permission from [16] / Elsevier.

percentage of Pd remaining in the solution largely increases along with the increase of O_2 pressure. A series of experimental exploration suggested that Pd(II) was reduced to Pd by CO and CH_3OSO_3H, which are the reaction intermediate products. The formation of Pd(0) makes the catalytic activity decay. The function of O_2 in sulfuric acid is to oxidize Pd(0) to $Pd(OSO_3H)_2$ through the following reaction: $Pd(0) + 1/2O_2 + 2\,H_2SO_4 \rightarrow Pd(OS_3H)_2 + H_2O$. For the synthesis of acetic acid through oxidation of CH_4 with H_2SO_4 with the assistance of O_2 catalyzed by Pd(II), a reaction pathway as Figure 13.4 was proposed [16]. Compared to

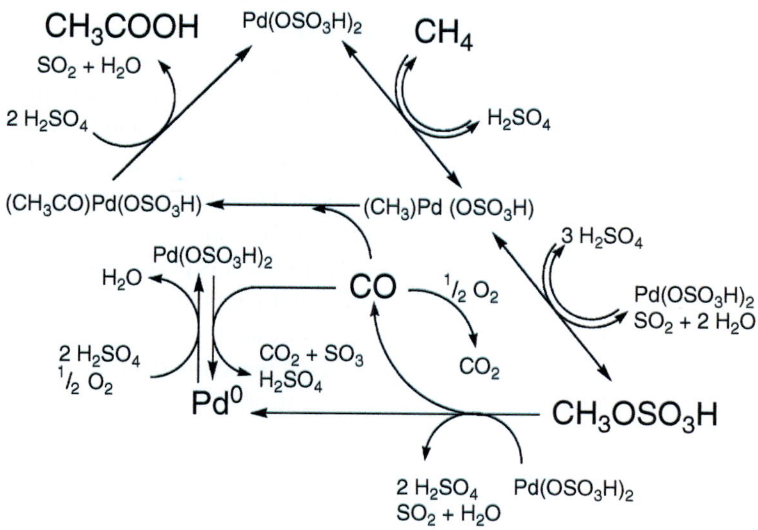

Figure 13.4 Reaction pathway. *Source:* Reproduced with permission from [16] / Elsevier.

the pathway proposed for synthesis of acetic acid through oxidation of CH_4 with H_2SO_4 without molecular O_2 catalyzed by Pd(II) in Figure 13.3, the mechanism in Figure 13.4 includes CO as a specific reaction intermediate, both CO and CH_3OSO_3H as reducing agents to reduce Pd(II) to Pd(0); then Pd(0) is oxidized to Pd(II) with O_2 and H_2SO_4 to restore the catalytically active species Pd(II) (Figure 13.4) [16].

Pd and Cu catalysis system. To prevent the potential formation of Pd black in the production of acetic acid through the oxidization of CH_4 with H_2SO_4 catalyzed by PdSO4, $CuCl_2$ can act as a cocatalyst [17]. The CO formed from the oxidation of CH_3OSO_3H, as shown in Figure 13.4, can reduce Pd(II) to Pd(0) in terms of black Pd. Other than CH_4, $PdSO_4$, and H_2SO_4, $CuCl_2$ and O_2 were introduced to reactors. The reaction is performed at 180°C. CO and CO_2 were formed at the headspace of the reactor and experimentally observed. The experimental observation of CO formation confirmed the deduction that CO was formed during the catalysis, and then the internally formed CO was inserted into Pd–CH_3 in terms of carbonylation. With $CuCl_2$ and O_2, Pd black was not observed, and the solution in the reactor remained light red [17]. Without $CuCl_2$ and O_2, Pd particles were readily found in the reactor. It suggests the important role of $CuCl_2$ in oxidizing the formed Pd(0) to Pd(II) to remain Pd as catalytically active species [17]. As shown in the following reaction 1, the Cu(II) was reduced to Cu(I), and then Cu(II) was regenerated through oxidizing Cu(I) using H_2SO_4 and O_2 in reaction 2. Reaction 1: $Pd(0) + 2Cu(II) \rightarrow Pd(II) + 2Cu(I)$ and Reaction 2: $2Cu(I) + 2H_2SO_4 + 1/2O_2 \rightarrow 2Cu(II) + H_2O + 2HSO_4^-$

Similar Pd and Cu catalysis systems can catalyze the oxidation of methane, ethane, and butane with the presence of O_2 and CO [18]. It is also active for catalyzing hydroxylation of primary C–H bonds of acids, alcohols, and aliphatic halides while CO and O_2 are involved [19].

13.5 Rh-based Molecular Catalysts for the Production of Acetic Acid with the Participation of External CO

Currently, the industrial production of acetic acid requests high-temperature methane steam reforming or coal gasification to produce syngas, high-temperature synthesis of methanol from syngas, and carbonylation of methanol to form acetic acid, typically through Monsanto process developed by 1970s′ [20]. It is reported that $RhCl_3$ can catalyze the oxidation of CH_4 to acetic acid through the participation of CO and O_2 at 100°C with a high yield [21]. The reaction is $2CH_4 + 2CO + O_2 \rightarrow 2CH_3COOH$. It is performed in an aqueous solution at 100 °C in a reactor with a headspace filled with a mixture of CH_4, CO, and O_2. The byproducts of this reaction are methanol and formic acid. Notably, different from the reaction, $2CH_4 + 4H_2SO_4 \rightarrow CH_3COOH + 4SO_2 + 6H_2O$ involving the inherently produced intermediate product CO as the carbonylation agent of CH_3– [16], here the source of CO in terms of carbonation agent in $2CH_4 + 2CO + O_2 \rightarrow 2CH_3COOH$ is external [21]. The C–H activation is performed through electrophilic attack to form $[Rh(CH_3)X_3]^-$, similar to the formation of Pd–CH_3 while Pd(II) salt was used in literature [10, 16]. The following carbonylation of Rh–CH_3 forms Rh(CO)CH_3 which is then hydrolyzed to form CH_3COOH (Figure 13.4).

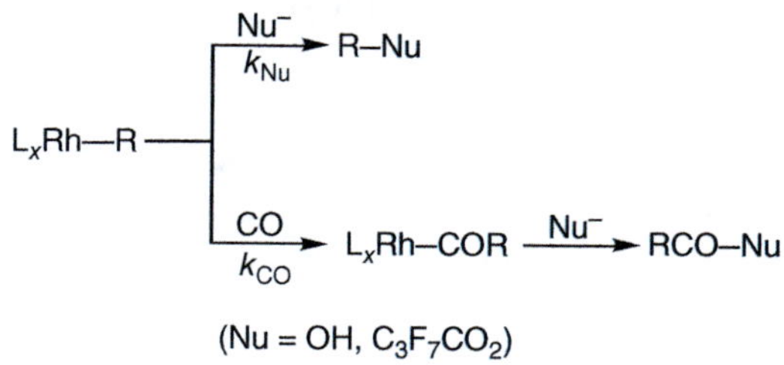

Figure 13.5 Schematic showing nucleophilic attack versus carbon monoxide insertion into an Rh-alky bond. *Source:* Reproduced with permission from [22] / American Chemical Society.

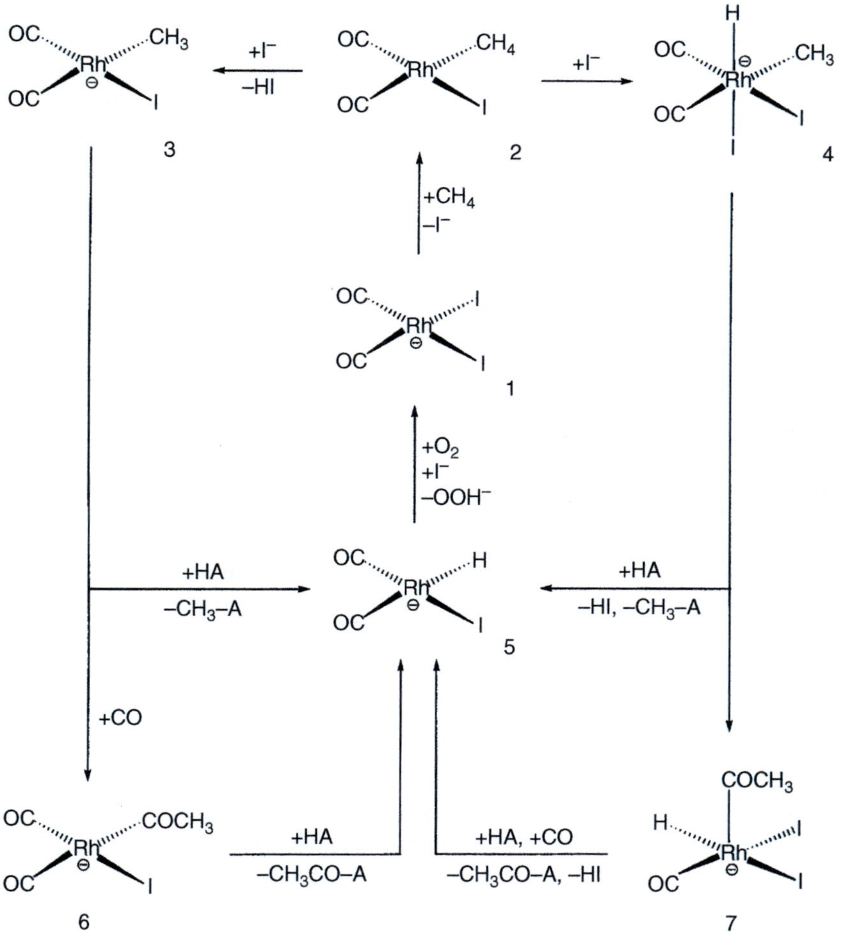

Figure 13.6 Proposed catalytic cycle for the Sen reaction. *Source:* Reproduced with permission from [23] / American Chemical Society.

Computational studies for the selective oxidation of methane with CO and O_2 catalyzed by $[Rh(CO)_2I_2]^-$ through the proposed mechanisms in Figure 13.6 [23]. As seen in Figure 13.6, a σ-complex, $Rh(CO)_2ICH_3$ (***3*** in Figure 13.6), can be formed through coupling CH_4 with a highly active $[Rh(CO)_2I_2]^{2-}$(***1***) or through a SN_2 substitution [23]. The σ-complex of CH_4, $Rh(CO)_2ICH_4$ (**2**), can perform a metathesis to form Rh(I)–CH_3 (**3**) or an oxidative addition of HI to form $Rh(CO)_2IHICH_3$ (**4**). In terms of **4**, one of the two CO molecule can be inserted into Rh-CH_3 of intermediate **4**, forming the intermediate, acyl complex *7*. Then, the CH_3CO^+ ligand is removed from *7* by nucleophilic attack from heptafluorobutyric anion to form $C_4F_7O_2COCH_3$ designated as CH_3CO–A at the bottom of Figure 13.6. $C_4F_7O_2^-$ is represented by A^- in Figure 13.6. Acetic acid is finally formed through hydrolysis of CH_3CO–A [23].

A parallel mechanism presented in Figure 13.6 is that intermediate **2** can transform to **3** by losing an HI and coordinating with an I^-. An external CO molecule can be inserted into the Rh–CH_3 (intermediate **3**), forming –CH_3CO– in intermediate **6**. Through a nucleophilic attack by $C_4F_7O_2^-$ (A^-), forming $C_4F_7O_2COCH_3$ (CH_3CO–A). A followed hydrolysis of CH_3CO–A forms acetic acid [23]. The selectivity for acetic acid was significantly improved by using a solution of perfluorobutyrate ($C_3F_7COO^-$) as a solvent and also a nucleophile since $C_3F_7COO^-$ is more nucleophilic than OH^- and thus it can readily attack L_xRh^- to form CH_3CO–$O_2CC_3F_7$ based on Figure 13.5 [22]. Notably, the addition of I^- anions or 5% metallic Pd supported on carbon largely increases the yield of acetic acid [21].

13.6 Hg-based Catalysts for Production of Methanol

Periana et al. reported a homogeneous catalysis system consisting of Hg(II) and sulfuric acid for the transformation of CH_4 to CH_3OH [24]. It is operated at 180°C. To prevent methanol from being oxidized into aldehyde, CO, formic acid, or even CO_2, the formed methanol was converted to ester, methyl bisulfate, and CH_3OSO_3H by H_2SO_4 [24]. Hg(II) was used to catalyze the oxidation of CH_4 to methyl bisulfate by H_2SO_4. Hydrolysis of methyl bisulfate produces methanol. The yield of methanol can reach 43% at 180 °C. The significance of this work is the use of Hg(II) as a catalyst to form methyl bisulfate as a primary product since methyl bisulfate is more insert than methane. If CH_3OH is the product, it must be further oxidized into byproducts. A later hydrolysis of methyl bisulfate and distillation readily transfer it to methanol. As reported in the literature, Hg(II) can activate C–H of arenes through electrophilic substitution of H of the C–H [24–27]. Periana et al. found that CH_4 can be oxidized by $Hg(CF_3SO_3)_2$ through the following reaction, $CH_4 + 2Hg(CF_3SO_3)_2 \rightarrow CF_3SO_3CH_3 + CF_3SO_3H + Hg_2(CF_3SO_3)_2$ [24]. When H_2SO_4 is used as a solvent, Hg(II) can catalyze the oxidation of CH_4 with H_2SO_4. The overall reaction is $CH_4 + 2H_2SO_4 \rightarrow CH_3OSO_3H + 2H_2O + SO_2$. It is reported that the conversion of CH_4 can reach 50% with a selectivity for methyl bisulfate at 84% [24]. The rate of the reaction catalyzed by Hg(II) is 10^{-3} s^{-1} in terms of 3.6 h^{-1} [24].

Even Pieriana et al. found that Tl(III), Pd(II), and cations of Pt and Au can also catalyze the oxidation of CH_4 in triflic and sulfuric acids to esters [24]. A reaction pathway of this catalysis is presented in Figure 13.7 [24]. In this transformation, $Hg(II)(OSO_3H)_2$ is the catalyst. The activation of CH_4 is performed through electrophilic displacement of a OSO_3H^- of $Hg(II)(OSO_3H)_2$ by CH_3, forming an intermediate, $Hg(CH_3)(OSO_3H)$ (***1***) in

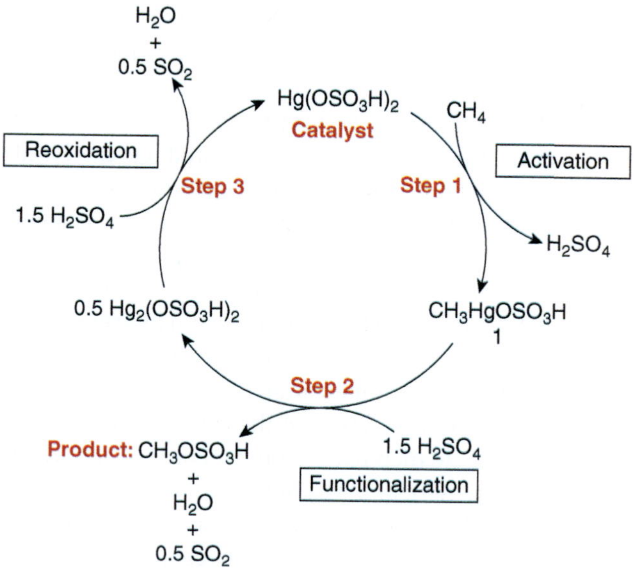

Figure 13.7 Schematic showing the three key steps in the catalytic sequence for oxidizing CH_4 through an Hg-based catalyst. *Source:* [24] / AAAS.

Figure 13.7. This is the first step. It forms the intermediate ***1*** in Figure 13.7. The second step was called the functionalization step, where two $(CH_3)Hg(OSO_3H)$ intermediates decompose to form two methyl bisulfates CH_3OSO_3H, two H_2O molecules, and one SO_2. The second step forms an intermediate, $Hg_2(OSO_3H)_2$ (Figure 13.7). Notably, the CH_3OSO_3H is the product in the second step. The third step is the regeneration of the catalyst, $Hg(OSO_3H)_2$, from the intermediate, $Hg_2(OSO_3H)_2$, through oxidation by H_2SO_4, as shown in the left-top corner in Figure 13.7. A free radical involved pathway was not considered in this work since the observed selectivity for the formation of CH_3OSO_3H is as high as 85% [24]. Thus, a free radical involved pathway is unlikely an option. They also excluded another possibility in terms of oxidative addition since the oxidation of Hg(II) $[5d^{10}4s^0]$ to Hg(IV) $[5d^8 4s^0]$ requests a significant activation barrier. The proposed pathway in Figure 13.7 was supported by their observation of $Hg(CH_3)(OSO_3H)$ (***1***) at a steady-state condition. This catalysis system has been a model system for understanding the mechanism of homogeneous catalysis at a molecular level since its key intermediates can be synthesized and studied readily. The proposed mechanism does not involve molecular O_2 since H_2SO_4 is the oxidant of this reaction.

CH_4 can undergo oxidation by SO_3 to produce methane sulfonic acid (MSA), H_3CSO_3H, a compound different from the previously mentioned product, methyl bisulfide (CH_3OSO_3H). The overall reaction is represented as $CH_4 + SO_3 \rightarrow CH_3SO_3H$, differing from the earlier reaction $CH_4 + 2H_2SO_4 \rightarrow CH_3OSO_3H + 2H_2O + SO_2$. The oxidation of CH_4 to CH_3SO_3H was catalyzed by $Hg(CF_3SO_3)_2$ at 148 °C when O_2 was included as a reactant in a reaction system containing SO_3 and H_2SO_4 [28]. The selectivity for CH_3SO_3H reached as high as 92%, with the primary byproduct being methyl bisulfate (CH_3OSO_3H), resulting in a CH_3SO_3H yield of 40% [28].

Bell et al. demonstrated the crucial role of introducing O_2 to enhance the selectivity for methane sulfonic acid. In the absence of O_2, the conversion of SO_3 to products catalyzed by $Hg(CF_3SO_3)_2$ at 148 °C was only 14% for methane sulfonic acid. In the presence of O_2 at 4 bar, the selectivity for methane sulfonic acid increased to 92%. The turnover number for this catalysis at 148 °C reached 118 [28]. The involvement of molecular O_2 is expected to suppress the proposed free radical mechanism for methane sulfonic acid formation, as O_2 can terminate the methyl radicals. Bell et al. suggested a nonradical pathway for the formation of methane sulfonic acid catalyzed by $Hg(CF_3SO_3)_2$ (Figure 13.8) [29]. This pathway involves three steps: activation of CH_4, functionalization of CH_4, and reproduction of the catalyst molecule. In the first step, CH_4 reacts with the catalyst, forming an intermediate, $CH_3HgOSO_2CF_3$. This intermediate then reacts with SO_3 or H_2SO_4 to yield the desired product, CH_3SO_3H, along with SO_2 and O_2, while reducing $Hg(CH_3)(OSO_2CF_3)$ to $Hg_2(OSO_2CF_3)_2$ as depicted at the bottom of Figure 13.8. In this step, the byproduct CH_3OSO_3H may also be formed. The final step involves the regeneration of the catalyst, $Hg_2(OSO_2CF_3)_2$, by oxidizing $Hg_2(OSO_2CF_3)_2$ with O_2. Notably, O_2 is produced during the formation of product CH_3SO_3H, eliminating the need for external O_2 supplementation.

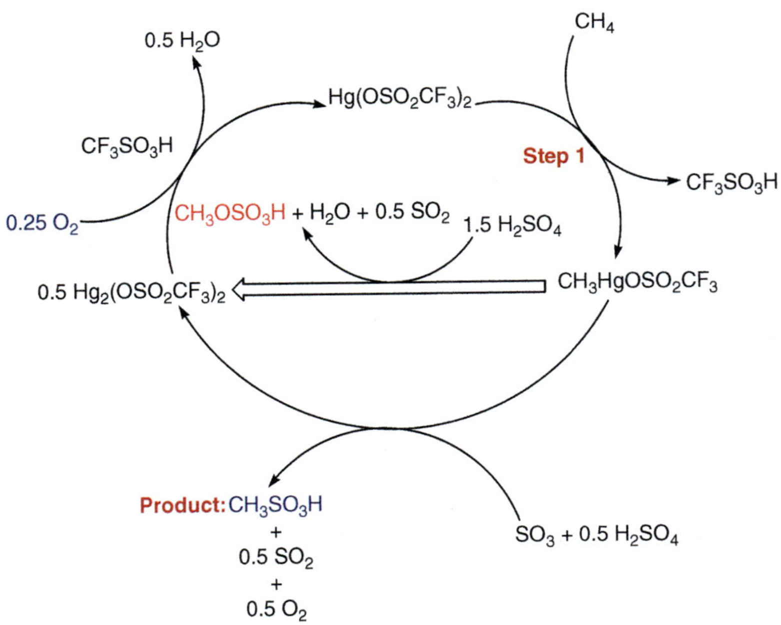

Figure 13.8 Proposed reaction mechanism of oxidizing CH_4 with H_2SO_4 to methane sulfonic acid (MSA), H_3CSO_3H through catalysis of $Hg(CF_3SO_3)_2$ at 148 °C in the presence of O_2. *Source:* Reproduced with permission from [28] / Elsevier.

13.7 Ru-based Catalysts for the Production of Methanol

In industrial oxidation of CH_4, molecular O_2 is not the most common oxidant, although CH_4 can be oxidized to methanol by molecular O_2 using the biocatalyst methane monooxygenase (MMO). MMO contains a binuclear μ-oxo(μ-hydroxo) iron complex. Notably, it has been reported that a salen-based binuclear ruthenium complex, analogous to the MMO complex, can catalyze the oxidation of CH_4 by molecular O_2 at 15 atm, producing CH_3OH at 30 °C [30]. The catalytic oxidation of CH_4 to CH_3OH by the binuclear bridged ruthenium complex, bis-(μ-acetato)(μ-oxo) bis-salen ruthenium (III), is illustrated in Figure 13.9 [30]. The proposed mechanism involves the binding of molecular O_2 to the Ru atom of the catalyst (1 in Figure 13.9), forming a peroxo complex (2). [30] This peroxo complex (2) reacts with a CH_4 molecule to generate a kinetic intermediate (3). Subsequently, the C–H and O–O bonds in the kinetic intermediate (3) undergo cleavage. An oxygen atom from the peroxo moiety transfers to the resulting carbonium ion center, yielding the first CH_3OH

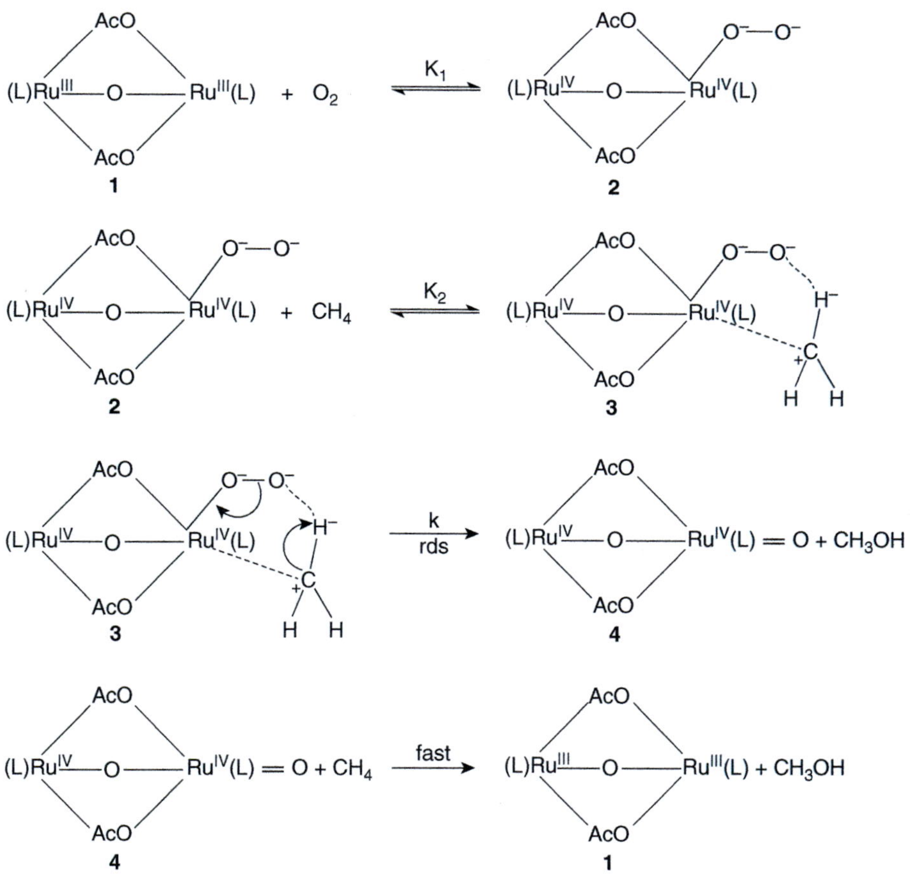

Figure 13.9 Proposed mechanism for the catalytic oxidation of CH_4 to CH_3OH on bis-(μ-acetato) (μ-oxo) bis-salen ruthenium (III). *Source:* Reproduced with permission from [30] / Elsevier.

molecule and forming an intermediate (4). Intermediate (4) then reacts with another CH_4 molecule to produce the second CH_3OH molecule, while the catalyst (1) is regenerated. Consequently, the overall reaction is $2CH_4 + O_2 \rightarrow 2CH_3OH$. The distinctive feature of this catalyst lies in its ability to activate both the O–O bond of the oxidant and the C–H bond of CH_4 before an oxygen atom inserts into the C–H bond of CH_4 at 30°C to form CH_3OH [30].

13.8 Peroxydisulphate for the Production of Acetic Acid without External CO

Sen et al. reported an oxidative carbonylation of methane utilizing $S_2O_8^{2-}$ at a temperature ranging from 105 to 115 °C. This process involves C–H activation of CH_4 through H-atom abstraction by an -oxo radical. Their finding represents an early example of oxidative carboxylation of hydrocarbons, including CH_4, through a one-pot synthesis. Figure 13.10 schematically illustrates a potential pathway involving the generation of an intermediate compound, CO, followed by the carbonylation of alkyl groups, such as methyl, to form carboxylic acids, such as acetic acid [31]. In this oxidative transformation, the SO_4^- derived from the O–O homolysis of $S_2O_8^{2-}$ serves as the oxidant. CO is formed through the sequential oxidation of methyl radicals to methanol, formic acid, and then carbon monoxide. The resulting carbon monoxide subsequently reacts with other methyl radicals, generating an acetyl radical. An acetyl radical combines with a H_2O molecule to produce acetic acid. The source of the carbon atom in the carbonyl group of acetic acid is carbon monoxide, confirmed through ^{13}CO-labeled experiments [31].

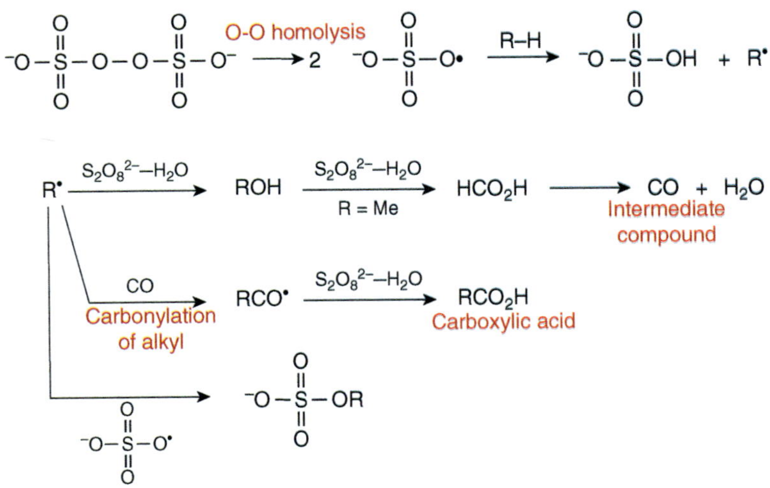

Figure 13.10 A proposed pathway which involves production of intermediate compound CO and then the carbonylation of alkyl such as methyl to form carboxylic acid such as acetic acid. *Source:* Reproduced with permission from [31] / Royal Society of Chemistry.

13.9 Polyoxometalates for the Production of Methanol

Mizuno et al. reported that $H_4PV_1Mo_{11}O_{40}$, a Keggin-type heteropolyacids, can catalyze the oxidation of CH_4 with H_2O_2 at 80°C for the synthesis of methanol. The pressure of CH_4 in the headspace of their reactor is 50 bar, and the solvent is the inert $(CF_3CO)_2O$. This catalysis utilized H_2O_2 up to 87% under an optimum condition, as described in the reference [32, 33]. It is suggested that this catalytic oxidation is performed through a radical path. This deduction was supported by a series of experiments in literature [33]. The oxidation to CH_4 by H_2O_2 was largely suppressed when a radical scavenger was added.

Molybdovanadophosphoric acid (HPA) is active in catalyzing the partial oxidation of methane by H_2O_2 or $K_2S_2O_8$ [32–34]. With the catalysis of $H_4PV_1Mo_{11}O_{40}$, a conversion of CH_4, 33% was achieved at 80°C in $(CF_3CO)_2O$ using H_2O_2 as the oxidant.

13.10 Ag-based Catalyst for Inserting CH_2

In contrast to the aforementioned oxidations of CH_4 to methanol or acetic acid catalyzed by various homogeneous catalysts, Perez et al. reported the transformation of CH_4 through the insertion of a carbene into a C–H bond of CH_4 to form a carbon-added organic molecule. This reaction is conducted at a temperature as low as 40 °C, with CH_4 being dissolved in the supercritical fluid of CO_2. The carbene, $CHCO_2CH_2CH_3$, is successfully inserted into a C–H bond of CH_4, resulting in the synthesis of ethyl propionate, $CH_3CH_2CO_3CH_2CH_3$, as illustrated in Figure 13.11a. The yield of ethyl propionate is 10% after catalysis at 40 °C for 19 hours [35]. Figure 13.11b schematically presents the hypothesized catalytic mechanism based on experimental observations from the literature. This catalytic mechanism is notably different from the metal-promoted C–H bond oxidative dissociation, followed by the insertion of a new functional group X into the formed (H···)M–C bond, leading to the intermediate (H···)M···X···C. Subsequent reductive elimination results in the product C–X–H through breaking the M···X bond. The challenge in utilizing this pathway for C–H bond activation lies in cleaving the robust M···X bond before the formation of C–X–H [36]. This type of C–H activation, catalyzed by a transition metal complex, has been reported in dehydrogenative silylation [37] or dehydrogenative borylation, [38] and dehydrogenation metathesis. However, the insertion of a new functional group X into the C–H bond of CH_4 had not been reported for C–H activation of CH_4 before these works, although electrophilic substitution-based oxidation has been reported by numerous groups [2, 10, 24, 39].

As illustrated in Figure 13.11, Perez et al. demonstrated that the C–H bonds of alkanes, R–H, can be activated by the insertion of an electrophilic X group into the C–H bond, catalyzed by a transition metal atom-based catalyst. In this mechanism, a polarized M–X bond is first formed in (1 in Figure 13.11), and then the X-group is inserted into the C–H bond to form a transition state (2). In this pathway, the metal atom M does not directly interact with the alkane, R–H. Release of the product C–X–H gives a product (3) in Figure 13.11, restoring the Ag catalyst as shown in Figure 13.11. The catalyst used is $T_p{}^xAg$, as shown in Figure 13.11a [35]. The overall reaction is presented in Figure 13.11a. In Figure 13.11b, an Ag atom of the catalyst bonds to the carbene, $C–CO_2CH_2CH_3$ formed upon the removal of N_2, resulting in the formation of the intermediate (1) in Figure 13.11b. Subsequently,

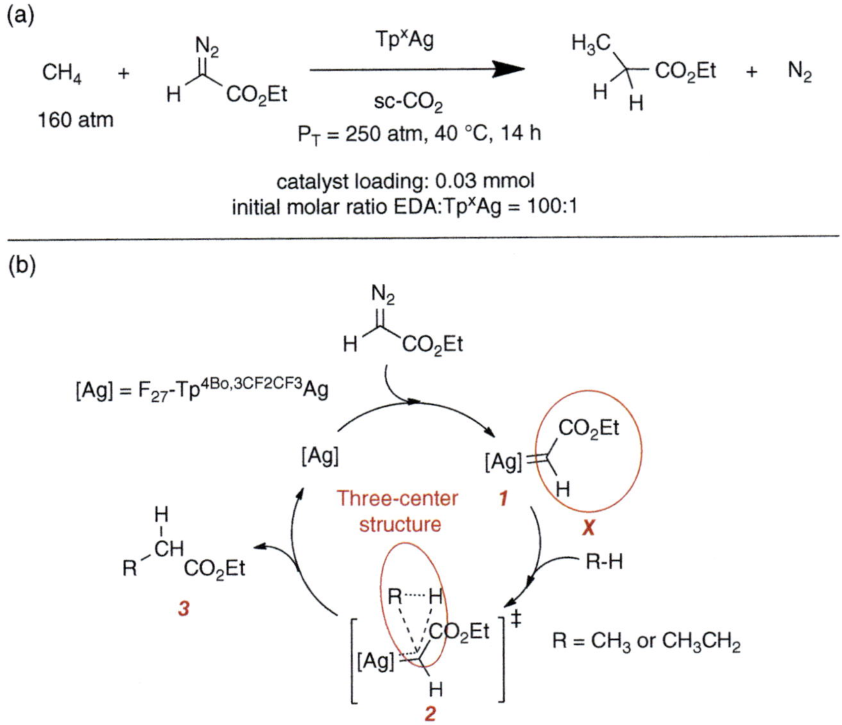

Figure 13.11 Mechanistic hypothesis for silver catalyzed alkane functionalization by ethyl diazoacetate. (a) Reaction equation. (b) Reaction mechanism. *Source:* [35] / AAAS.

intermediate (1) reacts with C-H of RH, such as CH_4, to form a three-center transition state (2 in Figure 13.11b) in which neither C nor H atoms bind to the metal atom Ag. In fact, the C and H of an alkane such as CH_4 bind to the carbene through a three-center structure shown in 2 in Figure 13.11b. The binding of the carbene to the C and H of CH_4 weakens the Ag=C interaction between the Ag atom and the carbene, ultimately cleaving Ag=C to form the product R-$CH_2COOCH_2CH_3$ (3 in Figure 13.11b).

13.11 Au-based Catalyst for the Production of Methanol

Periana reported that cationic gold (Au^{III}/Au^{I}) can catalytically transform CH_4 to CH_3OH at 180 °C [40]. It is deduced that this catalyst consists of both Au^{III} and Au^{I}. In this transformation, H_2SeO_4 was used as the oxidizing agent to oxidize Au to Au^{III}/Au^{I}. In the demonstrated catalysis, an aqueous solution consisting of 3M H_2SeO_3 and 96% H_2SO_4 oxidized metallic Au powder, producing Au^{III} and Au^{I}. Although it is not conclusive whether both Au^{III} and Au^{I} play a catalytic role in the production of methanol from methane, the DFT calculations performed by Periana et al. indicated that both Au^{III} and Au^{I} participate in this catalytic transformation. Figure 13.12 presents the enthalpy profile for the catalysis of Au^{III}

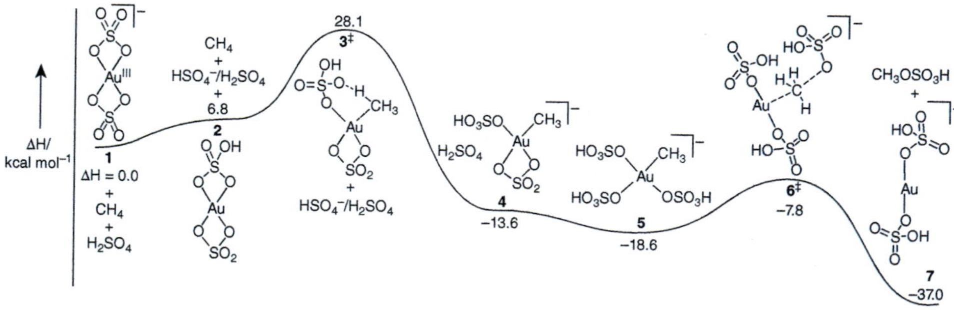

Figure 13.12 Calculated enthalpies (kcal mol^{-1}) relative to 1, including solvation by sulfuric acid (e = 50). Spectator sulfuric acid molecules have been omitted from some structures for clarity. *Source:* [40] / AAAS.

to the oxidation of CH_4 by H_2SO_4 to form CH_3OSO_3H [40]. In the Au^{III} pathway, $[Au^{III}(SO_4)_2]^-$ reacts with CH_4 to form the $[Au^{III}(CH_3)(HSO_4)_3]^-$ complex (5). C–H activation is performed through transition states, one of which is marked with $3^‡$ in Figure 13.12. In this transition state, the C–H bond of CH_4 is weakened through an intramolecular H-bond between O and Au of the catalyst and C and H of CH_4. The activation barrier of $3^‡$ is 28.1 kcal mol^{-1}, which is the highest one in the pathway for forming the product CH_3OSO_3H in Figure 13.12. Through this transition state, a C–H bond of CH_4 is cleaved, and then the left CH_3 bonds to Au through its C atom in complex (4). A rearrangement involving the weakly coordinated H_2SO_3 in complex (4) gives complex (5), which has an energy lower by 5.0 kcal mol^{-1}. At this point, CH_4 has been successfully activated, as a C–H bond of CH_4 has been broken. The formation of the final product CH_3OSO_3H is performed through the transition state marked as $6^‡$ in Figure 13.12. The activation barrier of $6^‡$ is 10.8 kcal mol^{-1}, which is the second-largest barrier in the Au^{III} pathway. In $6^‡$, the O* atom of an HO_3SO* switches its binding from O*–Au in (*6*) to O*–C of CH_3 in (*7*). The formation of $6^‡$ is a process of SN_2 where a HO_3SO* is detached from complex (5) and then attacks the CH_3 from the backside in $6^‡$. Then, the cleavage of the Au–C bond generates the final product of the Au^{III} pathway, CH_3OSO_3H (7 in Figure 13.12). This reductive elimination reduces Au^{III} to Au^I [40]. The coordination of Au^I is shown in complex (7) in Figure 13.12.

Based on the DFT calculation, [40] Au^I species adopt an electrophilic substitution mechanism, forming $[Au^I(CH_3)(H_2SO_4)]$ (10 in Figure 13.13) [40]. The activation barrier for forming the transition state ($9^‡$ in Figure 13.13) is 26.2 kcal mol^{-1}. This is the electrophilic substitution C–H activation pathway consisting of 7, 8, 9, and 10 on Au^I. Other than the electrophilic substitution, the complex (7) having Au^I can perform oxidative addition to form a complex (10). This pathway consists of 7, 11, 12, and 10 of Figure 13.13. From Au^I complex (7) without CH_4 or CH_3 to Au^I complex (10) with CH_3, both electrophilic substitution (7–8–9–10) and oxidative addition (7–11–12–10) are feasible. The oxidative addition pathway (7–11–12–10) is kinetically favorable since its activation barrier is 21.1 kcal mol^{-1}, which is lower than electrophilic substitution by 5.1 kcal mol^{-1}.

Notably, complex (10) of Au^I can be readily oxidized to complex (5) of Au^{III} by the sulfuric acid (1) or a much stronger oxidant, selenic acid. This oxidation reaction can be written as $Complex_{10}(Au^I) + Complex_1(Au^{III}) \rightarrow Complex_5(Au^{III}) + Complex_7(Au^I)$.

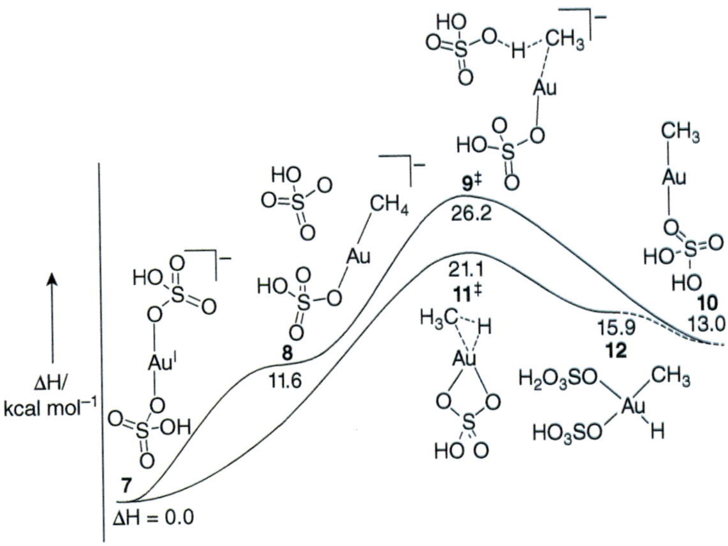

Figure 13.13 Calculated enthalpies (kcal mol^{-1}) at 175 °C relative to 7, including solvation by sulfuric acid (*e* = 50). For clarity, spectator H_2SO_4 groups have been omitted from some structures. The top curve illustrates electrophilic substitution, while the bottom curve illustrates oxidative addition. *Source:* [41] / AAAS.

Thermodynamically, this reaction is exothermic with $\Delta H_r = -32$ kcal mol^{-1}. Another option is oxidation of Complex$_1$(Au^{III}) to Complex$_7$(Au^{I}); ΔH_r of this option is −37 kcal mol^{-1} as seen in Figure 13.13. Notably, these calculations for Au^{III} and Au^{I} did not consider the unknown but real difference in concentrations of Au^{III} and Au^{I}. In summary, DFT calculations suggest both Au^{III} and Au^{I} can be possible active species of the observed catalytic transformation of CH_4 to CH_3OSO_3H [40].

13.12 Ir-based Catalyst for Borylation of Methane

Diverging from the extensively studied oxidation of methane to methanol or acetic acid, CH_4 can undergo transformation through borylation at 150 °C using an Ir-based catalyst [41]. Figure 13.14a illustrates the structure of this Ir-based molecular catalyst [41]. Under optimized conditions with a catalysis temperature of 150 °C and a pressure of 35 bar of CH_4, the yield of CH_3BR_2 is 52%, and the turnover number is nearly 100 [41]. The precatalyst [Ir(COD)(μ-X)]$_2$ (X^- = OME or Cl) evolves into the active catalyst (phen)Ir(Bpin)$_3$, also denoted as (phen)Ir(BR$_2$)$_3$. [41] Phenyl is the supporting ligand, 1,10-phenanthroline of Ir in this catalyst. Figure 13.14 presents the proposed reaction mechanism for monoborylation of methane. In the active catalyst, (phen)Ir(BR$_2$)$_3$ (Figure 13.14a), the central atom is Ir with a d^6 electron configuration and is in a pseudo-square-pyramidal coordination geometry. [41] The first step is a weak binding of CH_4 at the upper side of an octahedral coordination, forming an intermediate (Figure 13.14b), where an H atom of CH_4 binds to the metal center, Ir. A subsequent step is an oxidative addition to insert the metal center

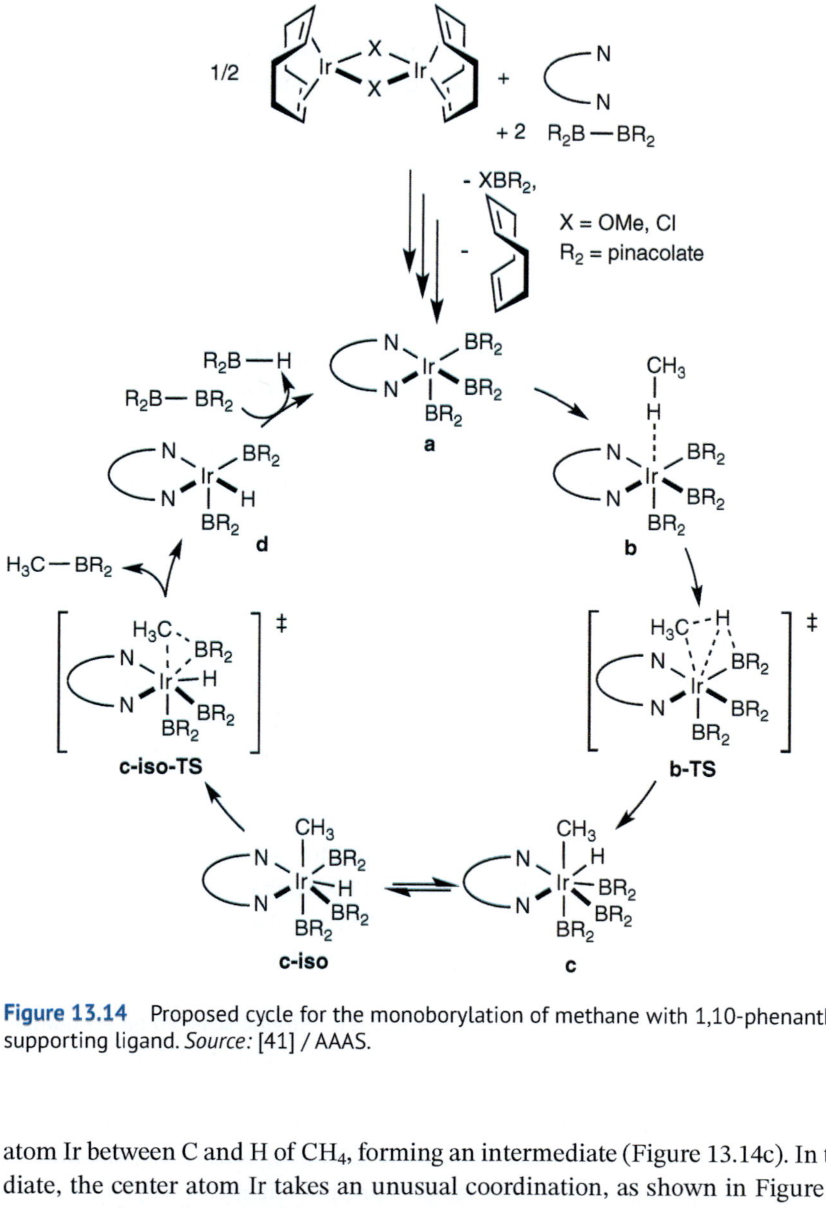

Figure 13.14 Proposed cycle for the monoborylation of methane with 1,10-phenanthroline as a supporting ligand. *Source:* [41] / AAAS.

atom Ir between C and H of CH_4, forming an intermediate (Figure 13.14c). In this intermediate, the center atom Ir takes an unusual coordination, as shown in Figure 13.14c. This step goes through a transition state with the highest activation barrier of 25.9 kcal mol^{-1} (Figure 13.14b-TS) and thus is the rate-determining step of this monoborylation [41]. The high energy of this transition state mainly results from the highly strained three-center structure of C, H, and Ir atoms in b-TS in Figure 13.14. The intermediate in Figure 13.14c performs an internal switch of the location of H and BR_2 groups, forming Figure 13.14c-iso. A reductive-elimination through the transition state in Figure 13.14c-iso-TS produces the product H_3C-BR_2 and intermediate (Figure 13.14d). In this intermediate, two BR_2 and one H atom coordinate with Ir, other than the bidentate, 1,10-phenanthroline. This intermediate (d) reacts with another R_2B-BR_2. The dissociation of the B–B bond of this reactant molecule regenerates the catalyst (Figure 13.14a). C–H of the CH_4 is broken in the transition state in Figure 13.14b-TS, performing oxidative addition. It oxidizes Ir(III) of (Figure 13.14a) to Ir(V) in Figure 13.14c. In the reductive elimination, Ir(V) of c-iso is reduced to Ir(II) in Figure 13.14d [41].

References

1 Shilov, A. E. 1984. *Activation of Saturated Hydrocarbons by Transition Metal Complexes.* Dordrecht, The Netherlands, D. Reidel Publishing Co.

2 Periana, R. A., Taube, D. J., Gamble, S., Taube, H., Satoh, T., Fujii, H. 1998. "Platinum catalysts for the high-yield oxidation of methane to a methanol derivative." *Science* 280, 560–564.

3 Mironov, O. A., Bischof, S. M., Konnick, M. M., Hashiguchi, B. G., Ziatdinov, V. R., Goddard, W. A., III, Ahlquist, M., Periana, R. A. 2013. "Using reduced catalysts for oxidation reactions: Mechanistic studies of the "Periana-Catalytica" system for CH_4 oxidation." *J. Am. Chem. Soc.* 135, 14644–14658.

4 Kurioka, M., Nakata, K., Jintoku, T., Taniguchi, Y., Takaki, K., Fujiwara, Y. 1995. "Palladium-catalyzed acetic acid synthesis from methane and carbon monoxide or dioxide." *Chem. Lett.* 24, 244–244.

5 Michalkiewicz, B., Kalucki, K., Sosnicki, J. G. 2003. "Catalytic system containing metallic palladium in the process of methane partial oxidation." *J. Catal.* 215, 14–19.

6 Zerella, M., Bell, A. T. 2006. "Pt-catalyzed oxidative carbonylation of methane to acetic acid in sulfuric acid." *J. Mol. Catal. A: Chem.* 259, 296–301.

7 Cheng, J., Li, Z., Haught, M., Tang, Y. 2006. "Direct methane conversion to methanol by ionic liquid-dissolved platinum catalysts." *Chem. Commun.* 4617–4619.

8 Periana, R., Taube, D. J., Gamble, S., Taube, H., Fuji, H. 1998. High yield, low temperature oxidation of methane to methanol. In: Derouane, E.G., Haber, J., Lemos, F., Ribeiro, F.R., Guisnet, M. (Eds.), *Catalytic Activation and Functionalisation of Light Alkanes. NATO ASI Series*, 44. Dordrecht, Springer.

9 Wu, X., Zhang, Q., Li, W., Qiao, B., Ma, D., Wang, S. L. 2021. "Atomic-scale Pd on 2D titania sheets for selective oxidation of methane to methanol." *Angew. Chem. Int. Ed.* 11, 14038–14046.

10 Periana, R. A., Mironov, O., Taube, D., Bhalla, G., Jones, C. J. 2003. "Catalytic, oxidative condensation of CH4 to CH3COOH in one step via CH activation." *Science* 301, 814–818.

11 Kao, L. C., Hutson, A. C., Sen, A. 1991. "Low-temperature, palladium(II)-catalyzed, solution-phase oxidation of methane to methanol derivative." *J. Am. Chem. Soc.* 113, 700–701.

12 Vargaftik, M. N., Stolarov, I. P., Moiseev, I. I. 1990. "Highly selective partial oxidation of methane to methyl trifluoroacetate." *J. Chem. Soc., Chem. Commun.*, 1049–1050.

13 Park, E. D., Choi, S. H., Lee, J. S. 2000. "Characterization of Pd/C and Cu catalysts for the oxidation of methane to a methanol derivative." *J. Catal.* 194, 33–34.

14 Sen, A., Benvenuto, M. A., Lin, M., Hutson, A. C., Basickes, N. 1994. "Activation of methane and ethane and their selective oxidation to the alcohols in protic media." *J. Am. Chem. Soc.* 116, 998–1003.

15 Tang, Y., Li, Y., Tao, F. 2022. "Activation and catalytic transformation of methane under mild conditions." *Chem. Soc. Rev.* 51, 376–423.

16 Zerella, M., Kahros, A., Bell, A. T. 2006. "Methane oxidation to acetic acid catalyzed by Pd2+ cations in the presence of oxygen." *J. Catal.* 237, 111–117.

17 Zerella, M., Mukhopadhyay, S., Bell, A. T. 2004. "Direct oxidation of methane to acetic acid catalyzed by Pd^{2+} and Cu^{2+} in the presence of molecular oxygen." *Chem. Commun.* 1948–1949.

18 Lin, M., Hogan, T., Sen, A. 1997. "A highly catalytic bimetallic system for the low-temperature selective oxidation of methane and lower alkanes with dioxygen as the oxidant." *J. Am. Chem. Soc.* 119, 6048–6053.

19 Shen, C., Garcia-Zayas, E. A., Sen, A. 2000. "A bimetallic system for the catalytic hydroxylation of remote primary C–H bonds in functionalized organics using dioxygen." *J. Am. Chem. Soc.* 122, 4029–4031.

20 Forster, D. 1979. "Mechanistic pathways in the catalytic carbonylation of methanol by rhodium and iridium complexes." *Adv. Organomet. Chem.* 17, 255–267.

21 Lin, M., Sen, A. 1994. "Direct catalytic conversion of methane to acetic acid in an aqueous medium." *Nature* 368, 613–615.

22 Lin, M., Hogan, T. E., Sen, A. 1996. "Catalytic carbon–carbon and carbon–hydrogen bond cleavage in lower alkanes. low-temperature hydroxylations and hydroxycarbonylations with dioxygen as the oxidant." *J. Am. Chem. Soc.* 118, 4574–4580.

23 Hristov, I. H., Ziegler, T. 2003. "Density functional theory study of the direct conversion of methane to acetic acid by RhCl3." *Organometallics.* 22, 3513–3525.

24 Periana, R. A., Taube, D. J., Evitt, E. R., Loffler, D. G., Wentrcek, P. R., Voss, G., Masuda, T. 1993. "A mercury-catalyzed, high-yield system for the oxidation of methane to methanol." *Science* 259, 340–343.

25 Aoyama, Y., Yoshida, T., Sakurai, K., Ogoshi, H. 1986. "Activation of arene carbon-hydrogen bonds. Highly regioselective, electrophilic aromatic metalation with rhodium(III) porphyrin and subsequent cleavage of carbon-rhodium bond." *Organometallics* 5, 168–173.

26 Sen, A. 1988. "Organometallic chemistry of electrophilic transition and lanthanide metal ions. The dominant pathways for reactions involving carbon-carbon double and single bonds and carbon-hydrogen bonds." *Acc. Chem. Res.* 21, 421–428.

27 Shilov, A. E. 1984. "*Activation of Saturated Hydrocarbons by Transition Metal Complexes.*" Reidel, Dordrecht, Springer.

28 Mukhopadhyay, S., Bell, A. T. 2004. "Catalyzed sulfonation of methane to methanesulfonic acid." *J. Mol. Catal. A: Chem.* 211, 59–65.

29 Bischof, S. M., Hashiguchi, B. G., Konnick, M. M., Periana, R. A. 2013. Designing molecular catalysts for selective CH functionalization. In: Gooßen, L. J. (Ed.), *Topics in Organometallic Chemistry: Inventing Reactions.*" Berlin, Springer-Verlag. 44, 195–232.

30 Khokhar, M. D., Shukla, R. S., Jasra, R. V. 2009. "Selective oxidation of methane by molecular oxygen catalyzed by a bridged binuclear ruthenium complex at moderate pressures and ambient temperature." *J. Mol. Catal. A: Chem.* 299, 108–116.

31 Lin, M., Sen, A., 1992. "Oxidation and oxidative carbonylation of methane and ethane by hexaoxo-μ-peroxodisulfate(2–) ion in aqueous medium. A model for alkane oxidation through the hydrogen-atom abstraction pathway." *J. Chem. Soc. Chem. Comm.* 892–893.

32 Seki, Y., Mizuno, M., Misono, M. 2000. "Catalytic performance of 11-molybdo-1-vanadophosphoric acid as a catalyst precursor and the optimization of reaction conditions for the oxidation of methane with hydrogen peroxide." *Appl. Catal. A. Gen.* 194–195, 13–20.

33 Seki, Y., Min, J. S., Misono, M., Mizuno, N. 2000. "Reaction mechanism of oxidation of methane with hydrogen peroxide catalyzed by 11-molybdo-1-vanadophosphoric acid catalyst precursor." *J. Phys. Chem. B* 104, 5940–5944.

34 Piao, D., Inour, K., Shibasaki, H., Taniguchi, Y., Kitamura, T., Fujiwara, Y. 1999. "An efficient partial oxidation of methane in trifluoroacetic acid using vanadium-containing heteropolyacid catalysts." *J. Organomet Chem.* 574, 116–120.

35 Caballero, A., Despagnet-Ayoub, E., Díaz-Requejo, M. M., DíazRodríguez, A., González-Núñez, M. E., Mello, R., Muñoz, B. K., Ojo, W.-S., Asensio, G., Etienne, M., Perez, P. J. 2011. "Silver-catalyzed C-C bond' formation between methane and ethyl diazoacetate in supercritical CO2." *Science* 332, 835–838.

36 Arndtsen, B. A., Bergman, R. G., Mobley, T. A., Petersen, T. H. 1995. "Selective intermolecular carbon-hydrogen bond activation by synthetic metal complexes in homogeneous solution." *Acc. Chem. Res.* 28, 154–162.

37 Sadow, A. D., Tilley, T. D. 2005. "Synthesis and characterization of scandium silyl complexes of the type Cp*2ScSiHRR'. σ-Bond metathesis reactions and catalytic dehydrogenative silation of hydrocarbons." *J. Am. Chem. Soc.* 127, 643–656.

38 Crabtree, R. H. 1985. "The organometallic chemistry of alkanes." *Chem. Rev.* 85, 245–269.

39 Shilov, A. E., Shul'pin, G. B. 1997. "Activation of C–H bonds by metal complexes." *Chem. Rev.* 97, 2879–2932.

40 Jones, C. J., Taube, D. Ziatdinov, V. R., Periana, R. A., Nielsen, R. J., Oxgaard, J., Goddard III, E. A. 2004. "Selective oxidation of methane to methanol catalyzed, with C-H activation, by homogeneous, cationic gold." *Angew. Chem. Int. Ed.* 43, 4626–4629.

41 Smith, K. T., Berritt, S., González-Moreiras, M., Ahn, S., Smith III, M. R., Baik, M. -H., Mindiola, D. J. 2016. "Catalytic borylation of methane." *Science* 351, 1424–1427.

14

Solid Organic Catalysts for the Selective Low-temperature Oxidation of Methane to Methanol

A molecular catalyst, the platinum bipyrimidine complex dissolved in concentrated H_2SO_4, has demonstrated activity for the catalytic oxidation of methane to methanol at a low temperature of 200 °C. Developed by Periana, as presented in Figure 13.1b, this catalyst can efficiently convert 90% of CH_4 with a selectivity of 81% to methylbisulfate complexes under the specified catalytic conditions [1–5]. However, a prominent challenge in molecular catalysis lies in the technical complexities associated with separating the molecular catalyst from reactants and products, especially when they coexist in the same liquid phase, and the substantial cost involved in the separation process. To address these challenges, Schuth et al. developed a solid catalyst that utilizes the immobilized molecular catalyst, the platinum bipyrimidine complex, as catalytic sites [1]. Figure 14.1 outlines the steps involved in preparing this solid organic catalyst. The catalyst, named Pt-CTF based on the literature, was synthesized through a reaction between a covalent triazine-based framework (CTF) and $K_2[PtCl_6]$ in water at 60 °C for four hours [1]. The CTF was synthesized from a monomer, 2,6-dicyanopyridine (DCP) in molten $ZnCl_2$ as shown in Figure 14.1 [1]. 2,6-dicyanopyridine (DCP) in molten zinc chloride was prepared through a unique method and proven stable up to 400 °C under oxidizing conditions [6, 7]. The resulting 2,6-dicyanopyridine was then polymerized into a covalent triazine-based framework (CTF), as shown in Figure 14.1 [1]. In the last step of preparing Pt-CTF, Pt atoms were coordinated into the CTF through a reaction between $K_2[PtCl_4]$ and CTF in water at 60 °C for four hours, followed by filtration, washing, and drying at 90 °C. Upon such a preparation, a Pt-CTF has been formed, as shown in Figure 14.1.

Other than Pt-CTF, another catalyst, designated as $K_2[PtCl_6]$-CTF, was formed in another preparation route. To prepare the $K_2[PtCl_6]$-CTF catalyst, both CTF and $K_2[PtCl_6]$ were mixed in a high pressure reactor and then CH_4 was introduced to the reactor until the CH_4 pressure reached 40 bar and then heated the sealed reactor to 215 °C for 2.5 hours and then cooling down to room temperature. Its structure is similar to Pt-CTF, as shown in Figure 14.1. Both Pt-CTF and $K_2[PtCl_6]$-CTF exhibited similar catalytic performances in transforming CH_4 to CH_3OH under the conditions (Table 14.1) [1]. The main reaction steps are proposed in Figure 14.2 [1]. Furthermore, Pt-CTF exhibits a remarkable reusability, maintaining stability through several catalytic runs. Figure 14.3a illustrates that the catalytic performance

Low-temperature Activation and Catalytic Transformation of Methane to Non-CO_2 Products,
First Edition. Franklin Tao.

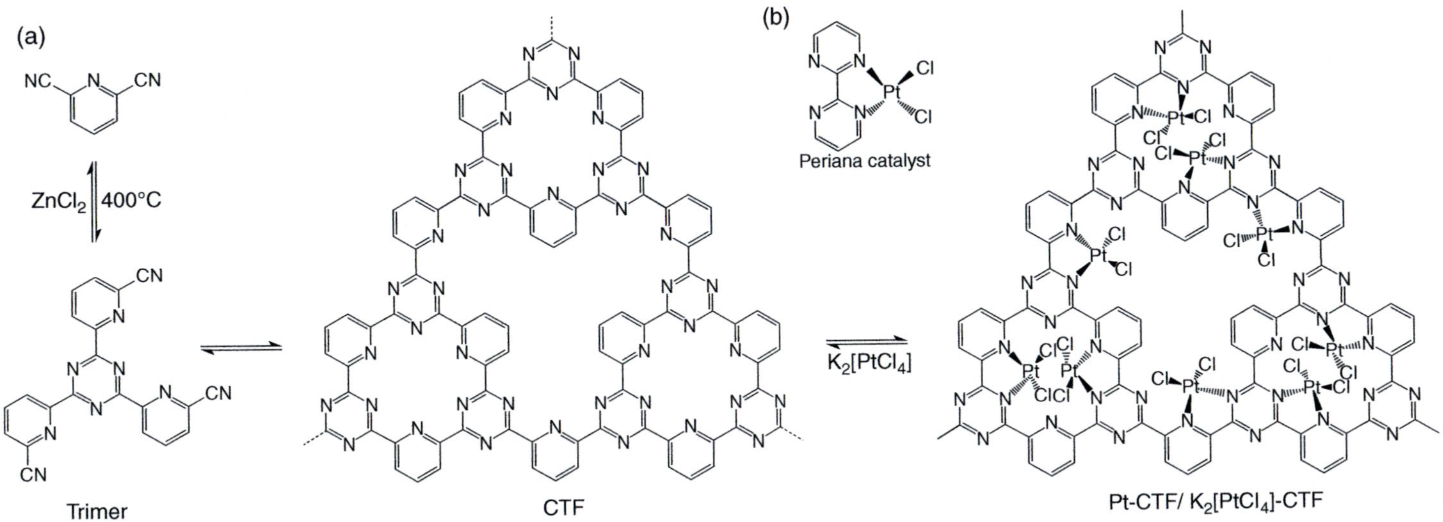

Figure 14.1 Preparation of a solid organic catalyst. (a) Trimerization of 2,6-dicyanopyridine (DCP) in molten $ZnCl_2$, conversion to a covalent triazine-based framework (CTF), and subsequent platinum coordination (Pt-CTF); (b) Periana's platinum bipyrimidine complex. *Source:* Palkovits et al. [1]/John Wiley & Sons.

Table 14.1 Catalytic activity of the molecular Periana catalyst and the heterogeneous Pt-CFT and $K_2[PtCl_4]$-CFT catalysts in methane oxidation.

Catalyst[a]	Final methanol cone ($mol\,L^{-1}$)	TON[b]
Periana catalyst[c]	1.65	158
Periana catalyst[d]	1.49	355
$K_2[PtCl_4]$-CTF[e]	1.54	201
Pt-CTF[f]	1.80	246

[a] Reaction conditions: 15 mL H_2SO_4 (30% SO_3), 40 bar CH_4 pressure (25 °C), 2.5 hours at 215 °C.
[b] TON based on the platinum content determined from SEM/EDX.
[c] 65 mg Periana catalyst.
[d] 26 mg Periana catalyst.
[e] 48 mg CTF with 92 mg $K_2[PtCl_4]$.
[f] Data from the second run with 62 mg Pt-CTF.
Source: Reproduced with permission from [1] / John Wiley & Sons.

$$CH_4 + H_2SO_4 + SO_3 \rightarrow CH_3OSO_3H + H_2O + SO_2 \quad (a)$$

$$CH_3OSO_3H + H_2O \rightarrow CH_3OH + H_2SO_4 \quad (b)$$

$$SO_2 + 1/2\,O_2 \rightarrow SO_3 \quad (c)$$

$$CH_4 + 1/2\,O_2 \rightarrow CH_3OH \quad (d)$$

Figure 14.2 Reaction steps performed on the solid organic catalyst. *Source:* Palkovits et al. [1]/John Wiley & Sons.

of Pt-CTF remained in a solid state after five catalytic runs, underscoring its enduring performance. The sustained reusability is attributed to the preservation of the platinum bipyrimidine complex, as suggested by the invariance of the N/Pt atomic ratio [1].

According to the literature [1], $K_2[PtCl_6]$-CTF is indeed a stable catalyst, exhibiting minimal deactivation in the test of Figure 14.3b. Even after five runs of catalytic tests, its turnover number remained around 300 in terms of TON. The preservation of $K_2[PtCl_6]$ within the CTF mesh, as shown in Figure 14.1 after multiple catalytic runs, was confirmed through characterizations using XPS and SEM. This catalyst demonstrated activity comparable to the molecular catalyst under the same catalytic conditions [1]. In contrast, Pt-CTF appears to require an incubation time, while $K_2[PtCl_6]$-CTF does not. The reason for the activation requirement in Pt-CTF is not entirely clear, but it is suggested to involve a rearrangement of the Pt-containing species [1]. It is possible that the activation process serves to remove solvent or contaminant molecules that may have remained during the Pt-CTF preparation step, allowing Pt atoms to interact effectively with CH_4 molecules.

These successful demonstrations of activity and selectivity in CH_4 oxidation using organic heterogeneous catalysts are significant as they demonstrated how to address the challenge of separating a molecular catalyst from the reactant and products in the same phase after catalysis. In this context, the CTF acts as a solid support with high stability for a molecular

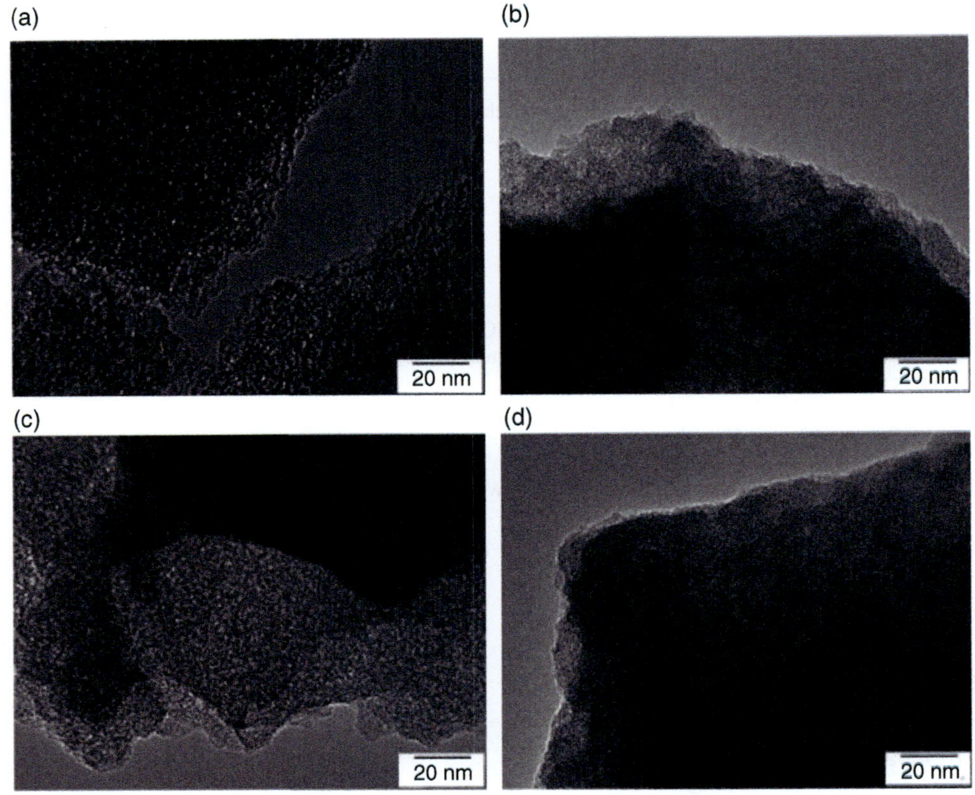

Figure 14.3 TEM micrographs of (a) pure CTF, (b) Pt-CTF from pre-coordination of platinum prior to catalysis, (c) Pt-CTF after the first catalytic run, and (d) Pt-CTF after utilization of the material in five catalytic runs. *Source:* Palkovits et al. [1]/John Wiley & Sons.

catalyst under the low-temperature catalytic condition. Notably, numerous publications have reported single-atom catalysts comprising single-atom metal sites anchored on a support [8–10]. In general, most supports of these single-atom catalysts are transition metal oxides, M_xO_y. Compared to transition metal oxides, an organic solid support may exhibit better durability in a solution of reactants and products. This is probably because the M–O bonds on the surface of the transition metal oxides, M_xO_y, can be sensitive to acidity or basicity. In other words, the acidic or basic solution of the reaction can readily etch the surface atomic layer of a transition metal oxide support. Thus, the single-atom catalytic sites anchored on the metal oxide support, M_xO_y, can be readily dissolved into the solution. This is the advantage of this type of solid organic catalyst. In addition, from a cost perspective, the cost of immobilizing metal atoms onto organic support through the formation of sigma bonds such as M_1–C bonds is generally higher than anchoring metal atoms on an oxide support for a single dispersion [8]. Therefore, the design of inexpensive organic supports and the development of low-cost immobilization of a molecular catalyst onto organic supports are expected to be crucial for the advancement of organic heterogeneous catalysts for the catalytic transformation of CH_4 in liquid at low temperatures.

References

1 Palkovits, R., Antonietti, M., Kuhn, P., Thomas, A., Schüth, F. 2009. "Solid catalysts for the selective low-temperature oxidation of methane to methanol." *Angew. Chem. Int. Ed.* 48, 6909–6912.

2 Periana, R. A. Taube, D. J., Gamble, S., Taube, H., Satoh, T., Fujii, H. 1998. "Platinum catalysts for the high-yield oxidation of methane to a methanol derivative." *Science* 280, 560–564.

3 Jones, C. J., Taube, D., Ziatdinov, V. R., Periana, R. A., Nielsen, R. J., Oxgaard, J., Goddard III, W. A. 2004. "Selective oxidation of methane to methanol catalyzed, with C-H activation, by homogeneous, cationic gold." *Angew. Chem. Int. Ed.* 116, 4726–4729.

4 Conley, B. L., Tenn III, W. J., Young, K. J. H., Ganesh, S. K., Meier, S. K., Ziatdinov, V. R., Mironov, O., Oxgaard, J., Gonzales, J., Goddard III, W. A., Periana, R. A. 2009. "Synthesis of highly substituted naphthalene and anthracene derivatives by rhodium-catalyzed oxidative coupling of arylboronic acids with alkynes." *Org. Lett.* 11, 5198–5201.

5 Periana, R. A., Bhalla, G., Tenn, W. J., Young, K. J. H., Liu, X. Y., Mironov, O., Jones, C. J., Ziatdinov, V. R. 2004. "Perspectives on some challenges and approaches for developing the next generation of selective, low temperature, oxidation catalysts for alkane hydroxylation based on the CH activation reaction." *J. Mol. Catal. A* 220, 7–25.

6 Kuhn, P., Antonietti, M., Thomas, A. 2008. "Porous, covalent triazine-based frameworks prepared by ionothermal synthesis." *Angew. Chem. Int. Ed.* 47, 3450–3453.

7 Kuhn, P., Thomas, A., Antonietti, M. 2009. "Toward tailorable porous organic polymer networks: A high-temperature dynamic polymerization scheme based on aromatic nitriles." *Macromolecules* 42, 319–326.

8 Yang, X. F., Wang, A., Qiao, B., Li, J.m., J. Liu, T. Zhang. 2013. "Single-atom catalysts: A new Frontier in heterogeneous catalysis." *Acc. Chem. Res.* 46, 1740–1748.

9 Zhang, X., Sun, Z., Wang, B., Tang, Y., Nguyen, L., Li, Y., Tao, F. 2018. "C-C coupling on single-atom based heterogeneous catalyst", *J. Am. Chem. Soc.* 140, 954–962.

10 Li, Y., Tang, Y., Tao, F. 2023. "C-N coupling through hydroaminoalkylation on a single-atom Rh heterogeneous catalyst." *Angew. Chem. Int. Ed.* 62, e202214332.

15

Confinement Effect in Micropores of Microporous Aluminosilicate

All the above activations of CH_4 or catalysis of CH_4 transformation were conducted using different oxidants or catalysts, respectively, including M^+, MO^+, $[ML_nX_a]$. A common feature of these activations and catalysis is that these oxidants, reactant(s), and catalysts are in an open environment. Different from these open environments, the activations and catalysis of CH_4 to be reviewed in this and the next several chapters, were mostly performed in a confined environment instead of an open one.

Micropores with a size up to a couple of nanometers could exhibit a confinement effect [1–6]. A confinement effect refers to unique chemical and physical properties resulting from the enhanced interaction between a molecule and the micropore (i.e., the wall of a micropore) where the molecule lands. Such a confinement effect is driven by the internal, concave surface in the micropore with a curvature at a couple of nm or less. The deep reason leading to the confinement effect is the constraint of the molecular orbitals in the micropore resulting from a "forced coupling between the frontier orbitals of a molecule and the concave surface." This chapter briefs the confinement effect of micropores in microporous aluminosilicate, as the low-temperature activation and catalysis of CH_4 oxidation in microporous aluminosilicate are expected to be correlated with this confinement effect although there are few studies of the confinement effect of a specific oxidation of methane at low temperature. To provide a complete picture for low-temperature activation and catalysis of CH_4 oxidation, this chapter is included.

15.1 Origin of Confinement: Elevation of Energy of Molecular Orbitals and Reduction of Gap of HOMO and LUMO

Based on literature, the energy levels and molecular orbitals of a sub-nanometer-sized molecule located in a micropore having dimensions similar to a molecule can be mathematically analyzed using quantum mechanics. More straightforwardly, the electrons of a molecule placed in a micropore can be considered as particles (electrons) in a three-dimensional (3D) well (the cavity of the micropore). As depicted in Figure 15.1a, the probability of finding the electrons of the molecules at the walls of a 3D well of a zeolite is zero. In an open environment without this wall (the cavity), the probability of finding electrons of the molecule at a

Low-temperature Activation and Catalytic Transformation of Methane to Non-CO_2 Products, First Edition. Franklin Tao.

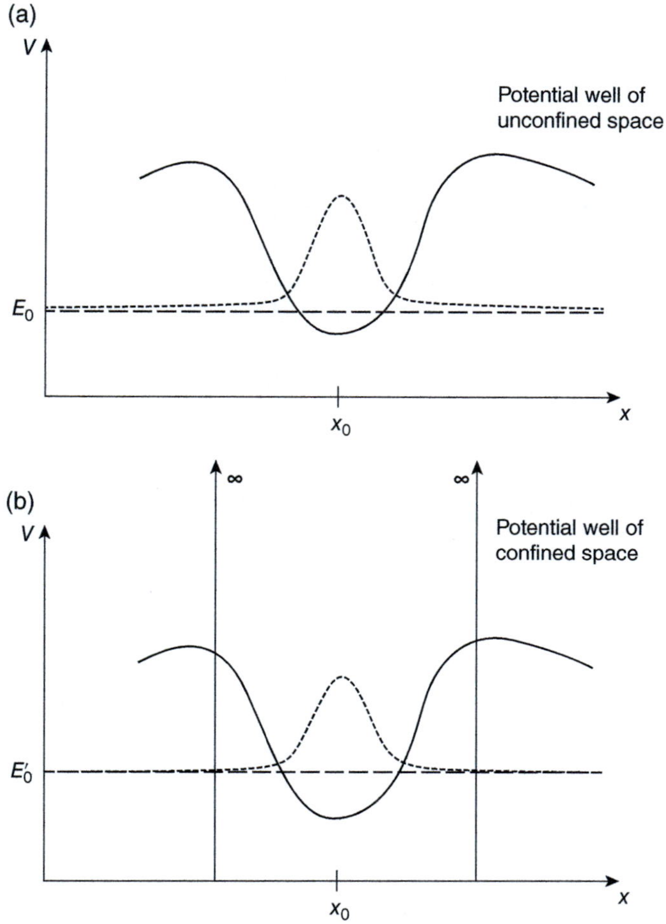

Figure 15.1 Microscopic particle in a potential well: (a) when being unconfined; (b) when being confined. The dashed curve depicts the probability function of the particle. *Source:* Reproduced with permission from [7] / American Chemical Society.

half of the well dimension is not zero. This difference predicted from this simulation result makes sense, as the electron density in zeolite bulk is homogeneously distributed. Thus, essentially, the edge of the cavity, in terms of the concave surface, is filled with electron density. On the contrary, the repulsion between the electron density of the cavity of zeolite and the electron density of the molecule in the cavity prevents the electron density of the molecule from penetrating into the wall of the micropore. This qualitative model can be described as a 3D structure with a wall that has an infinitely high potential to prevent electrons from flowing over the zeolite wall as schematically shown in Figure 15.1b [7].

Literature used ethylene as a molecular system. Its π electron was taken as the particle in a 3D well. The bonding π and anti-bonding π^* orbitals of the ethylene were described with Hückel molecular orbital (HMO) theory. Based on the literature, the bonding π and anti-bonding π^* orbitals of a free ethylene molecule can be expressed as a combination of $2p_z$ atomic orbitals of carbon atoms a and b. Thus, they are expressed as $|\pi> = |a> + |b>$ and

$|\pi^*>=|a>-|b>$. Energies of the two orbitals are $\epsilon\pi = \alpha+\beta$ and $\epsilon_{\pi^*} = \alpha - \beta$. Here, α and β are Coulomb integral and resonance integral for a free ethylene molecule, respectively. Similarly, the bonding π and anti-bonding π^* orbitals of an ethylene molecule in a micropore can be expressed with equations: $|\pi'>=|a'>+|b'>$ and $|\pi'^*>=|a'>-|b'>$. Energies of the two orbitals are $\epsilon_{\pi}' = \alpha'+\beta'$ and $\epsilon_{\pi'^*} = \alpha'-\beta'$. Correspondingly, the α' and β' are Coulomb integral and resonance integral for a confined ethylene molecule, respectively. Based on quantum mechanics, the energy states are not continuous. Assume ψ and ψ' are wavefunctions of a given electron state from the isolated ethylene and the confined one, respectively. The difference in energy of the electronic state from the free and confined molecule is $\Delta\varepsilon$. It can be expressed with the following equation: $\Delta\varepsilon = -\frac{h^2}{4m\pi^2}\cdot\frac{\int_S ds\psi^*(x)(\partial_N\psi')(x)}{\int_V d\tau\psi^*(x)\psi'(x)}$ [7]. Be specific, the difference in energy of one-electron state of ethylene in free state and confined state, $\Delta\varepsilon\pi$ can be expressed with $\Delta\varepsilon_\pi = \varepsilon_{\pi'} - \varepsilon_\pi = \Delta\alpha + \Delta\beta$ [7]. Similarly, the difference in energy of π' of a free state and a confined state is $\Delta\varepsilon_{\pi^*} = \varepsilon_{\pi'^*} - \varepsilon_{\pi^*} = \Delta\alpha - \Delta\beta$ based on the literature. Based on the literature [7], $\Delta\alpha$ and $\Delta\beta$ are expressed as:

$$\Delta\alpha = -\frac{1}{2}\frac{\int_{-\infty}^{\infty} dx\,dy\,a(x,y,d)\partial_z a'(x,y,d)}{\int_0^d dz\int_{-\infty}^{\infty} dx\,dy\,a(x,y,z)a'(x,y,z)}$$

$$\Delta\beta = -\frac{1}{2}\frac{\int_{-\infty}^{\infty} dx\,dy\,a(x,y,d)\partial_z b'(x,y,d)}{\int_0^d dz\int_{-\infty}^{\infty} dx\,dy\,a(x,y,z)a'(x,y,z)}$$

In the above two equations, $a(x, y, z)$ and $b(x, y, z)$ are the functions of $2p_z$ atomic orbitals on atoms a and b of a free ethylene, respectively [7]. As expected, $a'(x, y, z)$ and $b'(x, y, z)$ are the $2p_z$ atomic orbitals to atoms a and b of a confined ethylene, respectively. ∂_z suggests the partial derivative of $a'(x, y, d)$ with respect to z direction. d is the distance between the molecule and the border of the cavity [7]. Figure 15.2 schematically shows the wave function, $a(x, y, z)$ or $b(x, y, z)$ of a free ethylene molecule or $a'(x, y, z)$ or $b'^{(x, y, z)}$of a confined ethylene molecule as a function of z. $a(x, y, d)$ is $a(x, y, z)$ when z equals to d. Based on Figure 15.2, $a(x, y, d)$ is positive. From Figure 15.2, we can readily find that $\partial_z a'(x, y, d)$ is negative since $a'(x, y, d)$ decreases along the increase of z. In addition, based on the integration range of z, both $a(x, y, z)$ and $a'(x, y, z)$ are positive. Thus, based on the above expression of $\Delta\alpha$, $\Delta\alpha$ is positive. With a similar analysis, $\Delta\beta$ is positive as well. By combining the above three equations including $\epsilon_{\pi'^*} = \alpha' - \beta'$, $\varepsilon_{\pi'} - \varepsilon_\pi = \Delta\alpha + \Delta\beta$, and $\varepsilon_{\pi'^*} - \varepsilon_{\pi^*} = \Delta\alpha - \Delta\beta$, we obtain: $\alpha' = \alpha + \Delta\alpha$ and $\beta' = \beta + \Delta\beta$. Since both $\Delta\alpha$ and $\Delta\beta$ are positive, we can conclude α' and β' are larger than α and β, respectively. This is a significant conclusion. As defined in the literature, α, β, α', and β' represent the Coulomob internal of a free ethylene molecule, the resonance internal of a free ethylene molecule, the Coulombs internal of a confined ethylene molecule, and the resonance internal of a confined ethylene molecule, respectively.

As noted in the literature, both α and β are negative. In addition, there are the following reported relationship: $\alpha' > \alpha$, $\beta' > \beta$, $\Delta\alpha>0$, $\Delta\beta>0$. These relationships further

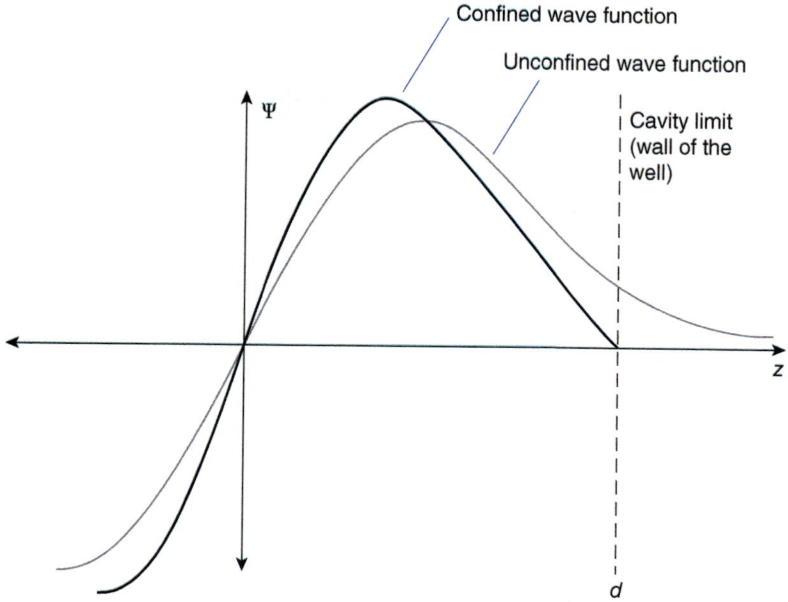

Figure 15.2 Schematic representation of the ψ function for *x* and *y* fixed at any point of space. The thin line depicts the unconfined wave function and the thick line the confined one. Dashed line represents the cavity limit. *Source:* Reproduced with permission from [7] / American Chemical Society.

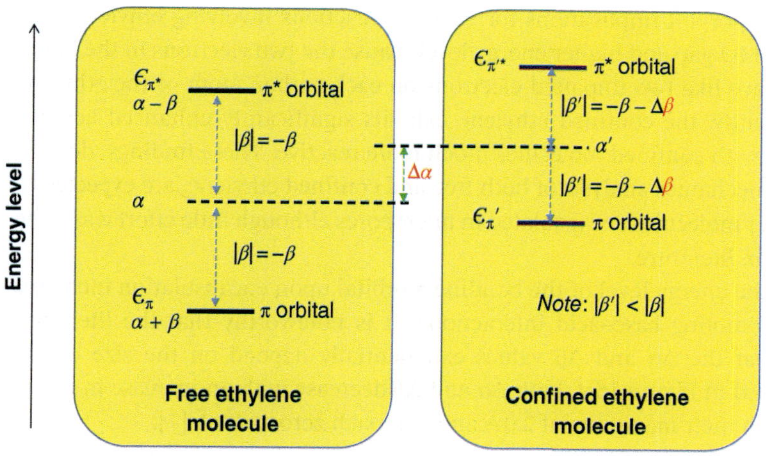

Figure 15.3 Representation of energy diagram of free ethylene molecules (left side) and confined ethylene molecule (right side) analyzed with quantum mechanistic theory. *Source:* Reproduced with permission from [7] / American Chemical Society.

suggest: $|\beta| > |\beta'|$. With this relationship, the energy levels of the $\epsilon_{\pi'^*}$, $\epsilon_{\pi'^*}$, $\epsilon_{\pi'^*}$, and $\epsilon_{\pi'^*}$ can be plotted in the same energy diagram as represented in Figure 15.3 [7]. Based on these mathematical and energy terms, it is concluded that both bonding π orbital and antibonding π* energy levels are elevated after a free ethylene molecule is confined in a cavity. Due

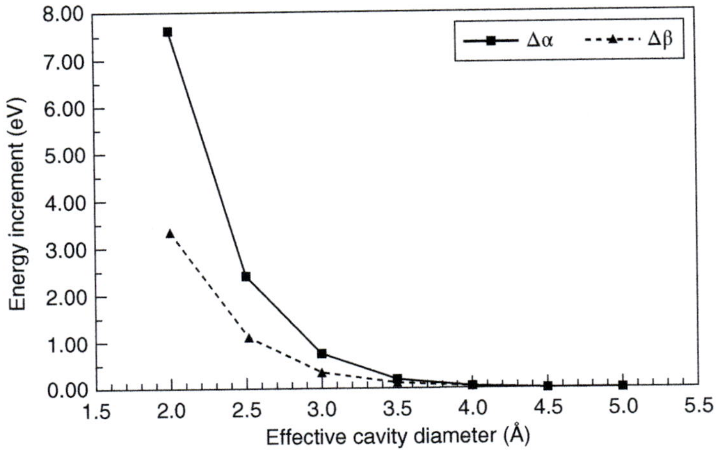

Figure 15.4 Variation of the Coulombic and resonance integrals ($\Delta\alpha$ and $\Delta\beta$, respectively) with the effective cavity size. Effective size should be 2.0–3.0 Å less than the crystallographic size. *Source:* Reproduced with permission from [7] / American Chemical Society.

to the confinement, energy levels of π and π^* orbitals of a free ethylene molecule move by $\Delta\alpha+\Delta\beta$ and $\Delta\alpha - \Delta\beta$, respectively. In addition, the gap between energy levels of π and π^* orbitals in a confined ethylene molecule were reduced by $2\Delta\beta$ in contrast to a free ethylene molecule in an open space [7].

The significant alteration of the energy diagram of ethylene upon encapsulation in a micropore carries important implications for catalytic reactions involving ethylene. The reduced HOMO-LUMO gap and higher energy levels cause the two electrons in the bonding π orbital to behave like two unpaired electrons on each carbon atom of the ethylene molecule. Consequently, the confined ethylene exhibits significantly enhanced activity, rendering reactions with confined molecules much more reactive. These findings, derived from the quantum mechanics analysis of both free and confined ethylene, are expected to be applicable to other molecules encapsulated in micropores although little effort was done on other molecules in literature.

The elevation of the energy level of the bonding π orbital upon encapsulation increases basicity, thereby promoting base–acid interactions. It is noteworthy that the literature further indicates that the $\Delta\alpha$ and $\Delta\beta$ values exponentially depend on the size of the cavity [7]. As depicted in Figure 15.4, both $\Delta\alpha$ and $\Delta\beta$ decrease with an increase in cavity size. $\Delta\alpha$ and $\Delta\beta$ reach their maximum at 2.0 Å and approach zero at 4.0 Å [7].

15.2 Relaxation of Atoms of the Concave Surface

One consequence of the confinement effect is that the zeolite exhibits behavior akin to a solid solvent, supported by computational studies [5, 8, 9]. Figure 15.5a and b display the optimized microporous wall and adsorbed dimethyl cyclohexane. In Figure 15.5a, the side view of the MOR micropore is nearly rectangular [5]. However, upon the adsorption of dimethyl cyclohexane, the shape of the micropore undergoes a noticeable change.

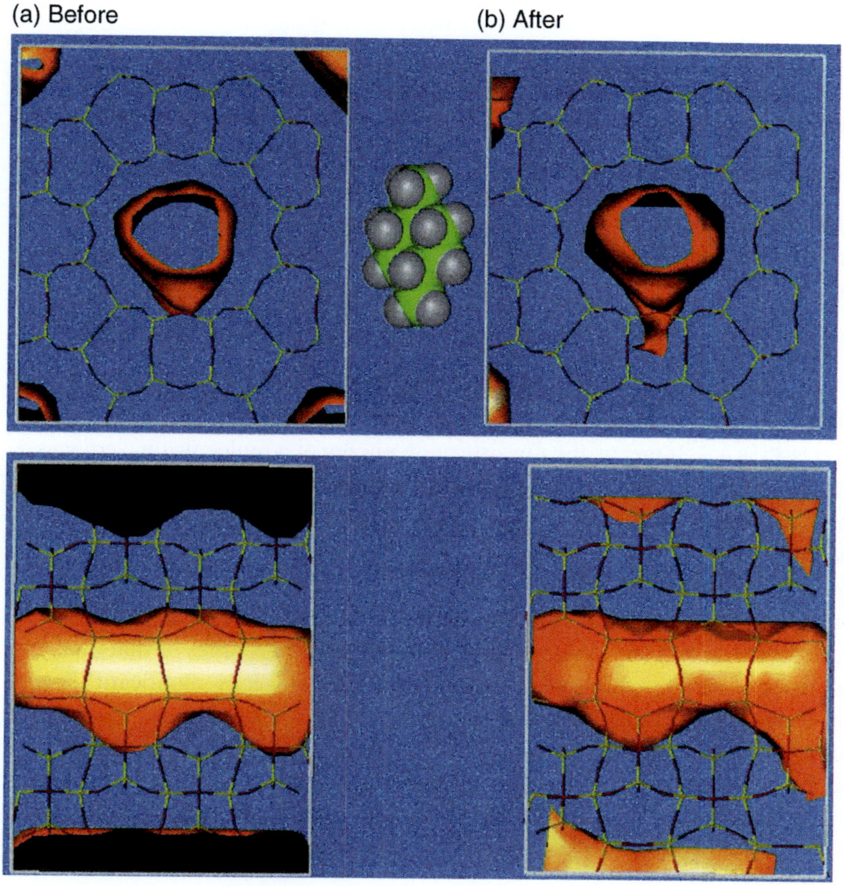

Figure 15.5 Relaxation of the MOR framework upon adsorption of 1,4-dimethyl-cyclohexane. (a) Before adsorption. (b) After adsorption. *Source:* Reproduced with permission from [5] / Elsevier.

Simultaneously, the edge of the micropores also transforms. Figure 15.5 suggests the relaxation of atoms on the concave surface induced by the adsorption of dimethyl cyclohexane, maximizing the interaction between the wall and the adsorbed molecules. This relaxation of the surface atoms on the microporous surface, prompted by the presence of adsorbed molecules, is similar to the restructuring observed in Pt(557) upon the adsorption of high-coverage CO [10]. The relaxation of the internal surface of the micropore is supported by experimental observations that ZSM-5 or ZSM-11, with a pore size of about 0.55 nm, can readily adsorb benzene molecules with a van der Waals cross-section of 0.60 nm. In simpler terms, the van der Waals cross-section of a molecule is a variable molecular parameter, contingent on the surrounding environment. The concave surface, in close proximity to the molecule with nearly zero distance, allows the benzene molecule to fit into the space of the ZSM-5. In addition, computational studies suggest that the internal surface of the micropore relaxes to some extent.

15.3 Quantification of the Confinement Effect

The van der Waals interaction between a molecule and the surface was explained with a model [2]. In this mode, a term, ratio of pore wall-molecule distance (d) to the cage or channel size (a), s was defined. It is expressed with eq. (1), $s = \frac{d}{a}$. In addition, $W(0)$ and $W(s)$ are defined to the adsorption energies for a flat surface with adsorbed molecule and a concaved surface with an adsorbed molecule, respectively. $W_r(s)$ represents the disperseion forces referenced to the energy of a flat surface. It is expressed with eq. (2), $W_r(s) = \frac{W(s)}{W(0)} = \frac{1}{(1-\frac{s}{2})^3}$ [5]. Based on eq. (1), s is 0 for a flat surface since $\boldsymbol{a}$ is infinite for a flat surface. Thus, for a flat surface $W_r(s)$ is 0. To make $W_r(s)$ meanful, s must be <2. For instance, $W_r(s)$ is 8 if s is 1. It means the physisorption energy of a molecule on a concaved internal surface of a micropore is 8 times of the physisorption energy of a flat surface to it. In the case of $s = 1$, the size of the molecule equals the radius of the micropore. If s is 0.5, the size of the molecules is 75% of the diameter of the micropore, $W_r(s)$ is 2.37. For a s at 0.25, $W_r(s)$ is 1.49. The $W_r(s)$ was plotted as a function of s in Figure 15.6. In general, if s is between 0.5 and 1.0, the molecular sizes are 75 and 50% of the pore size, respectively. Then, the physisorption energy of a molecule in the micropores are 8 or 3 times of that on a flat surface. This increase of physisorption energy is not negligible. It is significant. It can rationalize those observed chemical and physical properties of the micropores. Figure 15.6 is a plot based on a modified eq. (2) As suggested by Figure 15.6, the $W_r(s)$ reaches maxiumu when s is about 1.15. In other words, the physisorption energy is maximum when the

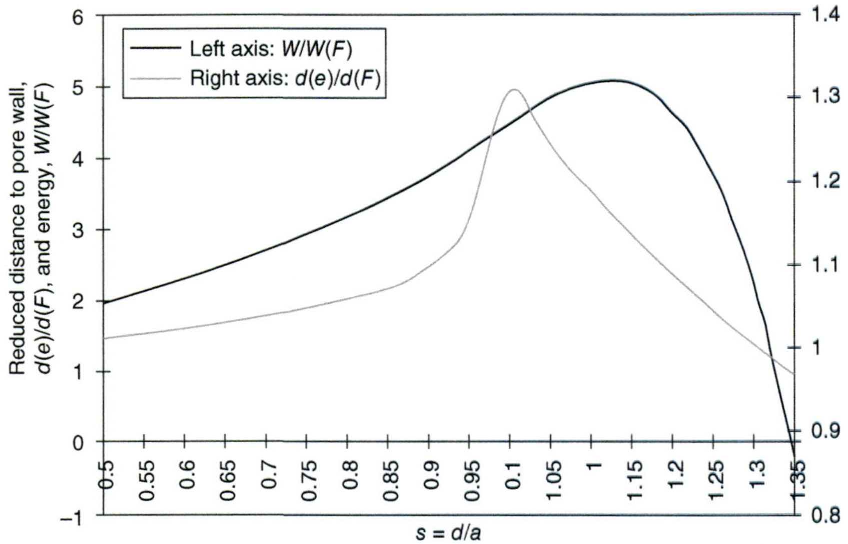

Figure 15.6 Variation at equilibrium of the reduced adsorption energy, *W/W(F)*, and reduced adsorption distance, *d(e)/d(F)*, as a function of the curvature parameter, *s* = *d/a*. *Source:* Reproduced with permission from [5] / Elsevier.

size of pore is 0.88 of the size of a molecule in a gas phase. This prediction suggests two important insights. One is that a molecule in an open space can be squeezed. Another is that a micropore with a diameter, ϕ can accommodate a molecule with a size of 1.15 ϕ. Certainly, the reason is the increased physisorption energy on the concave surface of a micropore in terms of the confinement effect of micropore.

15.4 Confinement-directed Catalytic Performance

Literature consistently highlights that the catalytic performance of a reaction conducted in micropores is highly contingent on the pore size of the zeolite. For instance, the turnover frequency (TOF) for *n*-pentane cracking in different zeolites exhibits a strong dependence on the pore size of various zeolites [11]. Illustrated in Figure 15.7, the TOF plot for *n*-pentane cracking across different zeolite pore sizes displays a volcano-like correlation. Compared to other zeolites, the MFI, in particular, demonstrates the highest TOF for pentane cracking, as its pore size optimally matches the molecular dimensions of *n*-pentane. The ensuing derivation provides a coherent explanation for the size-dependent catalytic activity.

For a reaction A → B catalyzed by a heterogeneous catalyst such as a zeolite, its reaction rate can be expressed with eq. (4), $r = k \times \theta_A$. Here, k is the rate constant. If we assume this reaction in a zeolite follows Langmuir–Hinshelwood, the reaction rate r can be expressed with eq. (5), $r = kK_A \frac{[A]}{1+K_A[A]+K_B[B]}$; here K_A and K_B are adsorption equilibrium

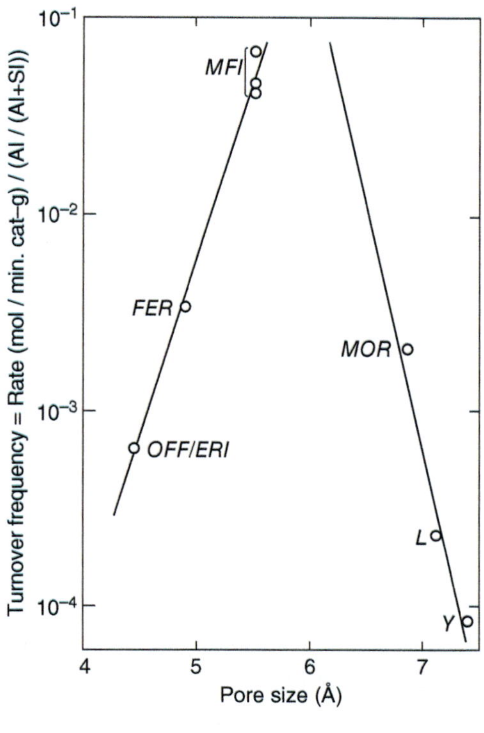

Figure 15.7 Dependence of turnover frequency for *n*-pentane cracking as a function of zeolite pore size (A). OFF/ERI = offretite-erionite; FER = ferrierite; MFI = ZSM-5; MOR = mordenite; L = Linde Type L; Y = Linde Type Y (faujasite). *Source:* Adapted from [11].

constants for the reactant and product [5]. For a reaction performed at high temperature and under kinetic-controlled regime, K_B is negligible. Thus, eq. (5) can be simplified to $r = k\ K_A$ [A]. By introducing energy term, r can be expressed as $r = Ake^{-E_A-\Delta H_A}$ [A], eq. (6) [5]. Here A is perfector; E_A is the activation barrier for the reaction A → B and ΔH_A is determined by physisorption heat of reactant molecules A (Q_A). It is expressed with eq. (7), $\Delta H_A = -Q_A$. Thus, eq. (6) can be rewritten as, $r = Ake^{-E_A+Q_A}$ [A], eq. (8) [5]. As eq. (2), described in the above section, the physisorption energy of a molecule in zeolite of a molecule in zeolite depends on the size of a micropore.. Thus, Q_A is dependent on the related size of the reactant molecules referenced to the pore size. This relevance is expressed with a term, s defined above, although this relevance of the two types of size can be expressed with other definition such as a simple ratio of molecular size to pore size. Based on the evolution of physisorption energy of a molecule as a function of s schematically presented in Figure 15.6, it is expected the physisorption heat Q_A increase along with the decrease of pore size and reach a maximum and then decrease. Thus, based on eq. (7), the reaction rate of A→B increase along the decrease of pore size and reach a maximum at certain size and then decease. As the dependence of rection rate, r on the physisorption heat Q_A is exponential, the increase or decrease of reaction rate r along decrease of size is exponential. This prediction is exactly the reported dependence of TOF of n-pentane cracking as a function of pore size of different zeolite as shown in Figure 15.7 [11].

Another crucial conclusion drawn from this confinement effect is that a more negative value, signifying a higher physisorption heat compared to that of adsorption on a flat surface, leads to a lower activation barrier for A → B. This explains why a reaction can take place on a site anchored in a micropore within a confined environment, while it might not occur on a site supported on a flat surface in an open environment.

References

1 Derouane, E. 1986. "Shape selectivity in catalysis by zeolites: The nest effect." *J. Catal.* 100, 541–544.

2 Derouane, E. G., Andre, J. M., Lucas, A. A. 1987. "A simple van der Waals model for molecule-curved surface interactions in molecular-sized microporous solids." *Chem. Phys. Lett.* 137, 336–340.

3 Derouane, E. G. 1987. "The energetics of sorption by molecular sieves: Surface curvature effects." *Chem. Phys. Lett.* 142, 200–204.

4 Derouane, E. G., André, J. -M., Lucas, A. A. 1988. "Surface curvature effects in physisorption and catalysis by microporous solids and molecular sieves." *J. Catal.* 110, 58–73.

5 Derouaner, E. G. 1998. "Zeolites as solid solvents." *J. Mol. Catal. A* 134, 29–45.

6 Sastre, G., Corma, A. 2009. "The confinement effect in zeolites." *J. Mol. Catal. A* 305, 3–7.

7 Zicovich-Wilson, C. M., Corma, A., Viruela, P. 1994. "Electronic confinement of molecules in microscopic pores. A new concept which contributes to the explanation of the catalytic activity of zeolites." *J. Phys. Chem.* 98, 10863–10870.

8 Derouane, E. G. 1996. "A molecular view of heterogeneous catalysis: Proceedings of the First Francqui Colloquium." In: Derouane, E. G. (Ed.), *Proceeding of the First Franequi Colloquium*, Brussels.

9 Derouane, E. G., Vanderveken, D. J. 1988. "Structural recognition and preorganization in zeolite catalysis: Direct aromatization of n-hexane on zeolite L-based catalysts." *Appl. Catal.* 45, L15–L22.

10 Tao. F., Dag, S., Wang, L. W., Liu, Z., Butcher, D. R., Bluhm, H., Salmeron, M., Somorjai, G. A. 2010. "Break-up of Pt catalyst surfaces by high CO coverage", *Science* 327, 850–853.

11 Kikuchi, E., Nakano, H., Shimomura, K., Morita, Y., Gakkaishi, S. 1985. "Cracking of hydrocarbons on zeolite catalysts (part 1) activity of various zeolites for *n*-pentane cracking." *J. Jpn. Petroleum Inst.* 28, 210–213.

16

Brief of Experimental Methods of Low-temperature Activation and Catalytic Conversion of CH_4 through M–O Clusters Anchored in Zeolite

The next chapters cover the activation and catalytic transformation of CH_4 through anchored metal sites in micropores. Chapters 17–19 discuss the activation and catalytic transformation of CH_4 using N_2O, H_2O_2, or O_2 as an oxidant, respectively. Chapter 20 reviews the synthesis of acetic acid through the catalytic oxidation of CH_4 and CO with O_2. These chapters provide insights into the experimental methods employed for activation and catalytic transformations of CH_4 and focus on reaction mechanisms at a molecular level, along with the identification of products and the measurement of product yields.

In general, CH_4 can be transformed into CH_3OH or other value-added chemicals through two types of processes known as low-temperature activation and low-temperature catalysis. Low-temperature activation by oxidizing CH_4 through anchored oxygen atoms/species in micropores of a zeolite is typically not a catalytic process. Its reaction has to be terminated for regenerating active sites before the reaction can be performed again. The limited conversion of CH_4 in the activation of CH_4 through oxygen atoms/species anchored in micropores has been suggested in several reports [1–8]. Strictly speaking, the activation of CH_4 by the oxygen atoms/species anchored in the micropores formed by pretreating a zeolite in N_2O or O_2 at a relatively high temperature is not a catalytic process since the formed product remains in the micropores and thus blocks the sites to be refilled with oxygen atoms/species in order to oxidize other CH_4 molecules continuously. In terms of low-temperature catalytic process, it is typically executed in a batch reactor within an enclosed system, such as a high-pressure reactor from Parr Instrument Co. Alternatively, it can be conducted through a reactor similar to a fixed-bed flow reactor exemplified by the catalytic oxidation of CH_4 to CH_3OH using Cu-encapsulated small-pore zeolite with a feed gas of CH_4, N_2O, and H_2O vapor or CH_4, O_2, and H_2O vapor [1–3]. In these examples, N_2O or O_2 serves as the oxidant. The catalyst for the two reactions is the Cu-encapsulated zeolite [1–3].

Differing from the above activation process of CH_4 in gas phase by the anchored oxygen atoms/species, catalytic oxidation of CH_4 by zeolite with anchored metal atoms can be accomplished at the solid–liquid interface in a liquid phase under a high-pressure gas phase sealed in a batch reactor. The oxidant can be O_2 for some reactions. The catalyst, H_2O_2 acting as an oxidant, and a solvent, typically H_2O, are mixed into an aqueous solution and placed in a batch reactor. CH_4, or a mixture of reactants such as CH_4, CO, and O_2, is

Low-temperature Activation and Catalytic Transformation of Methane to Non-CO_2 Products, First Edition. Franklin Tao.

introduced into the batch reactor, which is then sealed. The gas phase remains above the aqueous solution. The dissolved reactant molecules are the actual reactants as the catalytic reaction is performed at the solid–liquid interface. The dissolved gaseous reactants become a solute of this aqueous solution. They access the catalytic sites of heterogeneous catalyst particles dispersed in the solution. A Parr reactor can be efficiently heated to the optimal catalysis temperature. For instance, the dissolved CH_4 can access the catalytic sites in the micropores of zeolites dispersed in the solution under high-pressure gas phase sealed in a Parr reactor.

Compared to catalytic processes at solid–gas and solid–liquid interfaces, the formation of CH_3OH through partial oxidation can be achieved through an isothermal process of three key steps [1–3]. This process involves the oxidation of Cu-zeolite at a high temperature, exposing oxidized Cu-zeolite to CH_4 at a relatively low temperature or even room temperature, and extracting methanol by soaking it in water, an aqueous solution, or even water vapor at a relatively high temperature. Although the resulting products, such as CH_3OH, are adsorbed on the internal walls of the micropores, an extraction through co-diffusion with water vapor is the last but essential step before a zeolite containing oxygen atoms/species can be regenerated through oxidation by N_2O. Sheppard et al. conducted comparative studies on the two extraction methods, namely soaking in a liquid solution or diffusion in a gas phase. For the partial oxidation of CH_4 to CH_3OH by the oxygen species formed in ZSM-5, gas-phase diffusion was much more effective since the amount of methanol extracted from Cu-ZSM-5 through the gas-phase diffusion method is 2–6 times higher than that obtained through soaking in liquid water or a mixture of liquid H_2O and acetonitrile [1].

Measurement of the yields of products from the activation or catalysis of methane at a low temperature can be challenging, especially when the yield is at the level of a few micromoles or sub-micromoles. ^1H NMR is typically employed for the qualitative identification of products, including single-carbon products such as methanol, formaldehyde, and formic acid, as well as di-carbon products such as ethanol, acetic acid, aldehyde, and other oxygenates. ^1H NMR is also utilized for a quantitative analysis of these products. Since methanol is one of the ideal products of oxidative activation and catalysis of CH_4, the following paragraphs take the measurement of the amount of CH_3OH as one example to describe the quantitative analysis of these products. More details can be found in the literature, such as our work [8–13].

After low-temperature catalysis, the received solution in a reactor contains catalyst powder, solvent, and products. The catalyst powder is separated from the mixture through centrifugation, and the resulting solution is utilized for NMR analysis. The next step involves extracting a certain portion of the solution received from centrifugation. It needs to be diluted before analysis in some cases, depending on the concentration of the targeted product. For NMR testing, 3-(trimethyl-silyl)-1-propanesulfonic acid sodium salt (DSS) serves as a reference, with its peak in ^1H NMR set as zero chemical shift ($\delta = 0$). A 0.20 wt% DSS solution is prepared by diluting a commercial DSS with D_2O, and then 0.10 mL of this solution is added to the solution to be tested by NMR.

The concentration of a solution with an unknown amount of product, such as methanol, is determined by referring to a standard curve of methanol. To establish the standard curve, a series of standard solutions containing the same solvent but different amounts of methanol are prepared. These solutions have specific, known concentrations. The guessed concentration of methanol in the product solution to be analyzed should fall in the range of concentrations covered by the standard curve. NMR spectra of these solutions are collected under identical experimental parameters. For each spectrum, the ratio of the area of the methanol peak to the DSS peak is calculated. These ratios from each standard solution are plotted against the known concentrations of methanol. Fitting these data points yields a linear equation. This equation is a plot of the ratio of the area of the methanol peak to the DSS peak as a function of the methanol concentrations in these standard solutions. Then, the unknown concentration of methanol in the product solution, also called testing solution can be calculated using this linear equation by introducing the measured area ratio of the methanol peak of the testing solution to the DSS peak of the testing solution. Then, the amount of products formed in the batch reactor can be determined. In addition, if the CH_4 transformation occurs under a kinetics-controlled condition, the reaction rate can be calculated by dividing the number of methanol molecules by the time the transformation is processed. This rate is expressed in the unit of moles per hour. If the number of active sites of the catalyst is known or reasonably estimated, the turnover frequency, representing the number of methanol molecules formed on a site (e.g., Rh_1O_5 of Rh/ZSM-5 catalyst [11]) per hour, can be calculated by dividing the reaction rate by the number of active sites of the catalyst.

An isotope-labeled study is an important approach for identifying the origin of products and exploring catalytic mechanisms. It is widely used in the fundamental studies of low-temperature activation and catalytic transformation of CH_4. For instance, $^{13}CH_4$ was used to check whether the source of the carbon atom of the produced CH_3OH is CH_4. A mixture of $^{13}CH_4$ and $^{12}CH_4$ was introduced to the reactor containing the catalyst, Fe-ZSM-5, and aqueous H_2O_2. As $^{13}CH_3OH$ was observed in the product solution, the carbon source of CH_3OH is CH_4. In addition, isotope-labeled experiments can provide significant insights for understanding the catalytic mechanism in the oxidation of methane at a molecular level. For instance, acetic acid is industrially produced from methanol carbonylation catalyzed by an Ir-based molecular catalyst. To elucidate whether the high activity of the heterogeneous catalyst, Rh_1O_5 anchored in ZSM-15, is achieved from a similar pathway, methanol carbonylation, also called coupling of CH_3OH and CO or not, $^{13}CH_3OH$ was added as a potential intermediate compound to the catalysis system containing Rh_1O_5 anchored in ZSM-15, solvent H_2O, and reactants CH_4, CO, and O_2. It was found that no $^{13}CH_3COOH$ formed, although CH_3COOH was clearly observed. Thus, a pathway involving first formation of CH_3OH and then carbonylation of methanol was excluded. More details of this example can be found in the literature [8–13].

References

1 Sheppard, T., Hamill, C. D., Goguet, A., Rooney, D. W., Thompson, J. M. 2014. "A low temperature, isothermal gas-phase system for conversion of methane to methanol over Cu–ZSM-5." *Chem. Commun.* 50, 11053–11055.

2 Tomkins, P., Mansouri, A., Bozbag, S. E., Krumeich, F., Park, M. B., Alayon, E. M. C., Ranocchiari, M., van Bokhoven, J. A. 2016. "Isothermal cyclic conversion of methane into methanol over copper-exchanged zeolite at low temperature." *Angew. Chem. Int. Ed.* 128, 5557–5561.

3 Tomkins, P., Mansouri, A., Sushkevich, V. L., van der Wal, L. I., Bozbag, S. E., Krumeich, F., Ranocchiari, M., van Bokhoven, J. A. 2019. "Increasing the activity of copper exchanged mordenite in the direct isothermal conversion of methane to methanol by Pt and Pd doping." *Chem. Sci.* 10, 167–171.

4 Ravi, M., Ranocchiari, M., van Bokhoven, J. A. 2017. "The direct catalytic oxidation of methane to methanol—A critical assessment." *Angew. Chem. Int. ed.* 56, 16464–16483.

5 Starokon, E. V., Parfenov, M. V., Pirutko, L. V., Abornev, S. I., Panov, G. I. 2011. "Room-temperature oxidation of methane by α-oxygen and extraction of products from the FeZSM-5 surface." *J. Phys. Chem. C* 115, 2155–2161.

6 Groothaert, M. H., Smeets, P. J., Sels, B. F., Jacobs, P. A., Schoonheydt, R. A. 2005. "Selective oxidation of methane by the bis(μ-oxo)dicopper core stabilized on ZSM-5 and mordenite zeolites." *J. Am. Chem. Soc.* 127, 1394.

7 Smeets, P. J., Groothaert, M. H., Schoonheydt, R. A. 2005. "Cu based zeolites: A UV–vis study of the active site in the selective methane oxidation at low temperatures." *Catal. Today* 110, 303–309.

8 Shan, J., Huang, W., Nguyen, L., Yu, Y., Zhang, S., Li, Y., Frenkel, A. I., Tao, F. 2014. "Conversion of methane to methanol with a bent mono(μ-oxo)dinickel anchored on the internal surfaces of micropores." *Langmuir* 30, 8558–8569.

9 Tang, Y., Li, Y., Tao, F. 2022. "Activation and catalytic transformation of methane under mild conditions." *Chem. Soc. Rev.* 51, 376–423.

10 Huang, W., Zhang, S., Tang, Y., Li, Y., Nguyen, L., Li, Y., Shan, J., Xiao, D., Gagne, R., Frenkel, A. I., Tao, F. 2016. "Low-temperature transformation of methane to methanol on Pd1O4 single sites anchored on the internal surface of microporous silicate." *Angew. Chem. Int. Ed.* 55, 13411–13445.

11 Tang, Y., Li, Y., Fung, V., Jiang, D., Huang, W., Zhang, S., Iwasawa, Y., Sakata, T., Nguyen, L., Zhang, X., Frenkel, A. I., Tao, F. 2018. "Single rhodium atoms anchored in micropores for efficient transformation of methane under mild conditions." *Nat. Commun.* 9, 1231.

12 Li, Y., Tang, Y., Nguyen, L., Tao, F. 2018. "Catalytic oxidation of ethane to carboxylic acids in the liquid phase at near room temperature at ambient pressure." *ACS Sustain. Chem. Eng.* 7, 4707.

13 Zong, X., Chu, Y. C., Tang, Y., Li, X., Wu, X. P., Sun, Z., Tao, F. 2024. "Highly Selective Photocatalytic Synthesis of Acetic Acid at 0–25 °C." *Angew. Chem. Int. Ed.* 136, e202404598

17

Oxidation of Methane by N_2O through M–O Clusters Anchored in Zeolite in the Gas Phase at Low Temperature

17.1 Early Studies of Partial Oxidation of Methane

N_2O has demonstrated its effectiveness as an oxidant for activating CH_4 oxidation at either high temperature 600°C or relatively low temperature 100–200°C [1]. The decomposition mechanism of N_2O has been investigated since the 1950s, revealing a series of steps in its thermolysis: $N_2O \rightarrow N_2 + O$, $O + N_2O \rightarrow N_2 + O_2$, or even $O + N_2O \rightarrow 2NO$ [2–4]. The active species in this process is the atomic O, which can initiate the C–H activation of organic molecules through hydrogen atom abstraction, efficiently oxidizing organic molecules even at room temperature or low temperatures. For instance, ethane was catalytically partially oxidized by N_2O at 280–350°C, yielding C_2H_4 and CO_2 as the main products [5, 6]. The O-species initiated catalytic oxidation by abstracting H atoms from C_2H_6 [5]. Under identical conditions, N_2O proved much more effective than O_2 at 350°C. Figure 17.1 illustrates the N_2O oxidation of Mo (V) partially reduced by CO, forming Mo (VI)-O-, which, in turn, abstracts H atoms from C_2H_6 to create an ethoxide species [5]. This study even proposed the formation of surface species consisting of two adjacent Mo atoms bridged by an oxygen atom, suggesting a novel M–O–M bridge structure. This work marks the early observation of such a bridge structure [5]. The observed ethylene could be the initial product of the partial oxidation of ethane at 350°C [6].

The earliest instance of low-temperature oxidation of CH_4 involves the Mo_xO_y supported on silica [7]. N_2O served as the oxidant. The presence of water facilitated the partial oxidation of CH_4 to CH_3OH and HCHO. Figure 17.2 illustrates that, at 350°C, the yield of CH_3OH reached nearly 10% when supported MoO_x was used as a catalyst. Water played a crucial role in maintaining high selectivity for CH_3OH and HCHO. It was reported that water inhibited the oxidation of CH_3OH and HCHO. If 350°C is considered a relatively low temperature, this study, involving the partial oxidation of CH_4 to CH_3OH in H_2O through N_2O, is one of the earliest demonstrations of the feasibility of catalytic oxidation of CH_4 to CH_3OH at such temperatures [7]. Importantly, this process differs significantly from the oxidation of CH_4 with N_2O at 600°C to produce ethylene, where no catalyst was used, and the O atom generated from N_2O acted as the active species, abstracting H atoms from CH_4 in the gas phase [1].

Low-temperature Activation and Catalytic Transformation of Methane to Non-CO_2 Products,
First Edition. Franklin Tao.

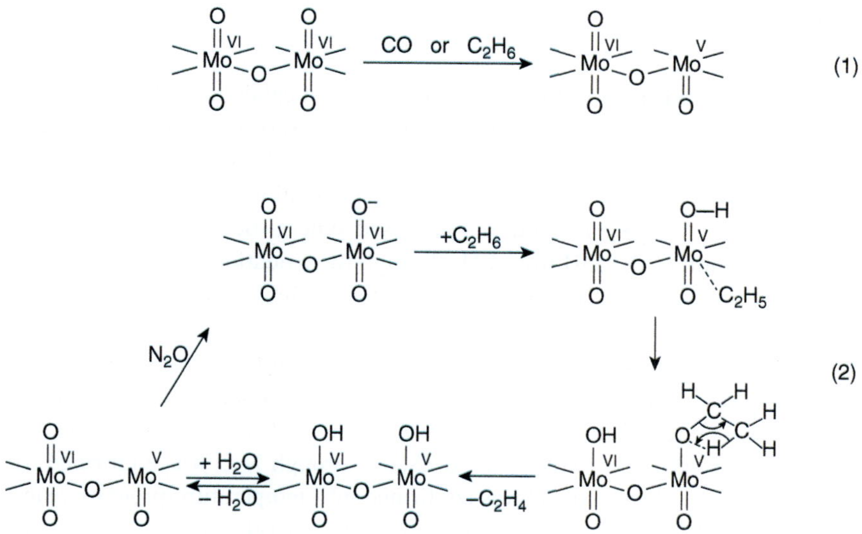

Figure 17.1 Schematic showing the capability of N_2O in oxidative dehydrogenation of C_2H_6 by Mo-based catalyst. (1) reduction of Mo_xO_y by C_2H_6 or CO and (2) oxidation of catalyst by N_2O (forming Mo_xO_y) and then reduction of the oxidized Mo_xO_y by C_2H_6 to form C_2H_4 and H_2O. The role of N_2O is to provide atomic O for oxidizing Mo(V) to Mo(VI) before the Mo(V)-O can activate the first C–H of C_2H_6. *Source:* [5] / U.S. Department of Energy / Public Domain.

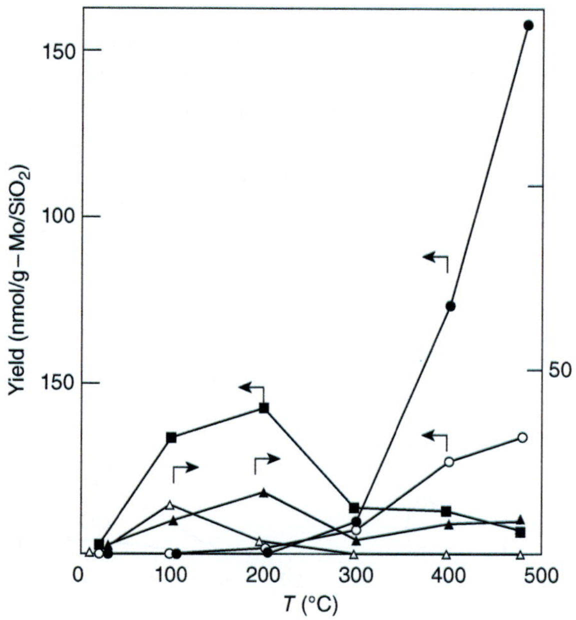

Figure 17.2 Yields of products as a function of the thermal desorption program following the reaction of CH_4 with O^- at 25°C. ● CH_3OH; o CO_2; ▪ CH_4; ▲C_2H_6; Δ C_2H_4. *Source:* Reproduced with permission from [7]/Royal Society of Chemistry.

17.2 Fe-ZSM-5

Noncatalytic partial oxidation. The first reported instance of catalytic oxidation of CH_4 to CH_3OH using a zeolite-based catalyst dates to 1987 [8]. N_2O served as the oxidant, and Fe^{3+} replaced Al^{3+} on the internal surface of the micropores of ZSM-5. This study demonstrated that Cu-Fe-ZSM-5 exhibited activity for the partial oxidation of CH_4 to CH_3OH at temperatures as low as 237°C [8]. The authors proposed synergistic roles for both Fe^{3+} and Cu^{2+} in the production of CH_3OH from CH_4 through the oxidation of N_2O. Since the catalysis temperature was sufficiently high, there is no need to include a solvation step to rinse the formed CH_3OH.

In 1990, the partial oxidation of CH_4 to CH_3OH at room temperature using Fe-ZSM-5 was reported for the first time [9]. The reaction involved the introduction of 108 Pa CH_4 to Fe-ZSM-5 at room temperature following treatment in 108 Pa N_2O at 250°C. The specific oxygen, designated "α-oxygen," introduced to Fe-ZSM-5 through N_2O was confirmed to play a crucial role. The role of "α-oxygen" was confirmed by a temperature-program reaction of CH_4 with Fe-ZSM-5 without α-oxygen (Figure 17.3a) [9] and Fe-ZSM-5 with α-oxygen (Figure 17.3b) [10]. The obvious decrease of partial pressure of CH_4 in Figure 17.3b but not Figure 17.3a confirms that the alpha-oxygen atoms are the oxidant of CH_4. The α-oxygen was generated on an α-Fe^{II} through $N_2O + (Fe^{II})\alpha \rightarrow N_2 + (Fe^{III}\text{–}O\cdot^{-})\alpha$ [11]. Thus, the room temperature reaction between α-oxygen and CH_4 via a hydrogen abstraction mechanism was expressed as $CH_4 + 2\,(Fe^{III}\text{–}O\cdot^{-})\alpha \rightarrow (Fe^{III}\text{–}OCH_3)\alpha + (Fe^{III}\text{–}OH)\alpha$ [12]. This reaction is called CH_4 activation sometimes.

The product was extracted by soaking the used Fe-ZSM-5 in a mixture of acetonitrile and H_2O, wherein the formed methoxy-like product underwent hydrolysis to produce methanol. This hydrolysis of methoxy groups was crucial. Panov concluded that

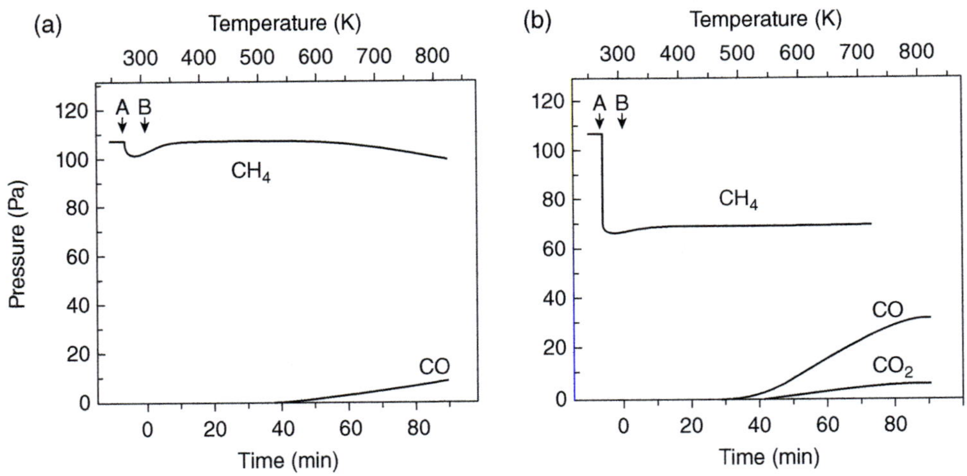

Figure 17.3 Temperature-programmed reaction of methane with Fe-ZSM-5 surface before a-oxygen loading (a) and after a-oxygen loading (b). The Y-axis is the pressure of CH_4. A is the time moment of opening the microreactor; B is the time moment of switching on the programmed heating ($6\,K\,s^{-1}$). *Source:* Reproduced with permission from [10] / Elsevier.

the α-oxygen is located in the micropores and bonded with Fe atoms. This di-iron complex structure was supported by the Mossbauer spectrum of ^{57}Fe-enriched Fe-ZSM-5 [13–26], highly similar to that of MMO observed previously [27]. As discussed in other sections of this chapter and the following chapters, numerous zeolites encapsulated with transition metals, including Cu, Fe, Co, Pd, Rh, Ir, and others, can carry highly active oxygen atoms to activate CH_4 or catalyze oxidation reactions of CH_4. These zeolite encapsulated systems serve as solid catalysts analogs of biomimetic systems.

The pivotal role of iron in Cu-exchanged Fe-ZSM-5 was subsequently confirmed in 1990s [28, 29]. Comparative investigations employing Fe_2O_3 and Fe-ZSM-5 revealed that the activity for producing CH_3OH per Fe atom in Fe_2O_3 was significantly lower than that in Fe-ZSM-5 by several orders of magnitude. The study proposed three chemical states of Fe: (1) Fe in the tetrahedral position of crystallite Fe_2O_3, (2) isolated Fe atoms anchored in the micropores of zeolite, and (3) small iron oxide clusters on the external surface of zeolite particles. However, the study could not conclusively identify which chemical state of Fe atoms was responsible for CH_3OH production. While the involvement of N_2O oxygen atoms in CH_4 oxidation was confirmed, the mechanism of how these oxygen atoms/species contributed to CH_3OH formation remained unclear in this work [30]. It suggests that the oxygen atoms generated from N_2O may participate in the oxidation through a radical-like O^-. This type of O^- could be produced upon the adsorption of N_2O to a partially reduced surface by H_2 or even CO. This radical-like O^- can abstract H atoms from CH_4. In fact, a study in 1992 confirmed that Fe atoms in Fe-ZSM-5 played a similar role to Fe_2O_3 supported on silica and ferrisilicate. Both are the sites of N_2O decomposition to generate O^- species for the oxidation of CH_4, although the specific coordination environments of Fe atoms in either Fe-ZSM-5 or Fe_2O_3 supported on silica were not clarified [30, 31]. In these studies [30, 31], this type of site responsible for N_2O decomposition was even called the α-site, and the oxygen atom originated at N_2O but formed on Fe before activating CH_4 is called α-oxygen.

The activity of Fe-ZSM-5 in the oxidation of CH_4 is temperature dependent. In Fe-ZSM-5, CH_3OH can be formed from the oxidation of CH_4 through α-oxygen in Fe-ZSM-5 originating at N_2O at 160°C. The formed CH_3OH can emigrate from the α-sites. This emigration of CH_3OH from an α-Fe site can vacate the site, allowing the refilling of the active α-oxygen to oxidize CH_4 with the α-oxygen to form the second CH_3OH molecule. This migration does not occur at a temperature lower than 160°C. Thus, at <160°C, the reaction is purely an oxidation of CH_4 to form methoxy on an α-Fe site without regeneration of the α-Fe site. For a run of 19 hours at 160°C, the turnover number of this oxidation is only 3.6 molecules per Fe α-site [32]. Thus, the turnover rate is only 0.2 h^{-1}. Certainly, this reaction rate is quite low due to the limited desorption of methoxy and methanol in the micropores at 160°C. Thus, it is not considered a typical catalytic event at ≤160°C. Through co-feeding H_2O and performing the reaction at a relatively high temperature, as reviewed in the next section, the partial oxidation of CH_4 with N_2O can be performed in a continuous catalytic manner since α-Fe sites can be regenerated along with the formation of CH_3OH and the subsequent emigration [11].

Catalytic oxidation. In addition to the previously discussed activation of CH_4 by Fe-ZSM-5 oxidized with N_2O, catalytic CH_3OH production from CH_4 on Fe-ZSM-5 by co-feeding N_2O and CH_4 at temperatures exceeding 200°C has been reported [11]. Below 200°C, CH_3OH does not readily desorb into the gas phase, although the co-feed H_2O can assist both the formation of CH_3OH and the followed emigration at 160–200°C [11]. Thus, the partial oxidation operates as a catalytic reaction only when the temperature of Fe-ZSM-5 is above 200°C. Figures 17.4 and 17.5 illustrate the conversion of CH_4 and the composition of products when the reaction is conducted at 300°C in the presence of a mixture of CH_4 and N_2O (Figure 17.4) or a mixture of CH_4, N_2O, and H_2O (Figure 17.5) [11]. Clearly, within 2.5 hours, the conversion of CH_4 and the composition of gas products remain relatively stable. Gas composition calculations exclude the amount of coke formed, indicating that the partial oxidation of CH_4 by N_2O at 300°C follows a classic catalytic mode. A distinction between the mixture of CH_4 and N_2O and that of CH_4, N_2O, and H_2O is the observed conversion and selectivity of CH_3OH. With the addition of water, the conversion of CH_4 slightly decreases. This is likely a result of the oxidation of some α-Fe sites, forming hydroxyl on the

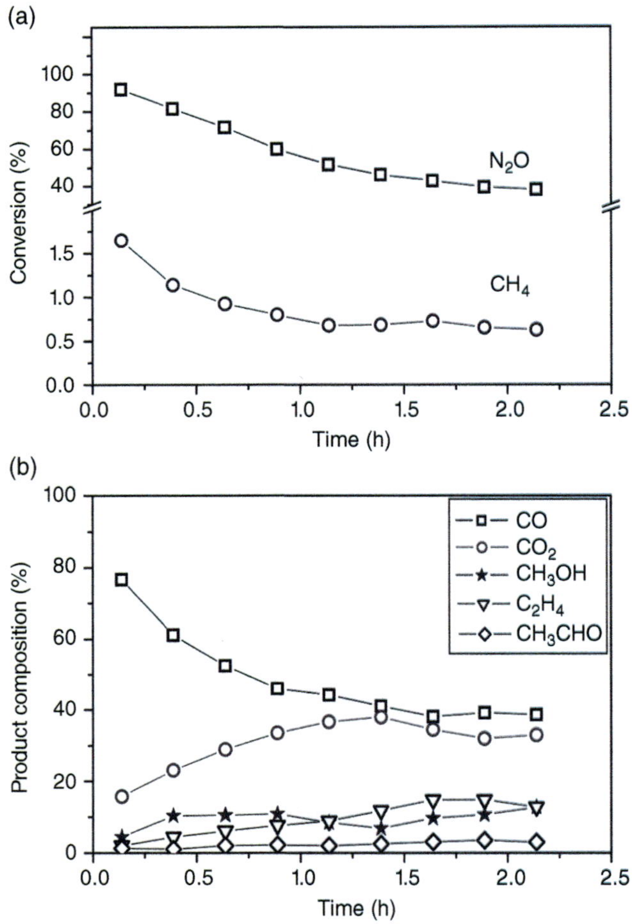

Figure 17.4 Catalytic oxidation of methane at 300°C: (a) X_{CH_4} (based on gas-phase products) and X_{N_2O}; (b) Product composition. Feed mixture: CH_4:N_2O: He = 20:2:78. No water was added to the flow of a mixture of CH_4, N_2O, and He. *Source:* Reproduced with permission from [11] / Elsevier.

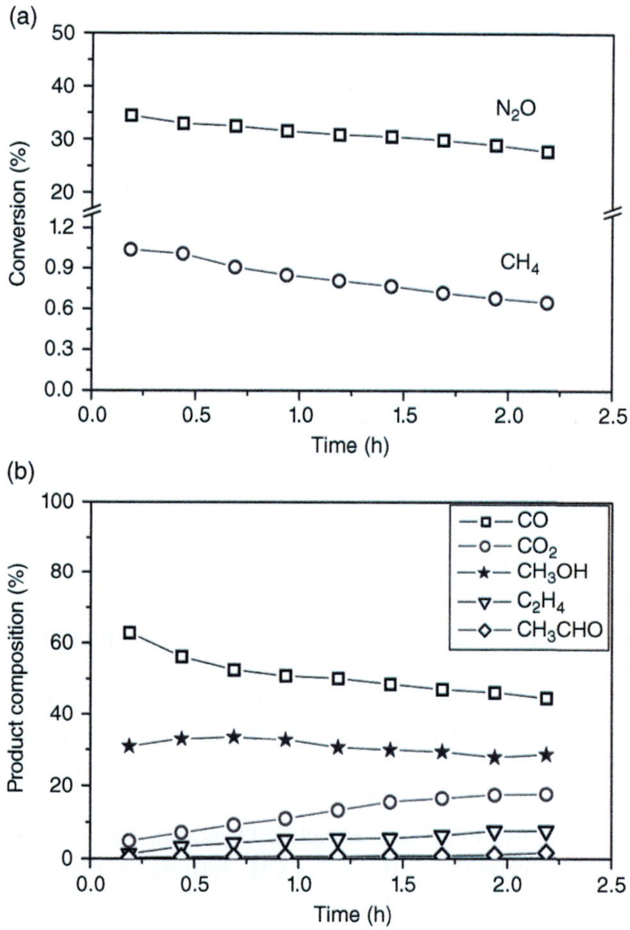

Figure 17.5 Catalytic oxidation of methane at 300°C with water addition: (a) X_{CH_4} (based on gas-phase products) and X_{N_2O}; (b) Product composition. Feed mixture: CH_4:N_2O:H_2O: He = 20:2:20:58. Water was added to the flow of a mixture of CH_4, N_2O, and He. *Source:* Reproduced with permission from [11] / Elsevier.

α-Fe sites and thus decreasing the number of the available sites for the partial oxidation of CH_4. Another noticeable difference is the higher concentration of CH_3OH in the downstream gas phase when water is added to the mixture of CH_4 and N_2O. According to the literature, H_2O prevents the formation of coke at 300°C. The addition of H_2O significantly reduces the amount of coke by 200 times at 275°C and 50 times at 300°C. The substantial decrease in selectivity for coke formation aligns with the observed increase in selectivity for methanol when H_2O is added to the reactants [11].

A comparison of product selectivity was conducted between quasi-catalytic reactions at 200°C and catalytic reactions at 300°C using the same mixture of CH_4 and N_2O. According to the literature, at 300°C without the addition of water, the selectivity for methanol is only 2%, with the majority of the products being coke. The literature suggests that the absence of dimethyl ether at 300°C is due to the decomposition of dimethyl ether to coke.

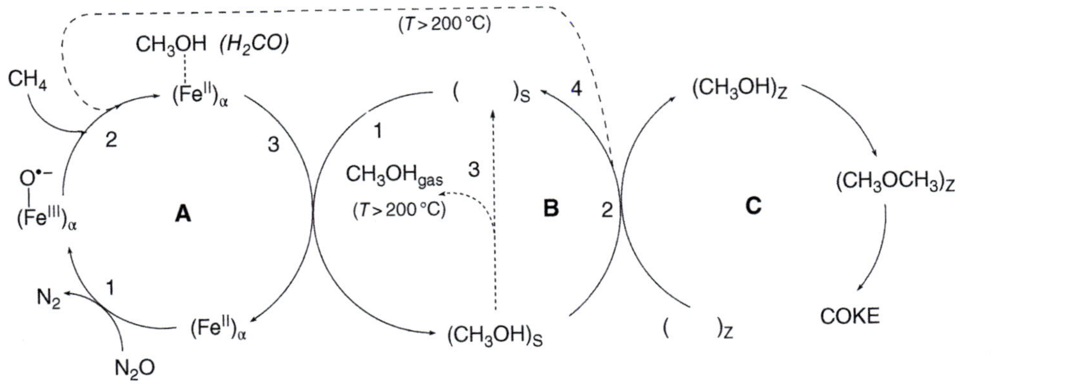

Figure 17.6 Mechanistic schemes of quasi-catalytic and catalytic oxidation of methane (A–C). Solid lines indicate the steps that are present in both the quasi-catalytic and catalytic modes of the reaction. Dotted lines display the steps that are present only in the catalytic mode. *Source:* Reproduced with permission from [11] / Elsevier.

A comparison of the partial oxidation of CH_4 under quasi-catalytic conditions at 200°C to catalytic conditions at 300°C enables the identification of intermediate compounds serving as precursors for the formation of final products at high temperatures. Thus, a proposed mechanism for the partial oxidation of CH_4 with N_2O at both 200 and 300°C is shown in Figure 17.6 [11]. In this figure, the solid line represents steps that can be performed at either quasi-catalytic reactions at 200°C or catalytic reactions at 300°C. The dashed line represents steps only performed as a part of the catalytic cycle at 300°C. The catalytic mechanism involves steps occurring on α-Fe sites (panel A), S sites (panel B), and Z sites (panel C). On an α-Fe site, N_2O dissociates to release an active O radical, forming an α-oxygen. This α-oxygen immediately activates the C–H bond of CH_4, forming methoxy or methanol on the α-Fe site. At temperatures >160°C, the methoxy or methanol molecule migrates to a neighboring α-Fe site, also called site S in reference [11]. Surface migration requires a low activation barrier, notably lower than the activation barrier for CH_3OH desorption to the gas phase. Thus, at temperatures lower than 200°C, the reaction is only quasi-catalytic since the formed CH_3OH could not desorb into the gas phase to vacate the sites at <200°C. At temperatures higher than 200°C, the formed CH_3OH can readily desorb to the gas phase, representing a major difference between the quasi-catalytic step at 200°C and catalysis at 300°C. However, at 300°C, formed CH_3OH can migrate to Z sites, an acidic site of ZSM-5 catalyzing the formation of dimethyl ether which eventually transforms to coke. Thus, at 300°C, at least three types of sites, including α-sites, sites S, and sites Z, participate in the catalytic reaction, leading to the formation of multiple products.

17.3 Small Pore Metallozeolite

While extensive studies have been conducted on the partial oxidation of CH_4 by medium-pore zeolites (primarily ZSM-5) with encapsulated Fe or Cu centers since the 1980s, small-pore zeolites with Fe for the partial oxidation of CH_4 have not been reported until recently. Only in 2015 was the first activity for the partial oxidation of CH_4 to CH_3OH through oxygen species in small pore metallozeolites reported by Wulfers et al. [33]. In this study, O species encapsulated in Cu-SSZ-13, Cu-SSZ-16, Cu-SSZ-39, and Cu-SAPO-34 can partially oxidize CH_4 to form CH_3OH. These O species formed through pre-oxidation of a zeolite with O_2 at high temperature. Most of these tests on small pore metallozeolites were performed with molecular O_2 [34–36]. The details of this type of work will be reviewed in one of the next chapters, discussing the oxidation of CH_4 through molecular O_2.

It was reported that oxygen species in Cu-SSZ-13 can oxidize CH_4 to CH_3OH upon the use of N_2O as the source of oxygen for small-pore Cu-SSZ-13 [37]. In this report, the oxidation of Cu-SSZ-13 to CH_4 was tested through an oxidation–extraction–regeneration three-step process. Additionally, CH_4, N_2O, and H_2O were co-fed to a microreactor maintained at 250–300°C for the catalytic oxidation of CH_4 to CH_3OH. It is a catalytic process as it continuously converts CH_4 without a separated re-activation of the catalyst. Cu atoms of copper acetate were introduced to SSZ-13 through ion exchange. The catalysis is performed at 250–300°C to avoid the formation of molecular O_2 during the decomposition of N_2O because the key step in this catalytic process is the oxidation of Cu atoms pre-encapsulated in SSZ-13 to a specific, active species by the atomic oxygen formed from N_2O decomposition. The gas feed mixture to the microreactor is 30% N_2O, 30% CH_4, and 3% H_2O vapor. Stable conversion of CH_4 and production of CH_3OH along with byproducts CO and CO_2 for 23 hours were confirmed. Selectivity for CH_3OH at 250–265°C remains at 20–27% [37].

A finding from this study is the limited yield of CH_3OH in this catalytic process. As shown in Figure 17.7a [37], selectivity for CH_3OH on Cu-SSZ-13 decreases along with the increase of CH_4 conversion, while the conversion of CH_4 was varied by setting the pressures of CH_4 between 20.3 and 40.5 kPa. It suggests the yield of CH_3OH in Figure 17.7a is nearly invariable. In addition, the invariance of the yield of CH_3OH was found when the catalysis temperature was changed between 250 and 270°C. As shown in Figure 17.7b, selectivity decreased along with an increase in the conversion of CH_4 while the catalysis temperature increased from 250 to 275°C [37]. The limit of the yield of CH_3OH results from the fact that the oxidation of the ideal product CH_3OH is faster than the oxidation of the reactant CH_4. The concern about the faster oxidation rate of the product such as CH_3OH than the reactant CH_4 was briefed in the earlier chapter on the oxidation of CH_4 catalyzed by a molecular catalyst in a liquid phase. Thus, to avoid the further oxidation of a product, CH_3OH regardless of a reaction in the gas or liquid phase, it is crucial to separate the product CH_3OH from the oxidant through a phase separation or remove the CH_3OH from the reaction system once it is formed. In addition, the found limit in CH_3OH yield suggests that CH_4 oxidation and CH_3OH decomposition or a further oxidation of CH_3OH are performed simultaneously. The deep reason for the yield limit is the activation of C–H of CH_4 is run through homolysis of C–H bonds [38].

17.4 A Comparison of Pore Size on Oxidation of Methane

A comparison was conducted between medium-pore metallozeolites, including Cu-ZSM-5 and Cu-MOE, and a small-pore metallozeolite, namely Cu-SSZ-13. The selectivities for producing CH_3OH on Cu-ZSM-5 (Cu/Al = 0.38) and Cu-MOR (Cu/Al = 0.33) are only

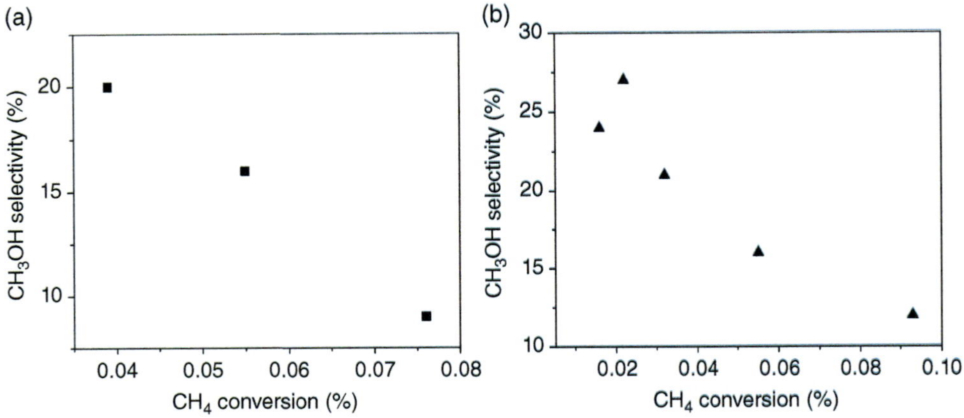

Figure 17.7 CH_4 conversion versus CH_3OH selectivity on Cu-SSZ-13. (a) The variation of CH_4 conversion is realized by changing CH_4 partial pressures between 20.3 and 40.5 kPa while the catalysis temperature is at 270°C and the gas pressure of N_2O and H_2O is 30.4 and 3.2 kPa, respectively. (b) The variation of CH_4 conversion is realized by changing catalysis temperature between 250 and 275°C while gas pressures of CH_4, N_2O, and H_2O are 30.4, 30.4, and 3.2 kPa, respectively. *Source:* Reproduced with permission from [37] / Royal Society of Chemistry.

0.3 and 1.1% at 270°C, but it is 39.7% on small-pore Cu-SSZ-13 (Cu/Al = 0.08) [37]. This difference suggests that a smaller size is favorable for this partial oxidation. Based on a quantum mechanics analysis of the confinement effect discussed in Chapter 15, the quantum effect of the small-pore size is larger than the medium-pore size. Thus, here the experimentally observed difference supports the analysis of the quantum mechanism in Chapter 15.

17.5 Isothermal Activation of Cu-ZSM-5, Partial Oxidation, and Gas Phase Extraction of Methanol

As documented in the literature [39], CH_4 can be catalytically converted to CH_3OH through the oxygen species in a bent mono(μ-oxo) di-copper structure formed in Cu-ZSM-5. This structure is generated through an activation by O_2, NO, or even N_2O. As CH_3OH decomposes at 200°C, typically, the selectivity for producing CH_3OH in the catalytic conversion of CH_4 at 200°C or a higher temperature is low [39]. Alternatively, CH_3OH can be extracted by soaking the Cu-ZSM-5 containing the formed methoxy in water or a mixture of water and acetonitrile at a relatively low temperature or even room temperature. This reaction, soaking, and extraction is neither a catalytic process as pointed out before, nor a continuous process.

Processes were developed for continuously producing CH_3OH [39–40]. Each cycle involves three steps: activation with NO or N_2O to form reactive species in the form of a bent mono(μ-oxo) di-copper structure in Cu-ZSM-5, partial oxidation of CH_4 by the oxygen atoms of the bent mono(μ-oxo) di-copper structure, and extraction of CH_3OH formed in the pore with water steam. All steps are performed at 150°C, making it an isothermal process. This choice of 150°C is a compromise between the ideal high activation temperature and the favorable low extraction temperature using steam to prevent CH_3OH decomposition. As depicted in Figure 17.8, a band at 450 nm contributed from a bent mono(μ-oxo) di-copper structure was observed from Cu-ZSM-5 activated by NO at 150°C [40]. Upon contact with water vapor at 150°C, this band disappeared, suggesting that the bent mono(μ-oxo) di-copper structure in Cu-ZSM-5 may not be stable and can be restructured by exposure to water vapor. After introducing NO, the band reappeared, indicating the regeneration of the bent mono(μ-oxo) di-copper structure (Figure 17.8). This regeneration was confirmed by the organic structure observed in Figure 17.8, where the regenerated bent mono(μ-oxo) di-copper structure reacted with CH_4 and then disappeared again. The spectrum collected while water vapor was flowing through for CH_3OH extraction is shown as the second to the top light blue line marked as H_2O 150 (2) in Figure 17.8. After this extraction, Cu-ZSM-5 was regenerated again in NO, as evidenced by the band at 450 nm in the orange spectrum (the top spectrum marked with NO 150 (s) in Figure 17.8). Figure 17.8 clearly shows that without annealing Cu-ZSM-5 to a temperature higher than 150°C, an introduction of NO can make the adsorbed water vapor desorb and then re-activate the Cu-ZSM-5 at 150°C to form a bent mono(μ-oxo) di-copper structure. Then, by switching NO to CH_4 at the same temperature (150°C), CH_4 can be oxidized to CH_3OH. Then, switching CH_4 to H_2O at 150°C, the formed CH_3OH can be extracted. This multi-step process can be repeated, and 0.589 mol g^{-1} of methanol can be sustainably extracted in each cycle [40].

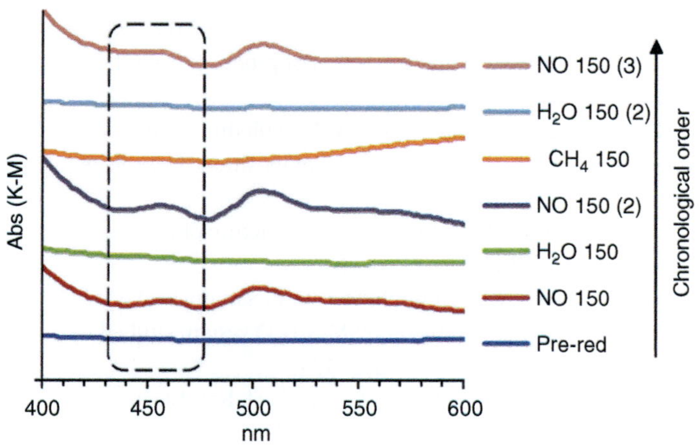

Figure 17.8 UV-vis spectra of Cu–ZSM-5 after pre-reduction at 500°C in Ar and treatment in NO, steam, and CH_4. A constant temperature of 150°C was used during treatment with each gas. Spectra were recorded in chronological order for a single sample *in situ* at room temperature. The band attributed to the di-copper core is highlighted. *Source:* Reproduced with permission from [40] / Royal Society of chemistry.

References

1 Falconer, J. W., Hoare, D. E., Overend, R. 1973. "The oxidation of methane by nitrous oxide." *Combust. Flame.* 21, 339–344.

2 Borisov, A. A., Zanianskii, V. M., Potmishil, K., Skachkov, G. I., Foteenkov, V. A. 1977. "Mechanism of methane oxidation with nitrous oxide." *Kinet. Katal.* 8, 307–315.

3 Robinson, P. L., Smith, E. J. 1952. "The kinetics of the oxidation of methane by nitrous oxide." *J. Chem. Soc.* 1952, 3895–3904.

4 Zhen, K. J., Khan, M. M., Mak, C. H., Lewis, K. B., Somorjai, G. A. 1985. "Partial oxidation of methane with nitrous oxide over V_2O_5-SiO_2 catalyst." *J. Catal.* 94, 501–507.

5 Yang, T. J., Lunsford, J. H. 1980. "The role of O^- ions in the oxidative dehydrogenation of ethane over molybdenum oxide supported on silica gel." *J. Catal.* 63, 505–509.

6 Ward, M. B., Lin, M. J., Lunsford, J. H. 1977. "The oxidative dehydrogenation of ethane by nitrous oxide over molybdenum oxide supported on silica gel." *J. Catal.* 50, 306–318.

7 Liu, R. S., Iwamoto, M., Lunsford, J. H. 1982. "Partial oxidation of methane by nitrous oxide over molybdenum oxide supported on silica." *J. Chem. Soc. Chem. Commun.* 1982, 78–79.

8 Anderson, J. R., Tsai, P. 1987. "Methanol from oxidation of methane by nitrous oxide over FeZSM5 catalysts." *J. Chem. Soc. Chem. Commun.* 1987, 1435–1436.

9 Pannov, G. I., Sobolev, V. I., Kharitonov, A. S. 1990. "The role of Iron in N_2O decomposition on Zsm-5 zeolite and reactivity of the surface oxygen formed." *J. Mol. Catal.* 61, 85–97.

10 Panov, G. I., Sobolev, V. I., Dubkov, K. A., KharitonovStudies, A. S. 1996. "Biomimetic oxidation on Fe complexes in zeolites." *Surf. Sci. Catal.* 101,493–502.

11 Parfenov, M. V., Starokon, E. V., Pirutko, L. V., Panov, G. I. 2014. "Quasicatalytic and catalytic oxidation of methane to methanol by nitrous oxide over FeZSM-5 zeolite." *J. Catal.* 318, 14–21.

12 Starokon, E. V., Parfenov, M. V., Pirutko, L. V., Abornev, S. I., Panov, G. I. 2011. "Room-temperature oxidation of methane by α-oxygen and extraction of products from the FeZSM-5 surface." *J. Phys. Chem. C* 115, 2155–2161.

13 Sobolev, V. I., Panov, G. I., Kharitonov, A. S., Romannikov, V. N., Volodin, A. M., Ione, K. G. 1993. "Catalytic properties of ZSM-5 zeolites in N_2O decomposition: The role of iron." *J. Catal.* 139, 435–443.

14 Malykhin, S., Zilberberg, I., Zhidomirov, G. M. 2005. "Electron structure of oxygen complexes of ferrous ion center." *Chem. Phys. Lett.* 414, 434–437.

15 Pirngruber, G. D., Grunwaldt, J. D., Roy, P. K., van Bokhoven, J. A., Safonova, O. V., Glatzel, P. 2007. "The nature of the active site in the Fe-ZSM-5/N_2O system studied by (resonant) inelastic X-ray scattering." *Catal. Today* 126,127–134.

16 Miranda, U., Varandas, A. J. C., Kaplan, I. G. 2014, "On the ferryl catalyst: Electronic structure and optimized ab initio geometry." *Chem. Phys. Lett.* 595–596, 175–179.

17 Zhidomirov, G. M., Shubin, A. A., Larin, A. V., Malykhin, S. E., Rybakov, A. A. 2012. In: J. Leszczynski, M.K. Shukla (Eds.), Practical Aspects of Computational Chemistry I: An Overview of the Last Two Decades and Current Trends, Dordrecht, Heidelberg, London, New York, Springer.

18 Li, G., Pidko, E. A., van Santen, R. A., Feng, Z., Li, C., Hensen, E. J. M. 2011. "Stability and reactivity of active sites for direct benzene oxidation to phenol in Fe/ZSM-5: A comprehensive periodic DFT study." *J. Catal.* 284, 194–206.

19 Guesmi, H., Berthomieu, D., Bromley, B., Coq, B., Kiwi-Minsker, L. 2010. "Theoretical evidence of the observed kinetic order dependence on temperature during the N_2O decomposition over Fe-ZSM-5." *Phys. Chem. Chem. Phys.* 12, 2873–2878.

20 Sazama, P., Wichterlova, B., Tábor, E., Stastny, P., Sathu, N. K., Sobalik, Z., Dedecek, J., Sklenak, S., Klein, P., Vondrova, A. 2014. "Tailoring of the structure of Fe-cationic species in Fe-ZSM-5 by distribution of Al atoms in the framework for N_2O decomposition and NH_3-SCR-NOx." *J. Catal.* 312, 123–138.

21 Taboada, J., Overweg, A., Kooyman, P. J., Arends, I. W. C. E., Mul, G. 2005. "Following the evolution of iron from framework to extra-framework positions in isomorphously substituted [Fe,Al]MFI with ^{57}Fe Mössbauer spectroscopy." *J. Catal.* 231, 56–66.

22 Rosa, A., Ricciardi, G., Baerends, E. 2010. "Is [FeO]2^+ the active center also in iron containing zeolites? A density functional theory study of methane hydroxylation catalysis by Fe-ZSM-5 zeolite." *J. Inorg. Chem.* 49, 3866.

23 Wood, B. R., Reimer, J. A., Bell, A. T., Janicke, M. T., Ott, K. C.2004. "Methanol formation on Fe/Al-MFI via the oxidation of methane by nitrous oxide." *J. Catal.* 225, 300–306.

24 Dubkov, K. A., Ovanesyan, N.S., Shteinman, A. A., Starokon, E. V., Panov, G. I. 2002. "Evolution of iron states and formation of α-sites upon activation of FeZSM-5 zeolites." *J. Catal.* 207, 341–352.

25 Kiwi-Minsker, L., Bulushev, D. A., Renken, A. 2003. "Active sites in HZSM-5 with low Fe content for the formation of surface oxygen by decomposing N_2O: is every deposited oxygen active?*J. Catal.* 219, 273–285.

26 Berrier, F., Ovsitser, O., Kondratenko, E. V., Schwidder, M., Grünert, W., Brückner, A. 2007. "Temperature-dependent N2O decomposition over Fe-ZSM-5: Identification of sites with different activity." *J. Catal.* 249 (2007), 67.

27 Dewitt, J. G., Bentsen, J. G., Rosenzweig, A. C., Hedman, B., Green, J., Pilkington, S., Papaefthymion, G. C., Dalton, H., Hodgson, K. O., Lippard, S. J. 1991. "X-ray absorption, Moessbauer, and EPR studies of the dinuclear iron center in the hydroxylase component of methane monooxygenase." *J. Am. Chem. Soc.* 113, 9219–9235.

28 Durante, V. A., Walker, D. W., Gussow, S. M., Lyons, J. E. 1990. US Patent 4918249.

29 Lyons, J. E., Ellis Jr. P. E., Durante, V. A. 1991. "Active iron oxo centers for the selective catalytic oxidation of alkanes." *Studies Surf. Sci. Catal.* 67, 99–116.

30 Ravi, M., Ranocchiari, M., van Bokhoven, J. A. 2017. "The direct catalytic oxidation of methane to methanol—A critical assessment." *Angew. Chem. Int. Ed.* 56, 16464–16483.

31 Polniser, R., Stolcova, M., Hronec, M., Mikula, M. 2011. "Structure and reactivity of copper iron pyrophosphate catalysts for selective oxidation of methane to formaldehyde and methanol". *Appl. Catal. A* 400, 122–130.

32 Starokon, E. V., Parfenov, M. V., Arzumanov, S. S., Pirutko, L. V., Stepanov, A. G., Panov, G. I. 2013. "Oxidation of methane to methanol on the surface of FeZSM-5 zeolite." *J. Catal.* 300, 47–54.

33 Wulfers, M. J., Teketel, S., Ipek, B., Lobo, R. F. 2015. "Conversion of methane to methanol on copper-containing small-pore zeolites and zeotypes." *Chem. Commun.* 51, 4447–4450.

34 Pappas, D. K., Borfecchia, E., Dyballa, M., Pankin, I., Lomachenko, K. A., Martini, A., Signorile, M., Teketel, S., Arstad, B., Berlier, G., Lamberti, C., Bordiga, S., Olsbye, U.,

Lillerud, K. P., Svelle, S., Beato, P. 2017. "Methane to methanol: Structure-activity relationships for Cu-Cha." *J. Am. Chem. Soc.* 139, 14961–14975.

35 Park, M. B., Ahn, S. H., Mansouri, A., Ranocchiari, M., van Bokhoven, J. A. 2017. "Comparative study of diverse copper zeolites for the conversion of methane into methanol." *ChemCatChem* 9, 3705–3713.

36 Narsimhan, K., Iyoki, K., Dinh, K., Roman-Leshkov, Y. 2016. "Catalytic oxidation of methane into methanol over copperexchanged zeolites with oxygen at low temperature." *ACS Cent. Sci.* 2, 424–429.

37 Ipek, B., Lobo, R. F. 2016. "Catalytic conversion of methane to methanol on Cu-SSZ-13 using N2O as oxidant." *Chem. Commun.* 52, 13401–13404.

38 Labinger, J. A. 2004. "Selective alkane oxidation: Hot and cold approaches to a hot problem." *J. Mol. Catal. A Chem.* 220, 27–35.

39 Alayon, E. M., Nachtegaal, M., Ranocchiari, M., van Bokhoven, J. A. 2012. "Catalytic conversion of methane to methanol over Cu–mordenite." *Chem. Commun.* 48, 404–406.

40 Sheppard, T., Hamill, C. D., Goguet, A., Rooney, D. W., Thompson, J. M. 2014. "A low temperature, isothermal gas-phase system for conversion of methane to methanol over Cu–ZSM-5." *Chem. Commun.* 50, 11053–11055.

18

Oxidation of Methane through M–O Sites Anchored in Zeolite or AuPd Nanoparticles by H_2O_2 at Low Temperature

18.1 Brief of the Difference between the Catalytic Oxidation of CH_4 with N_2O at a Relatively High Temperature and that with H_2O_2 in Aqueous Solution at a Low Temperature

Chapter 17 delved into the partial oxidation of CH_4 involving specific oxygen atoms generated through the reaction between N_2O and a particular site. This reaction can only occur again after the formed product (methoxy) desorbs from the specific site. However, the distinctive structural feature of microporous aluminosilicate hinders product desorption, typically making it a one-time reaction event at relatively low temperatures (<150 °C). In principle, at higher temperatures (150–200 °C), a portion of the product species can desorb into the gas phase. Only at a reasonably high temperature can the product species be desorbed into the gas phase, allowing the site to be oxidized again for reacting with the next CH_4 molecule, and the reaction may proceed in a catalytic manner. Unfortunately, the elevated temperature often triggers product decomposition and coke formation. For instance, at 300 °C, the partial oxidation of CH_4 involves at least three different sets of reactions, yielding several byproducts [1]. Despite following a routine catalytic cycle, the selectivity for CH_3OH at 300 °C is only 2%.

In parallel with the quasi-catalytic or catalytic oxidation of CH_4 in zeolites encapsulated with oxygen atoms/species, these catalysts can alternatively facilitate the oxidation of CH_4 by H_2O_2, a topic discussed in this chapter. When an H_2O_2 solution is used, the aqueous H_2O_2 serves as both an oxidant for the partial oxidation of CH_4 to methoxy in the solution and the aqueous solution for hydrolyzing the formed methoxy or other products. This feature allows the partial oxidation of CH_4 with H_2O_2 to operate in a closed catalytic cycle. Since this catalytic cycle is performed in an H_2O_2 aqueous solution at a relatively low temperature, the selectivity is relatively high. This contrasts sharply with the catalytic cycles of CH_4 oxidation with N_2O at high temperatures, as discussed in Chapter 17, where it exhibits a quite low selectivity for CH_3OH. In addition, compared to the low turnover rate of N_2O oxidation at less than $1\,h^{-1}$, the catalytic turnover rate by using H_2O_2 is higher by three orders of magnitude.

Low-temperature Activation and Catalytic Transformation of Methane to Non-CO_2 Products,
First Edition. Franklin Tao.

18.2 Fe-S-1 and Fe-ZSM-5

As documented in the literature, silicalite-1 (s-1) exhibits no activity in the partial oxidation of methane [2, 3]. In addition, the literature establishes that the Al and Ti sites are nonreactive for this reaction, as presented in Figure 18.1 [2]. However, the quantity of oxygenated products, including CH_3OH, CH_3OOH, and HCOOH, is directly proportional to the amount of introduced Fe ions within the micropores of silicalite-1 (Figure 18.1). Through ^{1}H NMR experiments examining the product of $^{13}CH_4$, the carbon source for these products was confirmed to be the carbon of CH_4 [2].

The Fe-ZSM-5 catalyst exhibits remarkable activity and selectivity in the oxidation of CH_4 by H_2O_2, yielding CH_3OH at low temperatures, as discussed below. The coordination environment of the impurity Fe, present at a concentration of 140 ppm in ZSM-5, was investigated using EXAFS. The first shell of an Fe atom contains approximately 4.9 ± 0.8 oxygen atoms, while the second shell consists of 1.6 ± 1.8 Fe atoms [4–6]. The Fe–O and Fe–(O)–Fe distances are measured at 2.04 ± 0.02 and 2.97 ± 0.04 Å, respectively. These values align with the di-iron structure reported in literature [4–6]. DFT calculations suggest that the structure labeled as **1** in Figure 18.2c closely matches the experimental data obtained from EXAFS studies of the as-prepared extra-framework di-iron species in ZSM-5.

Figure 18.2a depicts the time-stream selectivity of the main products, including methyl hydroperoxide (CH_3OOH), methanol (CH_3OH), formic acid (HCOOH), and carbon dioxide (CO_2). The evolutions of selectivities of the three products indicate that the oxidation of CH_4 proceeds through sequential reactions. The high selectivity of CH_3OOH, along with its sharp decrease coinciding with these simultaneous increases in selectivities of methanol, formic acid, and CO_2, suggests that CH_3OH, HOOCH, and CO_2 are formed from the

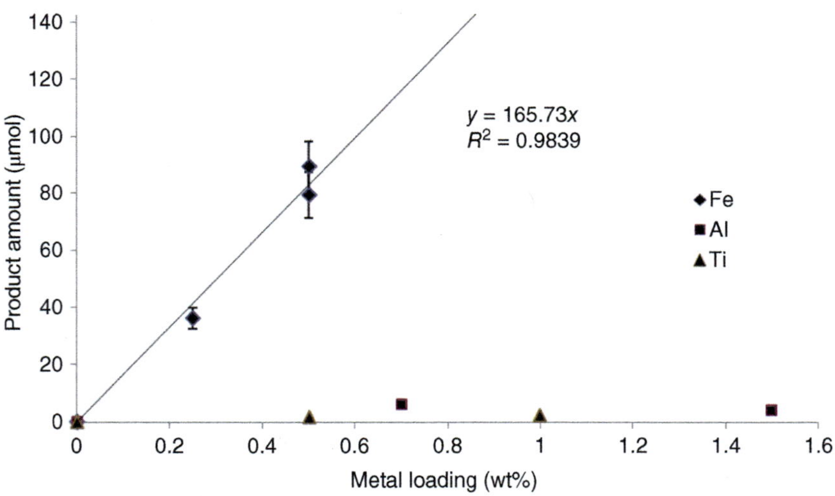

Figure 18.1 Effect of Fe, or Al, or Ti incorporation on the catalytic activity of silicalite-1. Reaction conditions; volume: 10 mL; time: 30 min; temp: 50 °C: P(CH_4): 30.5 bar; [H_2O_2]: 0.5 M; catalyst: 27 mg; stirring speed: 1500 rpm. Catalysts calcined prior to use (550 °C, 3 h, static air). *Source:* Reproduced with permission from [2] / John Wiley & Sons.

intermediate product, CH_3OOH. In other words, CH_3OOH serves as the intermediate for producing the other three products, CH_3OH, HOOCH, and CO_2 (Figure 18.2b). A proposed catalytic cycle for the formation of the intermediate compound, CH_3OOH, is presented in Figure 18.2c. In this cycle, an H_2O_2 molecule in an aqueous solution exchanges with a pre-coordinated H_2O molecule on the Feα atom of the di-iron structure (**1** in Figure 18.2c), forming structure **2**. Structure **2** acts as the catalytic site, facilitating the conversion of two H_2O_2 molecules and one CH_4 molecule to the initial product CH_3OOH through the reaction $CH_4 + 2\ H_2O_2 \rightarrow CH_3OOH + 2\ H_2O$. The specific steps are shown in Figure 18.2c. The first step of the catalytic cycle involves the rearrangement of the bound H_2O_2 molecule (H–O–O–H) on the Feα atom of structure **2** to form a HOO group on the Feα atom in structure **3**, where the H left from HOOH on Feα atom binds to OH on Feα atom, creating a bound H_2O on the Feα atom in structure **3** of Figure 18.2c. In the formed structure **3**, the Feα and Feβ atoms are Fe^{4+} and Fe^{2+}, respectively. The second step (from structure **3** to **4**) involves the exchange of an H_2O_2 molecule with the H_2O bound to Feα, releasing the H_2O from Feα into the solution. This interaction between H_2O_2 and Fe^{2+} forms a ferryl ion of Fenton's reagent in structure **4** in Figure 18.2c. Furthermore, another H_2O is released in the second step after the rearrangement of H_2O_2 to form a ferrite ion. In Structure **4**, both Feα and Feβ atoms are Fe^{IV}, making it a unique bifunctional oxidation center. This bifunctional oxidation center exhibits high activity for activating C–H bonds of CH_4 to form the precursor (5 in Figure 18.2c). This activation requests a barrier of only 50 kJ mol^{-1}. The

(a)

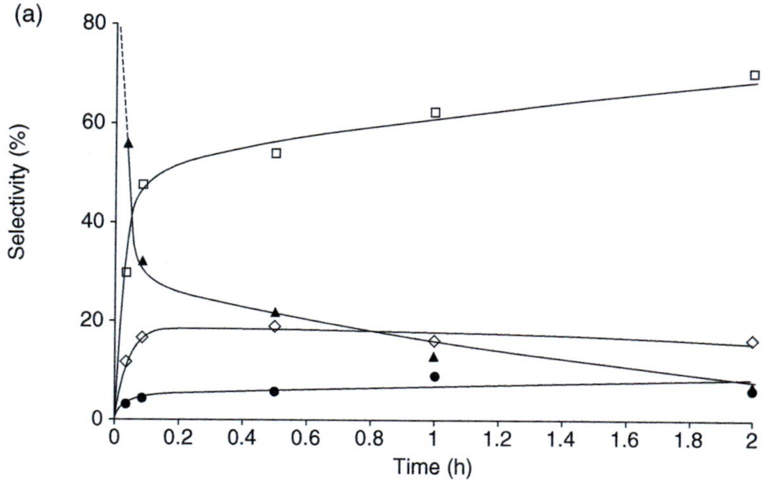

Figure 18.2 Catalytic performance and proposed pathway for oxidation of CH_4 by H_2O_2 catalyzed by anchored Fe_1 in the pore of ZSM-5. (a) Time-on-line profile for the liquid-phase oxidation of methane with hydrogen peroxide and ZSM-5. Key: ▲ = methyl hydroperoxide (MeOOH); ◆ = methanol (MeOH); □ = formic acid (HCOOH); ● = carbon dioxide (CO_2). (b) A potential reaction scheme for the oxidation of methane based on the time-on-line profile. Methanol is formed through the conversion of the methyl hydroperoxide intermediate over the Fe sites present in the catalyst.·OH radicals produced during the reaction are later responsible for the over-oxidation of methanol. (c) Catalytic cycle for the oxidation of methane to CH_3OOH using H_2O_2, catalyzed by a binuclear Fe species in ZSM-5. The overall charge in each case is formally +2 as the species acts as an extra-framework cation within the zeolite. *Source:* Reproduced with permission from [2] / John Wiley & Sons.

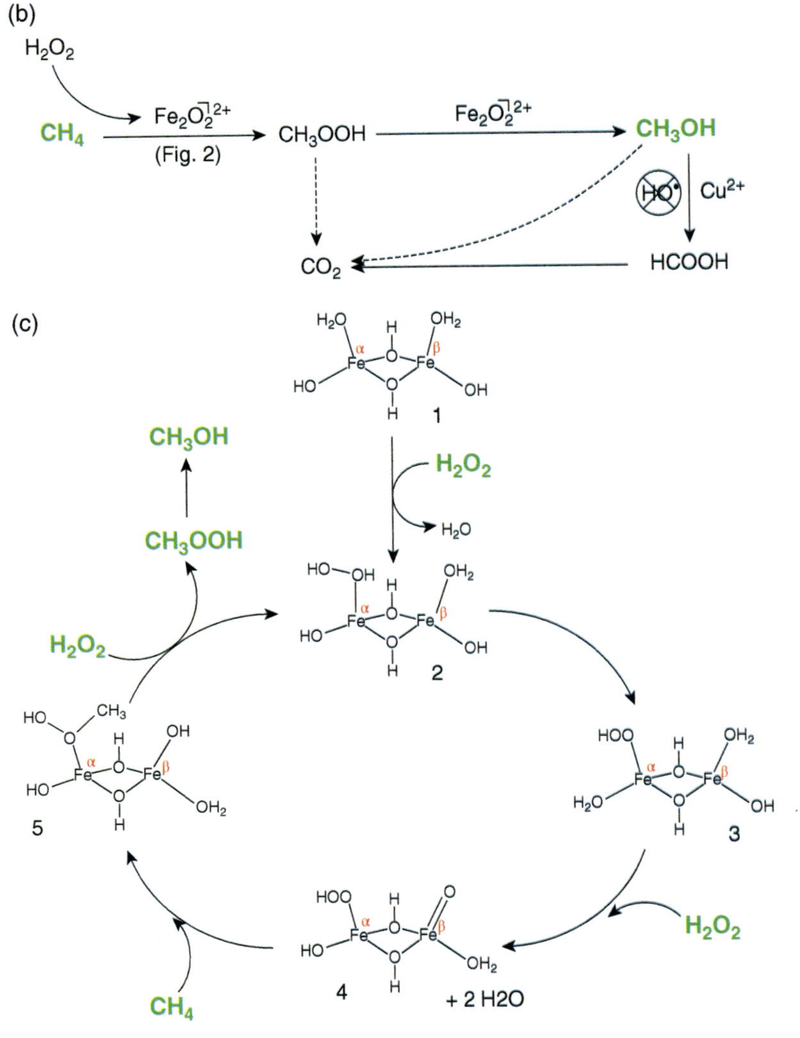

Figure 18.2 (Continued)

mechanism proposed in Figure 18.2c rationalizes the effective oxidation of CH_4 by H_2O_2 in Fe-ZSM-5 at near-room temperature.

Figure 18.3 illustrates the reaction coordination from intermediate 4 to **5** of Figure 18.2c. The pathway of transforming intermediate 4 to 5 depicted in Figure 18.2c is optimized with spin-unrestricted DFT and is validated by the high consistency between the theoretically predicted barrier and the experimentally measured apparent barrier (Figure 18.3). The top panel of Figure 18.3 presents the optimized structure of the pathway, mapping from intermediate 4 with an adsorbed CH_4 molecule to intermediate 5 of Figure 18.2c. The lower panel in Figure 18.3 is the energy profile for the reaction coordination from intermediate 4

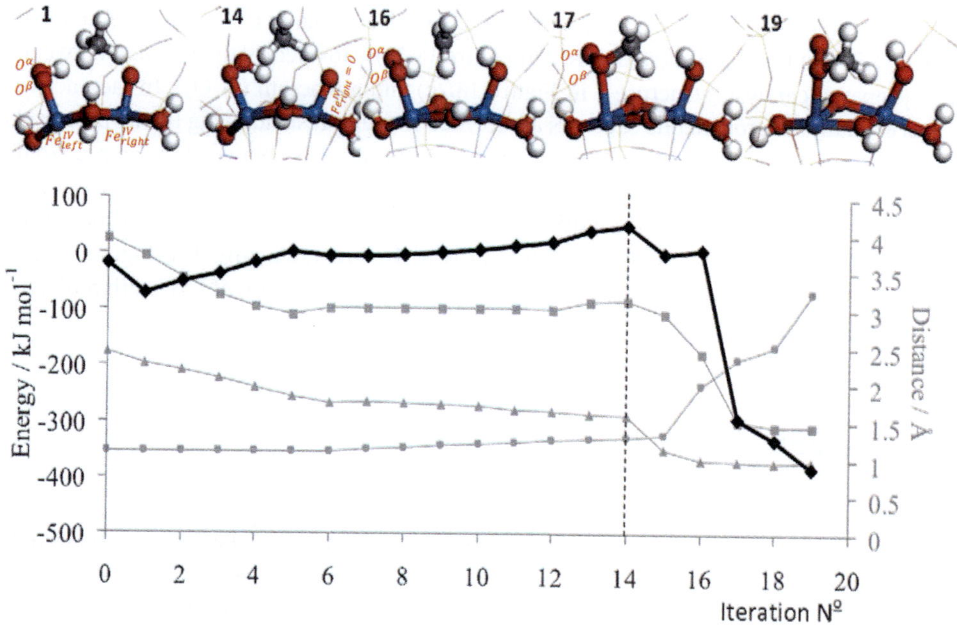

Figure 18.3 Spin-unrestricted DFT-optimized (PBE, $U_{Fe} = 4.0$) pathway for methane to methyl hydroperoxide showing energy relative to the free methane, hydroperoxide and resting state of the catalyst (in black referred to left-hand axis, diamonds) and (in gray referred to the righthand axis) the key atom distances along the pathway, with $H_3C \cdots OOH$ as triangles, $(H_3C)H \cdots O{=}Fe$ as squares and $H_3C \cdots H$ as diamonds. The *x*-axis is the iteration number along the reaction coordinate. Top: graphical representations of the active-site region at the points along the reaction coordinate indicated by iteration number are given; in each case, the reactants and molecules coordinated to Fe centers are shown in ball-and-stick representations while the zeolite framework is wireframe, most of the periodic structure of the MFI unit cell is omitted for clarity. Atoms are colored according to C gray, O red, H white, Si yellow(wire), Al pink(wire), and Fe mauve. Further details of the method are deposited in the Supporting Information. *Source:* Reproduced with permission from [2] / John Wiley & Sons.

to 5 of Figure 18.2c. In this pathway, the interaction between the H atom of a C–H bond of CH_4 and $HO^{\beta}O^{\alpha}Fe^{IV}_{left}$ (Structure **1** in Figure 18.3) weakens the C–H bond of CH_4, causing an increase in the C–H bond length (Structure **14** in Figure 18.3). When the C–H bond length reaches 1.30 Å, the H–C bond is broken, and the hydrogen atom is released to the O of $Fe^{IV}_{right} = O$ (structure **16** in Figure 18.3). Consequently, a CH_3 radical is formed and immediately bonds to the Oβ atom of $HO^{\beta}O^{\alpha}Fe^{IV}_{left}$ (17 and 19 in Figure 18.3, also 5 in Figure 18.2c). Importantly, this step does not release any species into the aqueous solution. This oxidation step of CH_4 by H_2O_2 through a Fe-based site (from 4 to 5 in Figure 18.2c or 18.3) differs from the oxidation of CH_4 by α-oxygen directly bonded to α-Fe of Fe-ZSM-5 as reviewed in Chapter 17 where CH_3 is formed, and that mechanism involves the oxidation of the $CH_3\cdot$ radical by a terminal oxidant in Structure A in Figure 17.6.

Following the desorption of the formed CH_3OOH in structure **5** in Figure 18.2c, the catalytic site (structure **2** in Figure 18.2c) is regenerated and ready for the next catalytic cycle.

An important distinction in this heterogeneous catalysis, oxidizing CH_4 with H_2O_2 to form CH_3OOH, is that the catalyst (structure **2** in Figure 18.2c) is formed by the reaction between an as-prepared catalyst (structure **1** in Figure 18.2c) and a reactant (H_2O_2). In fact, active phases of many heterogeneous catalysts are formed through restructuring the as-prepared catalysts during pretreatment with H_2, O_2, or during a reaction between an as-prepared catalyst and one or two reactants of a reaction. The formation of an active phase from the reaction between an as-prepared catalyst and a reactant in the catalytic reaction is not unusual. If the oxidant H_2O_2 is replaced with O_2, the active site (structure **2** in Figure 18.2c) could not form. This is why this catalyst is not active for the oxidation of CH_4 with molecular O_2 dissolved in water.

18.3 Pd-ZSM-5

The preceding section highlighted the robust activity in generating CH_3OOH in an H_2O_2 aqueous solution. The pivotal active site in ZSM-5 containing 0.14% Fe is the bifunctional di-iron center, $HO\beta O\alpha Fe_{left}$. . .. $Fe_{right} = O$ (structure **4** in Figure 18.2c). The roles of the $HO\beta O\alpha$ and $Fe_{right} = O$ to activate the H–C bond of CH_4 and receive the H atom of the first C–H in CH_4, respectively. In contrast to ZSM-5 containing 0.14% Fe, another catalyst, Pd@ ZSM-5 containing 0.01 wt% Pd, has been recognized for its high activity and selectivity in producing CH_3OH through the initial formation of CH_3OOH [7]. The key active site in this catalyst is the Pd=O, which activates the C–H bond of CH_4, leading to the formation of CH_3OOH, subsequently transformed into CH_3OH [7]. This mechanism of Pd@ZSM-5 proposed in reference [7] stands in stark contrast to the distinctive bifunctional di-iron center [4–6].

This catalyst was prepared through a modified incipient wetness impregnation method, and detailed information can be found in literature [7]. The actual loading of Pd is 0.01 wt%. Neither Pd or Pd_xO_y nanoparticles were formed on 0.01wt%Pd/ZSM-5 (Figure 18.4a). As depicted in Figure 18.4c, XPS studies indicate no Pd atoms in the surface region of the zeolite particles, suggesting that the loaded 0.04 wt% Pd is within the micropores. The presence of Pd in the micropores is confirmed by EXAFS studies of Pd K-edge, as illustrated in Figure 18.5. According to the EXAFS studies, a Pd atom is bonded to 4.1 ± 0.5 oxygen atoms (Table 18.1). Throughout the subsequent discussion, 0.01 wt% Pd/ZSM-5 and Pd_1O_4@ ZSM-5 are used interchangeably [7].

The HZSM-5 used in the preparation of Pd_1O_4@ZSM-5 exhibits no activity for the partial oxidation of CH_4 under the conditions of 50 °C, 30 bar CH_4, 5 mmol H_2O_2, and 10 mL deionized H_2O (Figure 18.6). Upon the addition of 0.01 wt% Pd to HZSM-5, 60 μmol HCOOH, 40 μmol CH_3OOH, and 8 μmol CH_3OH were produced within 30 minutes (green pillars in Figure 18.6). However, no activity was observed when 0.01 wt% Pd was loaded on either SiO_2 (blue pillars) or Al_2O_3 (dark green). Hence, the activity must originate from the Pd-based structure formed in the micropores of 0.01 wt% Pd/ZSM-5. This origin was further confirmed by the fact that the formed palladium oxide nanoclusters (Figure 18.4b) are not active for CH_4 partial oxidation at all. This is because the activity on 28 mg of 2 wt% Pd/ ZSM-5 is lower than 28 mg of 0.01 wt% Pd/ZSM-5 based on contrasting experiments in

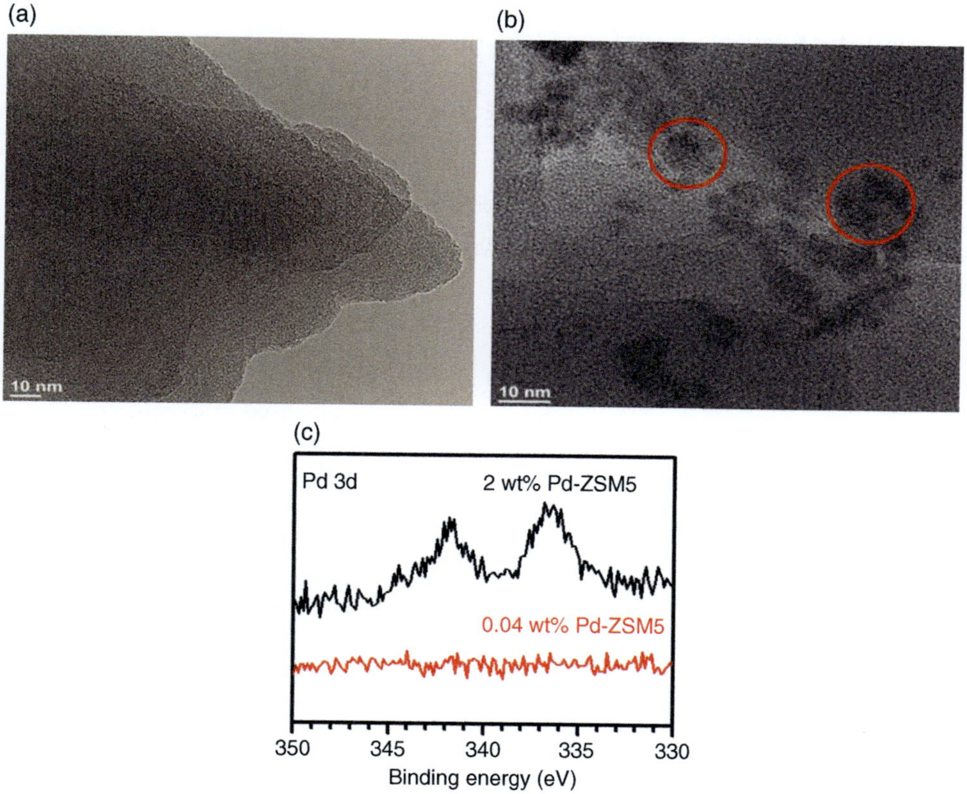

Figure 18.4 TEM and XRD studies of the as-prepared catalysts of 0.04 wt% Pd/ZSM-5 and 2.0 wt% Pd/ZSM-5. (a) TEM image of 0.01 wt% Pd/ZSM-5. (b) TEM image of 2.0 wt% Pd/ZSM-5; Representative PdO nanoparticles were circled. (c) Photoemission features of Pd 3d of as-synthesized 0.04 wt% Pd/ZSM-5 (red spectrum) and 2.0 wt% Pd/ZSM-5 (black spectrum). *Source:* Reproduced with permission from [7] / John Wiley & Sons.

Figure 18.5 EXAFS studies of Pd K-edge of 0.04 wt% Pd/ZSM-5 and relevant reference samples. Fourier transform magnitudes of k^2-weighted EXAFS data of 0.04 wt% Pd/ZSM-5 (blue) and reference samples, including Pd foil (red) and PdO nanoparticles (black). *Source:* Reproduced with permission from [7] / John Wiley & Sons.

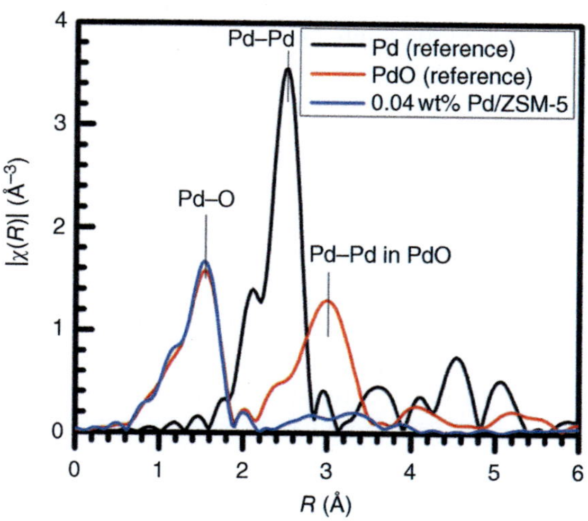

Table 18.1 Coordination number of O or Pd atoms to a Pd atom, and bond lengths of O–Pd or Pd–Pd in 0.04 wt% Pd/ZSM-5 studied with EXAFS.

Sample	N (Pd–Pd)	N (Pd–O)	R(Pd–Pd) (Å)	R (Pd–O) (Å)
Pd foil	12	0	2.740 ± 0.002	–
0.04 wt% Pd/ZSM-5	0	4.12 ± 0.49	–	2.001 ± 0.009

Source: Reproduced with permission from [7] / John Wiley & Sons.

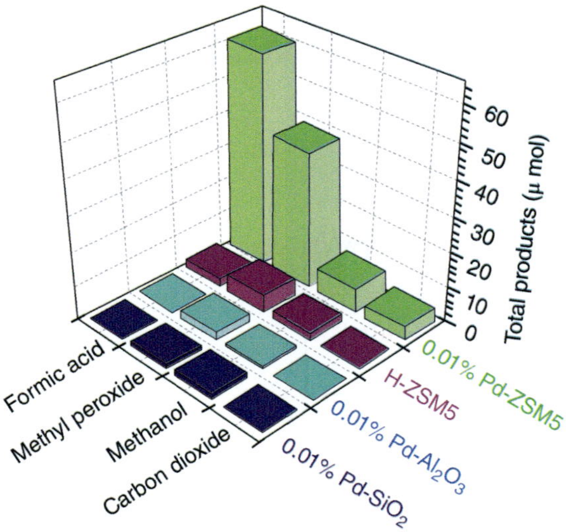

Figure 18.6 Catalytic performance of 0.01 wt% Pd loaded on different supports. Yields of products of methane partial oxidation on 0.01 wt% Pd/ZSM-5, ZSM-5, 0.01 wt% Pd/Al$_2$O3, 0.01 wt% Pd/SiO$_2$ at 50 °C for 30 minutes in 30 bar CH_4, 5 mmol H_2O_2 and 10 mL deionized water. *Source:* Reproduced with permission from [7] / John Wiley & Sons.

Figure 18.7. Once again, the activity of 0.01 wt% Pd@ZSM-5 must stem from the encapsulated Pd_1O_4 species.

It was found that the selectivity for CH_3OH can be significantly promoted by adding Cu^{2+}. Due to the limited solubility of CH_4 in H_2O, the concentration of H_2O_2 in aqueous solution of the work in reference [7] is much higher than CH_4. Unfortunately, the heavy excess of H_2O_2 readily oxidizes the formed CH_3OH to formic acid and even CO_2. Further oxidation to the formed CH_3OH gives a low selectivity for the ideal product, CH_3OH. As Cu^{2+} can catalyze the decomposition of H_2O_2 to H_2O and O_2, the excess H_2O_2 can be consumed by the added Cu^{2+}. In other words, the addition of Cu^{2+} to the solution can suppress the further oxidation of CH_3OH, thus promoting the selectivity for CH_3OH. As illustrated in Figure 18.8, the catalyst 0.01 wt% Pd/ZSM-5 loaded with 2.0 wt% CuO exhibits a much higher selectivity for methanol (Figure 18.8b) compared to 0.01wt%Pd/ZSM-5 without any CuO (Figure 18.8a) under the same catalytic condition.

Based on the coordination environment of the Pd_1 atom measured with EXAFS, the geometry of Pd_1O_4 was optimized with DFT, as shown in structure **1** in Figure 18.9 [7]. Structure **1** in Figure 18.9 represents Pd_1O_4 of 0.01 wt% Pd/ZSM-5. Structure **1** loses one H atom, forming structure **2** containing a Pd=O. This Pd=O is the center for activating C–H

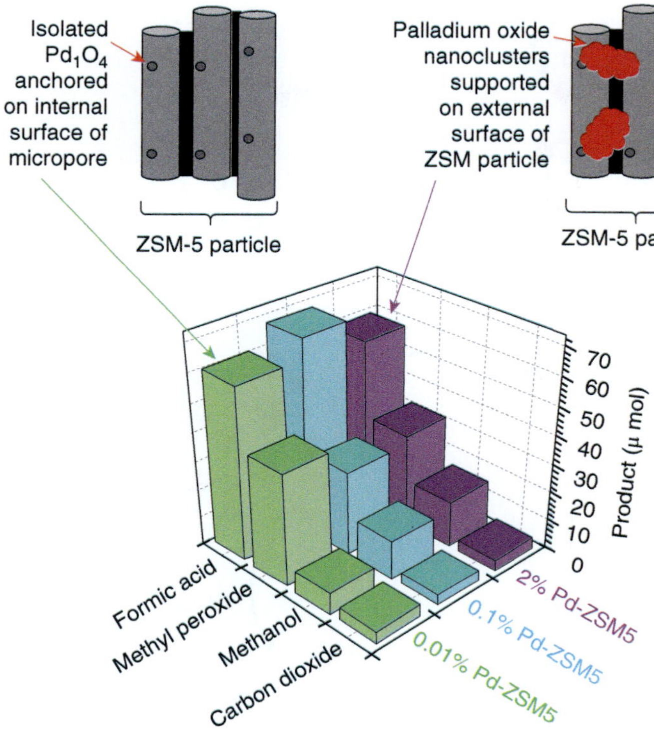

Figure 18.7 Yields of products of transformation of CH_4 catalyzed by 0.01, 0.10, and 2.0 wt% Pd/ZSM-5 for methane partial oxidation at 50 °C. *Source:* Reproduced with permission from [7] / John Wiley & Sons.

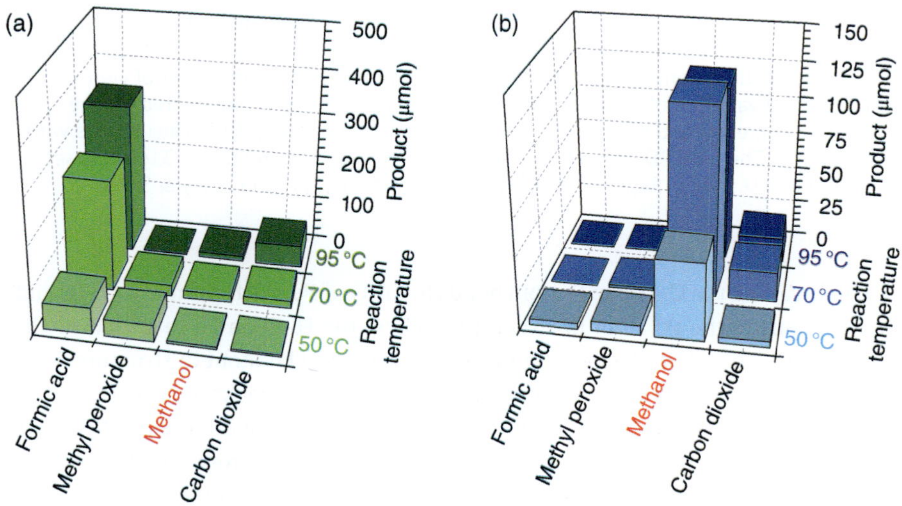

Figure 18.8 Catalytic performance of (a) 0.01 wt% Pd/ZSM-5 at 50, 70, and 95 °C, and (b) 0.01 wt% Pd/ZSM-5 loaded with 2 wt% CuO at 50, 70, and 95 °C. The plotted yields of products formed on 28 mg 0.01 wt% Pd/ZSM-5 or 0.01 wt% Pd/ZSM-5 loaded with 2.0 wt% CuO are the values after subtraction of the yields of the products formed on 28 mg ZSM-5 (without Pd sites) under the same catalytic conditions. *Source:* Reproduced with permission from [7] / John Wiley & Sons.

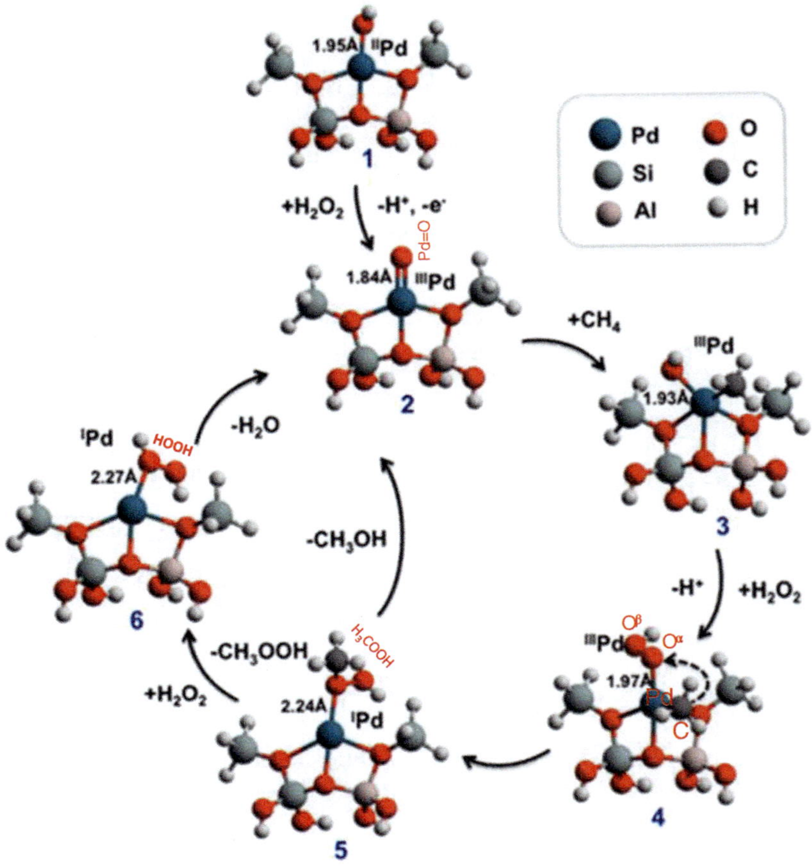

Figure 18.9 The proposed mechanism for the conversion of CH_4 to methanol is based on the optimized structural model (1) representing the active site Pd_1O_4 anchored on the micropores of ZSM-5. The reaction pathway and these intermediates were proposed from the point of view of thermal catalysis. The variation of the Pd–O distance on ZSM-5 surfaces is highlighted. All the chemical structures here are optimized using the density functional theory (DFT) calculations with the B3LYP functional. The LANL2DZ basis set was used for Pd, Si, and Al, and the 6-31g*basis set was used for C, O, and H. *Source:* Reproduced with permission from [7] / John Wiley & Sons.

of CH_4. Upon breaking the C–H, the H atom bonds to the O of Pd=O, forming H–O–Pd and the C of CH_3 bonds to the Pd atom in structure **3**. Then, the H–O on Pd in structure **3** is replaced by HOO–. Thus, both HOβOα– and CH_3– bond to the Pd atom in structure **4**. As shown with the black arrow in structure **4**, the C atom of the CH_3 attacks the Oα of HOβOα, forming $HOOCH_3$ in structure **5**. Then, the $HOOCH_3$ is replaced with a HOOH (H_2O_2), forming HOO-Pd in structure **6** along the desorption of CH_3OOH from structure **5**. The HOOH binding to the Pd atom in structure **6** morphs to an O=Pd along the desorption of an H_2O to the aqueous solution, forming the catalytic site, structure **2**, for the next catalytic cycle. Alternatively, CH_3 of Structure **5** couples with OH, forming CH_3OH. Once the formed CH_3OH desorbs, a Pd=O structure is formed to perform a next catalytic cycle.

18.4 AuPd Supported on ZSM-5

AuPd bimetallic nanoparticles supported on TiO_2 are active for the partial oxidation of CH_4 to form CH_3OH, utilizing either preexisting aqueous H_2O_2 or on-site synthesized H_2O_2 as an oxidant [8]. Prepared through the incipient wetness method, these nanoparticles have a size ranging from 5 to 20 nm (Figure 18.10a, b, d, and e), with the lattice fringe of an individual particle clearly depicted in Figure 18.10b. The partial oxidation was conducted under the main condition, where H_2O_2 was added as a preexisting reactant, and CH_4 was introduced to the AuPd nanoparticle surface by dissolving it into H_2O under the high-pressure CH_4 gas phase. The catalyst was dispersed in the H_2O_2 aqueous solution, and all reactants, solvents, and gas phases were sealed in a batch reactor. Catalysis was performed at temperatures ranging from 25 to 90 °C. As AuPd bimetallic nanoparticles are reported to be active catalysts for H_2O_2 synthesis from H_2 and O_2 at near-room temperature [9], the H_2O_2 can be produced on-site from H_2 and O_2. The observed activity for CH_4 partial oxidation was confirmed by the formation of CH_3OOH, CH_3OH, and CO_2 (Figure 18.11) [9]. Importantly, a control experiment using pure TiO_2 showed no product formation under the same catalytic condition. Clearly, AuPd nanoparticles played a catalytic role in the partial oxidation to form the oxygenates presented in Figure 18.11. Notably, only CH_3OOH was initially formed ($t < 2$ min), followed by a rapid increase in the selectivities of CH_3OH and CO_2, while the selectivity of CH_3OOH decreased simultaneously (Figure 18.11a). This suggests that the formation of CH_3OH proceeds through a sequential reaction involving the initial formation of CH_3OOH (marked with triangles in Figure 18.11a) and its subsequent transformation into CH_3OH. This pathway is similar to that observed from Fe-ZSM-5 [2]. However, in contrast to the formation of formic acid in the case of Fe-ZSM-5, a significant difference is that no formic acid formed by AuPd/TiO_2. This indicates that AuPd nanoparticles participate in both the formation of CH_3OOH and the transformation of CH_3OOH into CH_3OH and CO_2 but not the formation of formic acid. The involvement of Au-Pd/TiO_2 in the transformation of CH_3OOH to CH_3OH was confirmed by isotope-labeled experiments (Figure 18.11b), where preexisting $^{13}CH_3OOH$ was transformed to $^{13}CH_3OH$ in aqueous H_2O_2 solution under high-pressure CH_4 conditions [8].

The partial oxidation of CH_4 with H_2O_2 to CH_3OH proceeds through a radical mechanism. To confirm the presence of radicals such as HOO·, HO·, H_3C·, H_3CO·, and O^{2-}·, a trapper of radicals, 5,5′-Dimethyl-1-pyrrolidine-N-oxide (DMPO), was added to the reaction system under the catalytic condition described above. DMPO can trap these radicals to form adducts such as DMPO-OH and DMPO-CH_3 that can be examined by electron paramagnetic resonance (EPR). Both DMPO-OH and DMPO-CH_3 formed by trapping HO· and H3C· on DMPO, respectively, were detected by EPR (Figure 18.12), showing the formation of H_3C· and HO· radicals during partial oxidation of CH_4 to CH_3OH. This is distinctly different from the partial oxidation of CH_4 to form CH_3OH through the catalysis of Fe-ZSM-5, where neither HO· nor CH_3· was formed during catalysis [2]. In the radical mechanism performed on 1.0wt%AuPd/TiO_2 (IW) catalyst, HO· is formed from the dissociation of H_2O_2, and CH_3· is produced through the absorption of an H atom of CH_4 by HO· [9]. Subsequently, CH_3· can couple with an HOO· bound to the catalyst surface, forming the

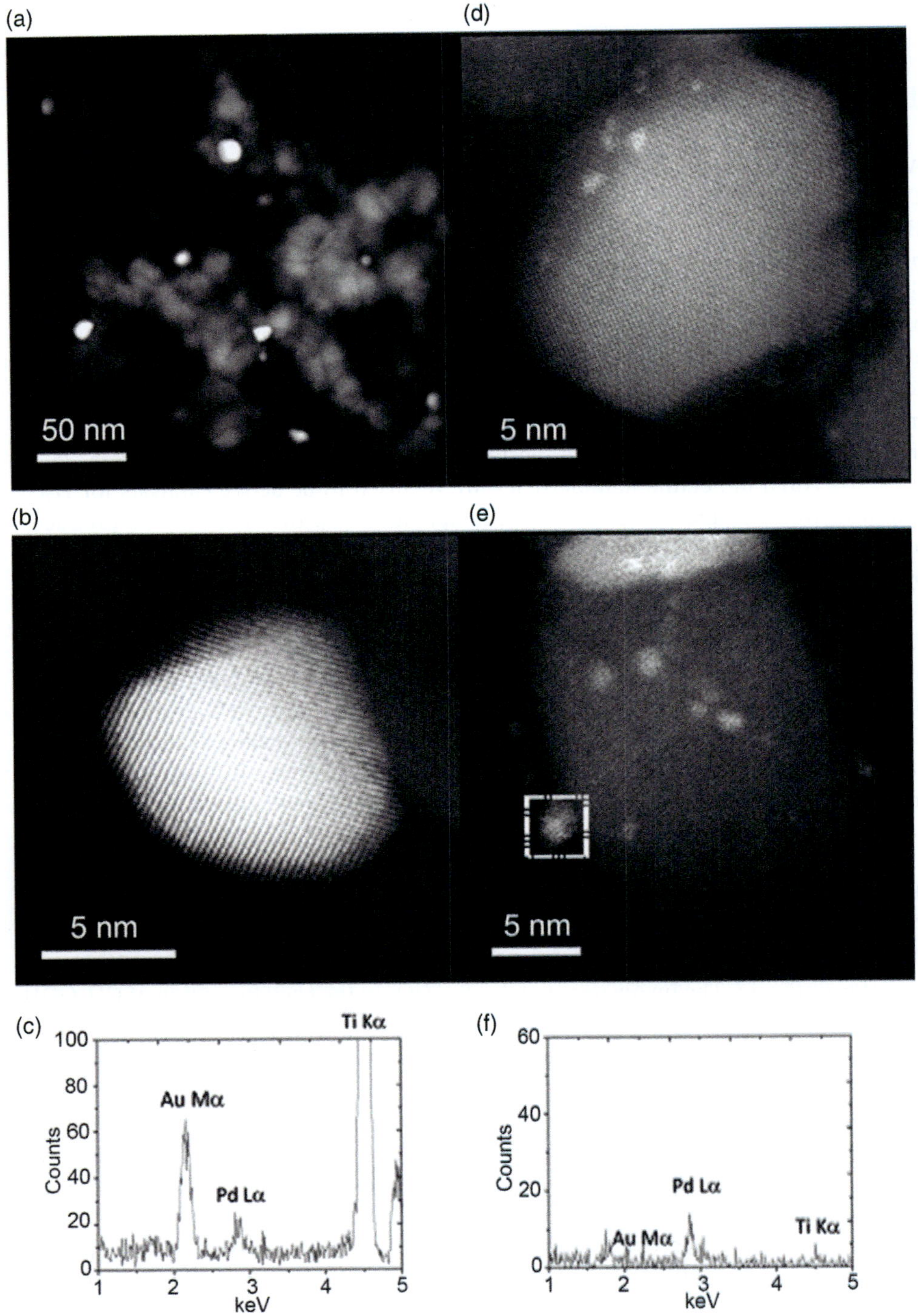

Figure 18.10 Representative scanning transmission electron microscopy (STEM) HAADF images and corresponding STEM–XEDS spectra from the 1.0 wt% AuPd/TiO_2 (IW) sample. (a, b) Low and higher magnification HAADF images show metal particles in the 5–20 nm size range. (c) The XEDS spectrum obtained from the 10 nm particle shown in (b) indicates that it is an Au-Pd alloy that is enriched in Au compared to the nominal composition. A distinct Au-rich core/Pd-rich shell morphology can also be deduced from image (b) as the particle exhibits significantly fainter contrast at its extremity. (d, e) Even higher magnification HAADF images show a significant population of 1–2 nm particles and subnanometer clusters co-existing on the TiO_2 support surface. (f) The XEDS spectrum of the particle highlighted in (e) suggests that these smaller species are definitely Pd-rich. *Source:* Reproduced with permission from [8] / John Wiley & Sons.

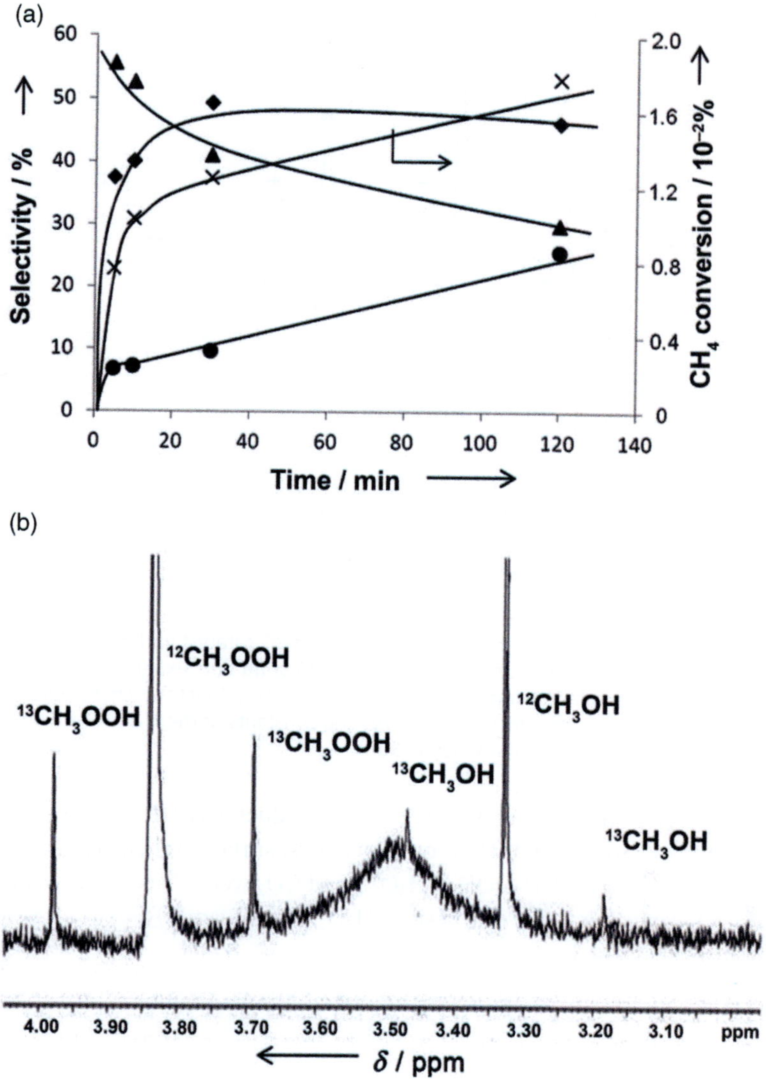

Figure 18.11 Catalytic performance of 1.0 wt% AuPd/TiO_2 (IW) catalyst. (a) The time-on-line plot of methane oxidation with the addition of H_2O_2 in the presence of a 1.0 wt% AuPd/TiO_2 (IW) catalyst. Key: ▲ selectivity to methyl hydroperoxide, ◆ selectivity to methanol, ● selectivity to carbon dioxide, *x* methane conversion. Reaction conditions: p_{CH_4} = 30.5 bar, [H_2O_2] = 0.5m, *T*=50°C, stirring rate = 1500 rpm, and catalyst mass = 10 mg. (b) ^{1}H NMR spectrum of the reaction filtrate collected when using a solution initially spiked with $^{13}CH_3OOH$ in a typical reaction using 5 wt% Au-Pd/TiO_2 (IW) and pure $^{12}CH_4$. *Source:* Reproduced with permission from [8]/John Wiley & Sons.

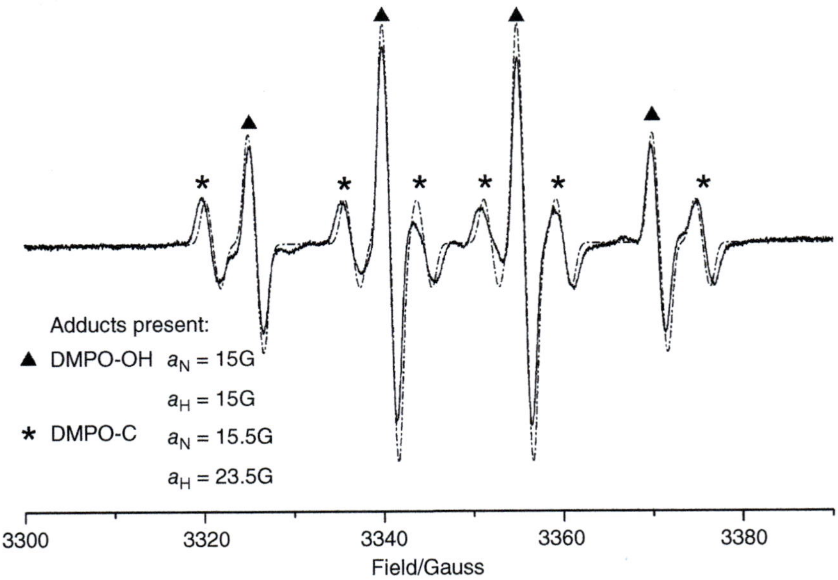

Figure 18.12 Electron paramagnetic resonance (EPR) spectrum showing radical species detected during the reaction of methane and H_2O_2 over a 5.0 wt% Au–Pd/TiO_2 (IW) catalyst with DMPO added to the reaction mixture as the radical trapping agent. Key: black solid line: experimental signal, black dashed line: combined simulated signal for both ·OH and ·CH_3 adducts, ▲: DMPO-OH adduct, and *: DMPOCH_3 adduct (α_N and α_H are the hyperfine coupling constants of the N and H atoms). *Source:* Reproduced with permission from [8]/John Wiley & Sons.

initial product, CH_3OOH, before transforming into the ideal product, CH_3OH. The bound HOO· is formed through the dissociation of H_2O_2 on the catalyst surface. When H_2O_2 was replaced with a mixture of H_2 and O_2, CH_3OOH, CH_3OH, and CO_2 were observed as well, indicating that AuPd/TiO_2 can catalyze the synthesis of H_2O_2 from H_2 and O_2.

References

1 Parfenov, M. V., Starokon, E. V., Pirutko, L. V., Panov, G. I. 2014. "Quasicatalytic and catalytic oxidation of methane to methanol by nitrous oxide over FeZSM-5 zeolite." *J. Catal.* 318, 14–21.

2 Hammond, C., Forde, M. M., Rahim, M. H. A., Thetford, A., He, Q., Jenkins, R. L., Dimitratos, N., Lopez-Sanchez, J. A., Dummer, N. F., Murphy, D. M., Carley, A. F., Taylor, S. H., Willock, D. J., Stangland, E. E., Kang, J., Hagen, H., Kiely, C. J., Hutchings, G. J. 2012. "Direct catalytic conversion of methane to methanol in an aqueous medium by using Copper-Promoted Fe-ZSM-5." *Angew. Chem. Int. Ed.* 51, 5129–5133.

3 Xu, J., Armstrong, R. D., Shaw, G., Dummer, N. F., Freakley, S. J., Taylor, S. H., Hutchings, G. J. 2016. "Continuous selective oxidation of methane to methanol over Cu- and Fe-modified ZSM-5 catalysts in a flow reactor." *Catal. Today* 270, 93–100.

4 Battiston, A. A., Bitter, J. H., Heijboer, W. M., de Groot, F. M. F., Koningsberger, D. C. 2003. "Reactivity of Fe-binuclear complexes in over-exchanged Fe/ZSM5, studied by in situ XAFS spectroscopy: Part 1: Heat treatment in He and O_2." *J. Catal.* 215, 279–293.

5 Jia, J., Sun, Q., Wen, B., Chen, L. X., Sachtler, W. M. H. 2002. “Identification of highly active iron sites in N_2O-activated Fe/MFI.” *Catal. Lett.* 82, 7–11.

6 Battiston, A. A., Bitter, J. H., Koningsberger, D. C. 2003. “Reactivity of binuclear Fe complexes in over-exchanged Fe/ZSM5, studied by in situ XAFS spectroscopy 2. Selective catalytic reduction of NO with isobutane.” *J. Catal.* 218, 163–177.

7 Huang, W., Zhang, S., Tang, Y., Li, Y., Nguyen, L., Li, Y., Shan, J., Xiao, D., Gagne, R., Frenkel, A. I., Tao, F. 2016. “Low-temperature transformation of methane to methanol on Pd_1O_4 single sites anchored on the internal surface of microporous silicate.” *Angew. Chem. Int. Ed.* 55, 13441–13445.

8 Rahim, M. H. A., Forde, M. M., Jenkins, R. L., Hammond, C., He, Q., Dimitratos, N., Lopez-Sanchez, J. A., Carley, A. F., Taylor, S. H., Willock, D. J., Murphy, D. M., Kiely, C. J., Hutchings, G. J. 2012. “Oxidation of methane to methanol with hydrogen peroxide using supported gold–palladium alloy nanoparticles.” *Angew. Chem. Int. Ed.* 52, 1280–1284.

9 Edwards, J. K., Solsona, B. E., Landon, P., Carley, A. F., Herzing, A., Kiely, C. J., Hutchings, G. J. 2005. “Direct synthesis of hydrogen peroxide from H_2 and O_2 using TiO_2-supported Au–Pd catalysts.” *J. Catal.* 236, 69–79.

19

Noncatalytic and Catalytic Oxidation of Methane with O_2 through M–O Clusters Anchored in Zeolite in Liquid at Low Temperature

In contrast to the low-temperature oxidation of CH_4 with H_2O_2 or N_2O catalyzed by zeolite-based catalysts, the oxidation of CH_4 by O_2 has garnered significant attention due to the lower cost associated with O_2. This chapter reviews the catalytic and noncatalytic oxidation of CH_4 by O_2 at low temperatures, focusing on M–O clusters encapsulated in zeolites.

19.1 Cu-ZSM-5

19.1.1 Identification of Reactive Oxygen Species for Oxidizing CH_4 in Cu-ZSM-5 through O_2 Treatment

Driven by the aim to mimic the Cu sites in particulate CH_4 monooxygenase (pMMO) [1], Cu-ZSM-5 activated with O_2 was selected as an example for exploration. Extensive studies were conducted on the partial oxidation of CH_4 to CH_3OH using Cu-ZSM-5 activated with O_2. In 2005, it was reported that reactive oxygen species in Cu-ZSM-5, formed by activating Cu-ZSM-5 in O_2, could partially oxidize CH_4 to produce CH_3OH. Upon oxidation, a band centered at 22 700 cm^{-1} was observed in oxidized ZSM-5 [3]. This peak was attributed to a charge transfer from O_{bridge} to Cu [4]. As illustrated in Figure 19.1, the active oxygen species contributing to the 22 700 cm^{-1} band exhibited decreasing intensity with the rise in temperature when exposed to CH_4 [2]. Therefore, it was established that the oxygen species associated with the 22 700 cm^{-1} band served as the oxidant for CH_4 in the formation of CH_3OH. Generating reactive oxygen species through activation in O_2 requires a critical temperature range. Figure 19.2 demonstrates that a minimum temperature of 350–400 °C is necessary for the formation of these species featured at 22 700 cm^{-1} in the UV–vis spectrum, reaching maximum intensity at 650 °C [2]. Surprisingly, the band nearly disappeared at 750 °C, indicating restructuring or decomposition of the formed reactive oxygen species upon overheating. The optimal temperature for generating these species is within the range of 450–650 °C. Following annealing at 750 °C, Cu-ZSM-5, when annealed again at 450 °C in O_2, could not restore the reactive oxygen species, as evident from the top spectrum marked at 450 °C in Figure 19.2. This suggests that annealing at extremely high

Low-temperature Activation and Catalytic Transformation of Methane to Non-CO_2 Products, First Edition. Franklin Tao.

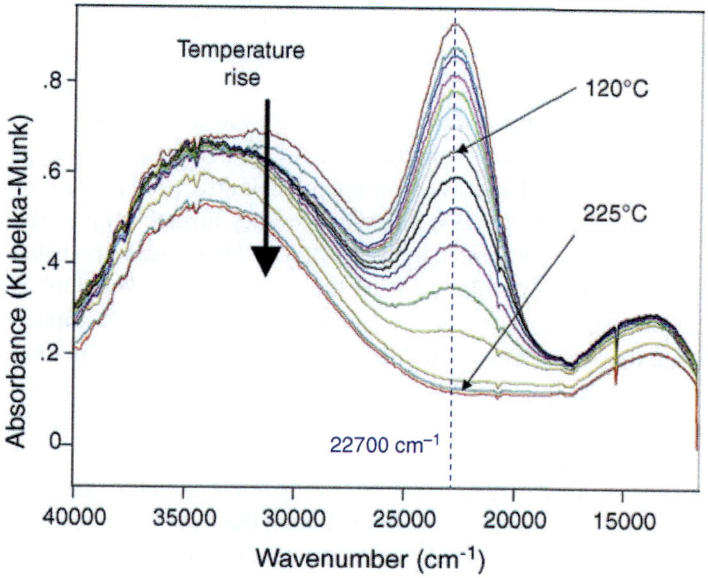

Figure 19.1 Fiber-optic UV–vis spectra of a calcined CZ-12-0.54 sample during reaction with CH_4 at a heating rate of 10 °C min^{-1} from 25 up to 225 °C. The time interval between two spectra is 1.5 min or a temperature difference of 15 °C. The arrow represents the temperature evolution. *Source:* Reproduced with permission from [2] / Elsevier.

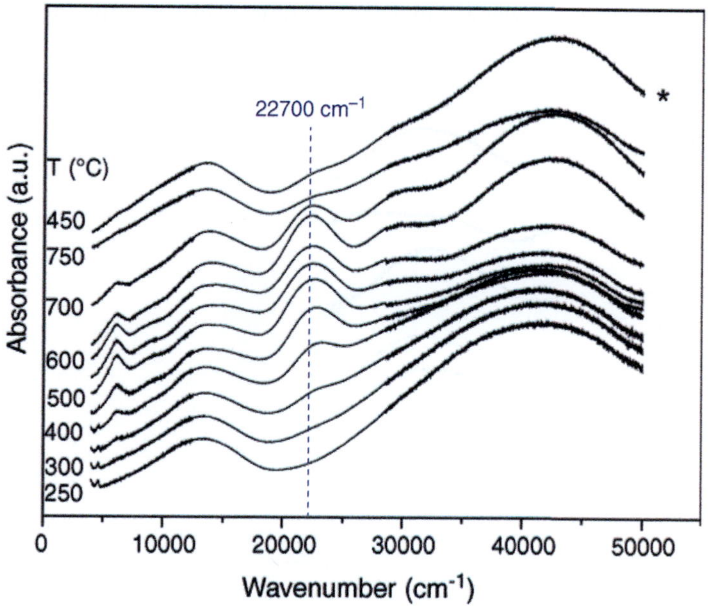

Figure 19.2 Diffuse reflectance spectroscopy (DRS) spectra of the CZ-12-0.54 sample were taken at RT after calcination in O_2 at different temperatures ranging from 250 up to 750 °C. The temperature difference between the two spectra is 50 °C. The spectrum marked with an asterisk (*) is the DRS spectrum of CZ-12-0.54 calcined at 450 °C after calcination at 750 °C. *Source:* Reproduced with permission from [2] / Elsevier.

temperatures renders Cu-ZSM-5 incapable of being re-oxidized, making it impractical to be oxidized by O_2 to form reactive oxygen species or oxidize CH_4 with its oxygen atoms anymore.

Through EXAFS studies, it was determined that the average Cu–Cu distance is 2.87 Å, with the number of Cu atoms in the second coordination shell of Cu being 1 [4]. A bis-(μ-oxo)di-copper core, $[Cu_2(\mu\text{-}O)_2]^{2+}$, was deduced as one of the possible structural models for the reactive oxygen species bonded to Cu in ZSM-5. UV–Vis studies further support the notion that the reactive oxygen species represented by the 22 700 cm^{-1} band are responsible for the partial oxidation of CH_4 to CH_3OH (Figure 19.1) [2]. Here "partial oxidation" is broadly defined to an oxidation of CH_4 to a product except CO_2, the product of a complete oxidation of CH_4. In addition, the Si/Al ratio of Cu-ZSM-5 is significant for the formation of the reactive oxygen species, suggesting that these species are formed between two Cu atoms located on two adjacent Al sites. Figure 19.3 presents the UV–Vis spectra recorded in diffuse reflectance spectroscopy (DRS) mode of Cu-ZSM-5 with different Si/Al ratios: 12 in CZ-12-0.54, 25 in CZ-25-0.51, 30.3 in CZ-30.3-0.47, 77.5 in CZ-77.5-0.55, and CZ-120-0.88. An interesting finding is that at a high Si/Al ratio, the intensities of the band of the reactive oxygen species at 22 700 cm^{-1} are weak. Conversely, at Si/Al ratios as low as 12–30.3, the intensities at 22 700 cm^{-1} are noticeably higher. This suggests that the number of reactive oxygen species strongly depends on the number of Al sites. Furthermore, it implies that the number of reactive oxygen species is closely related to the concentration of Cu atoms in the micropores, given that Cu atoms are introduced to the micropores through ion exchange. There are very few pairs of two adjacent Cu atoms at high Si/Al ratios since the concentration of Cu atoms in the pore is lower at a higher Si/Al ratio. Consequently, only a

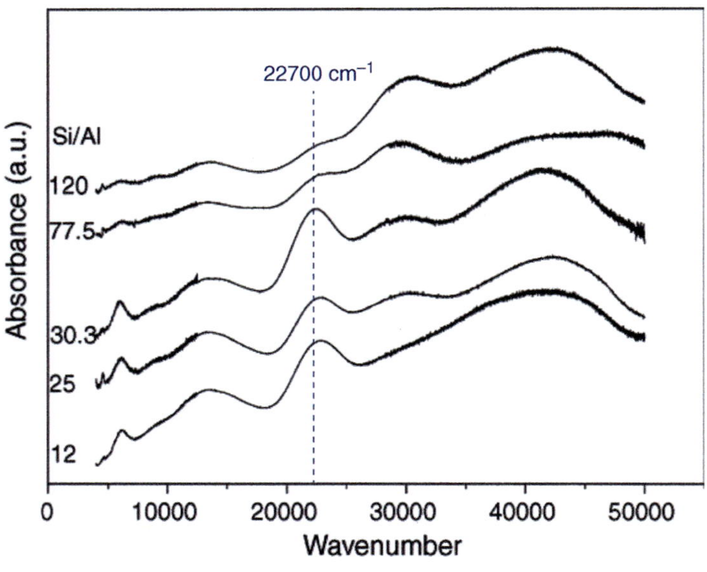

Figure 19.3 Diffuse reflectance spectroscopy (DRS) spectra at room temperature of a Cu-ZSM-5 series with increasing Si/Al ratio (CZ-12-0.54, CZ-25-0.51, CZ-30.3-0.47, CZ-77.5-0.55, and CZ-120-0.88) after calcination in O_2 at 450 °C. *Source:* Reproduced with permission from [2] / Elsevier.

small number of reactive oxygen species could be formed, as their formation requires two adjacent Cu atoms. At low Si/Al ratios, the number of pairs of two adjacent Cu atoms is higher, leading to a higher intensity of the 22 700 cm^{-1} band representing the reactive oxygen species formed between adjacent Cu species, as shown in Figure 19.3 [2].

19.1.2 Confirmed Reactivity of the Formed Oxygen Species in Oxidizing CH_4

In 2005, Cu-ZSM-5 activated under the same conditions as reported in the literature exhibited high activity for oxidizing CH_4 to CH_3OH at 175 °C [4, 5]. A notable yield of 8.2 μmol of CH_3OH per gram of Cu-ZSM-5 was reported. CH_3OH was identified as the predominant product through various analytical techniques, including GC, GC–MS, ^{13}C NMR, and ^{1}H NMR. The observation of ^{18}O and ^{13}C-labeled CH_3OH, resulting from the use of ^{18}O-labeled O_2 and ^{13}C-labeled CH_4, respectively, confirmed that the molecular O_2 used in the step of activation provided the active oxygen atoms, and CH_4 served as the source of carbon atoms in the production of CH_3OH. Remarkably, when Cu-ZSM-5 activated at 25 °C was exposed to CH_4, only 0.81 μmol g^{-1} could be extracted from the reacted Cu-ZSM-5. The appearance of a band centered at 22 700 cm^{-1} in the UV–vis spectrum upon activation by O_2 suggested that the 22 700 cm^{-1} band is contributed by the formed reactive oxygen species in the activation step [4]. This assignment was further supported by the time-dependent weakening of the band at 22 700 cm^{-1}, which completely disappeared within three minutes when the activated Cu-ZSM-5 was exposed to CH_4 at 175 °C, as shown in Figure 19.4 [4]. The assignment of the 22 700 cm^{-1} band in UV–Vis

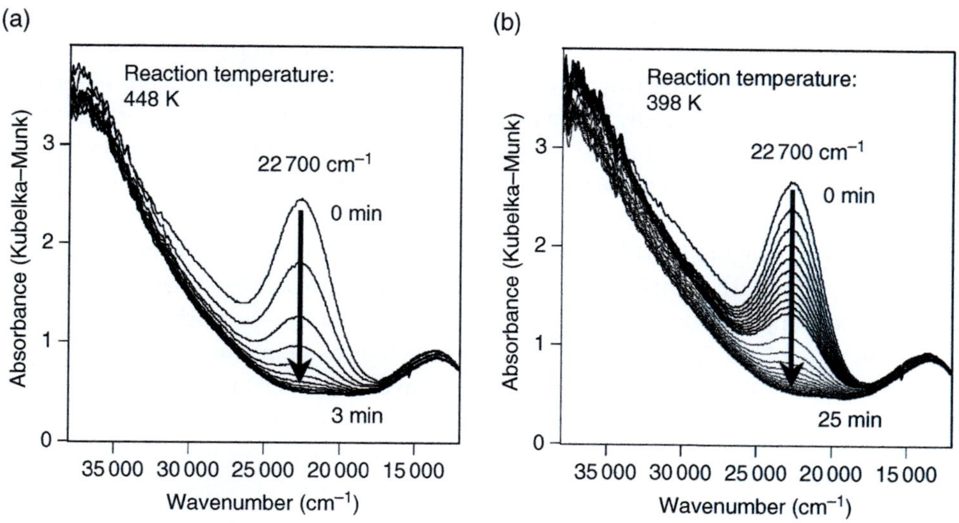

Figure 19.4 Fiber-optic UV–vis spectra of O_2-activated CZ-12-0.58 during reaction between CH_4 (5% in N_2; 25 mL min^{-1}) and oxygen species formed in CZ-12-0.58 at 448 K (a) and at 398 K (b). Time interval between two spectra in full line = 25 s; time interval between two spectra in dashed line = 125 s. Kubelka–Munk = $f(R) = (1 - R)^2/2R$ with R the absolute reflectance of the sample. *Source:* Reproduced with permission from [4] / American Chemical Society.

spectroscopy was corroborated by other reports, where the band was definitively attributed to the charge transfer from Cu to its neighboring O atom(s).

19.1.3 Characterization of the Formed Oxygen Species with Resonance Raman Spectroscopy

In 2009, a comprehensive study employing experimental resonance Raman (rR) spectroscopy and computational studies was conducted [5]. It concluded that the reactive site for oxygen species involved in the partial oxidation of CH_4 to CH_3OH is a bent mono-(μ-oxo)di-cupric complex. This study highlighted the challenging task of distinguishing reactive oxygen species from spectator oxygen species coexisting in the micropores of ZSM-5, making it difficult to achieve a conclusive assignment of the reactive oxygen species. The integrated experimental observations using rR and DFT calculations, however, provided a clear identification of the reactive oxygen species responsible for the formation of CH_3OH.

Figure 19.5a presents rR spectra collected from Cu-ZSM-5 under different treatments: pure $^{16}O_2$ (red in Figure 19.5a), pure $^{18}O_2$ (blue in Figure 19.5a), and a mixture of $^{16}O_2$ and $^{18}O_2$ [5]. This Raman spectroscopy (Figure 19.5a) features the highest peak at 456 cm^{-1} in the red spectrum, attributed to the reactive oxygen species. This assignment is supported by rR spectra collected with different excitation wavelengths (Figure 19.5b), from ZSM-5 with various Cu/Al ratios treated in O_2 (Figure 19.5c), and under different pretreatment conditions (Figure 19.5d). The peak at 456 cm^{-1} was consistently observed for Cu-ZSM-5 with Cu/Al ratios of 0.31, 0.42, and 0.54 in Figure 19.5c, suggesting it is a common feature of these reactive oxygen species in different Cu-ZSM-5 samples. Figure 19.5d conclusively shows that the peaks at 456 and 514 cm^{-1} must be contributed by the reactive oxygen species, as they disappear when activated Cu-ZSM-5 reacts with CH_4 or is heated to 450 °C in He for decomposition. Upon replacing $^{16}O_2$ with $^{18}O_2$, corresponding shifts of these vibrational peaks in the Raman spectrum in Figure 19.5a (red versus blue) were clearly observed, indicating that these peaks truly originate from the reactive oxygen species formed from external O_2 rather than any intrinsic oxygen species of Cu-ZSM-5 [5]. Here, intrinsic oxygen species refer to the oxygen atoms and related species of Cu-ZSM-5 before treatment in O_2 at a certain temperature.

As confirmed in Figure 19.5a, the absence of a peak at about 600 cm^{-1} characteristic of a bis(μ-oxo)di-copper(III) complex (Table 19.1) shows that the reactive oxygen species is not a bis(μ-oxo)di-copper(III) complex. In addition, the observed Cu–O stretching frequencies at 456 and 870 cm^{-1} in Figure 19.5a are distinct from the stretching frequencies of Cu–Cu at 270 cm^{-1} and O–O at 750 cm^{-1} of μ-(η^2: η^2) peroxo di-copper (II) reported in the literature [7], as well as the stretching frequency of O–O at 1050–1100 cm^{-1} of η^2-superoxo copper(II) [8, 9]. To determine whether the reactive oxygen species is a μ-1,2-peroxo or hydroperoxo di-copper(II) species, a mixed isotope $^{16,18}O_2$ was introduced. If the formed reactive oxygen species contains an –O–O– structure, such as the η1-hydroperoxo copper(II) and trans-μ-1,2-peroxo di-copper(OO) in Table 19.1, the Cu-ZSM-5 activated by $^{16,18}O_2$ should have three peaks of stretching mode contributed from –^{16}O–^{16}O–, –^{16}O–^{18}O–, and –^{18}O–^{18}O–, which are 870, 845, and 830 cm^{-1}, respectively, based on the reference frequencies listed in Table 19.1. However, as shown in the green

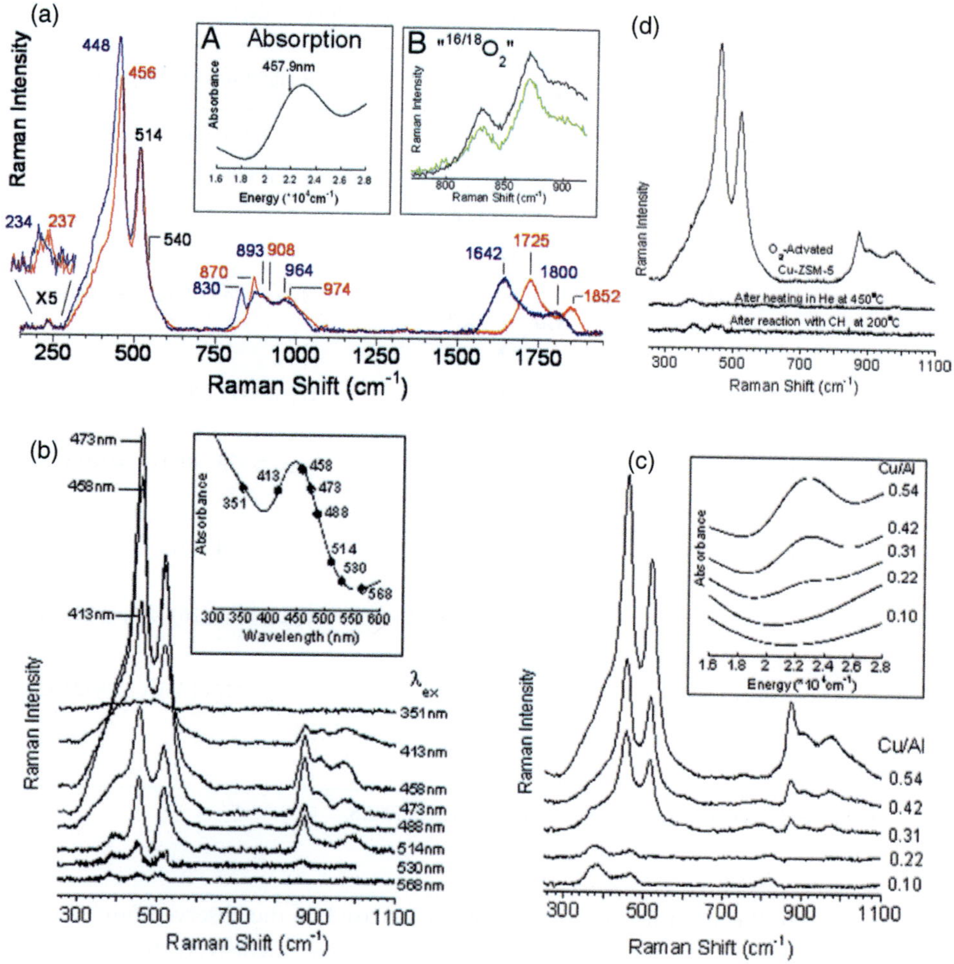

Figure 19.5 Resonance Raman (rR) spectra (λ_{ex} = 457.9 nm) of Cu-ZSM-5 activation by O_2. (a) Resonance Raman (rR) spectra (λ_{ex} = 457.9 nm) of Cu-ZSM-5+ $^{16}O_2$ (red), $^{18}O_2$ (blue). *Inset A*: Absorption spectrum of oxygen-activated Cu-ZSM-5. *Inset B*: "$^{16,18}O_2$" (green), and 1:1 normalized sum of $^{16}O_2$ and $^{18}O_2$ (black). (b) rR spectra of O_2-activated Cu-ZSM-5 (Cu/Al = 0.54) collected at eight λ_{ex} from 351 to 568 nm with corresponding absorption spectrum *Inset*. (c) rR spectra (λ_{ex} = 457.9 nm) of O_2-activated Cu-ZSM-5 with varying Cu/Al ratios from 0.10 to 0.54 with corresponding absorption spectra *Inset*. (d) rR spectra (λ_{ex} = 457.9 nm) of Cu-ZSM-5 (Cu/Al = 0.54) pretreated in O_2 at 450 °C, recorded before and after heating in He at 450 °C and after reaction with CH_4 at 200 °C. *Source:* Reproduced with permission from [5] / Proceedings of the National Academy of Sciences PNAS.

spectrum in inset B of Figure 19.5a, peaks at 870 and 830 cm^{-1} were clearly observed after $^{16,18}O_2$ was used as the oxidant to activate Cu-ZSM-5. However, there is a lack of the peak at 845 cm^{-1} of –^{16}O–^{18}O– in the green spectrum. It suggests there is no –O–O– structure in the formed reactive oxygen species. In fact, the 830 and 870 cm^{-1} peaks in the green spectrum are attributed to the stretching frequencies of Cu–^{18}O and Cu–^{16}O in Cu–O–Cu, respectively. The peak at 1729 cm^{-1} in the red spectrum of Figure 19.5a is the second

Table 19.1 Spectroscopically characterized mononuclear and binuclear Cu/O_2 species.

Cu/O_2 species		rR vibrations ($\Delta^{18}O_2$)/cm^{-1}
O_2 activated Cu-ZSM-5		*456 (8) 870 (40)*
bis(μ-oxo) dicopper(III)	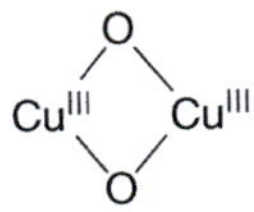	ν_{Cu-O}=606 (23)
μ-(η^2:η^2) peroxo dicopper(II)	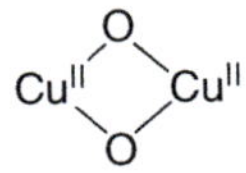	ν_{Cu-Cu}=284 (0) ν_{O-O}=763 (40)
η^1-superoxo copper(II)	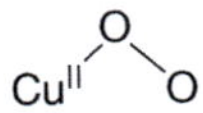	ν_{Cu-O}=472 (20) ν_{O-O}=1121 (63)
η^2-superoxo copper(II)	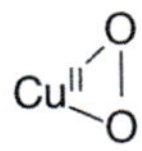	ν_{Cu-O}=554 (20) ν_{O-O}=1043 (59)
η^1-hydroperoxo copper(II)	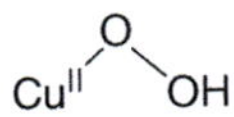	ν_{Cu-O}=624 (17) ν_{O-O}=843 (44)
trans-μ-1,2-peroxo di-copper(II)	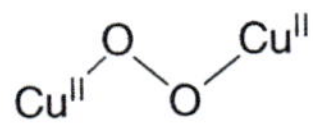	ν_{Cu-O}=561 (26) ν_{O-O}=832 (44)
μ-1,1-hydroperoxo di-copper(II)	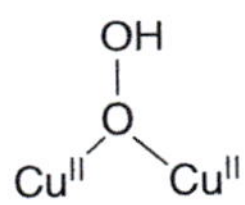	ν_{Cu-O}=322 (10) ν_{O-O}=892 (52)

Source: [5] / Proceedings of the National Academy of Sciences PNAS.

quantum ($2\nu_{as}$) of the asymmetric vibration of Cu–O based on the observed fundamental frequency at 870 cm^{-1}. These detailed rR spectroscopy studies conclusively suggest that the reactive oxygen species formed in the micropores of Cu-ZSM-5 is a mono-μ(oxo)-di-copper complex, $[Cu-O-Cu]^{2+}$ [10].

The geometric structure of the Cu–O–Cu arrangement was investigated using normal coordinate analysis (NCA). A triatomic Cu–O–Cu mode was employed to determine the ∠Cu–O–Cu angle in the mono-μ(oxo)-di-copper complex. The variation in ∠Cu–O–Cu results in changes in symmetric and asymmetric stretching frequencies; Correspondingly, the gap between ν_s and ν_{as}, denoted as an $\Delta^{18}O_2$, varies. The angle corresponding to the observed gap $\Delta^{18}O_2$, ∠Cu–O–Cu is calculated as 140° based on Figure 19.4. In contrast, the previously proposed bis-μ-oxo di-copper complex has a ∠Cu–O–Cu angle of 100°. NCA analysis suggests that the expected frequency for the ∠Cu–O–Cu at 100° for the bis-μ-oxo di-copper complex is approximately 600 cm^{-1} [10]. However, the absence of a peak at 600 cm^{-1} in Figure 19.5a implies that the reactive oxygen species should be assigned to a different structure instead of a bis-μ-oxo di-copper complex, as previously proposed. Based on the confirmed mono-μ-oxo, $[Cu-O-Cu]^{2+}$, a structural model for Cu-ZSM containing the reactive oxygen species was proposed in the literature (Figure 19.6) [5].

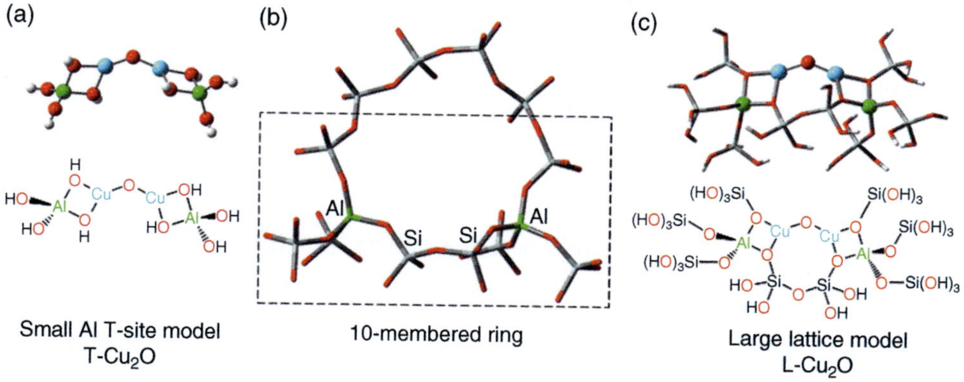

Figure 19.6 Structural models of ZSM-5 and the Cu_2O intermediate used for DFT calculations. (a) Small model, T-Cu_2O. (b) 10-membered ring of ZSM-5 containing two Al-sites separated by two Si-sites (c). Large model constructed from part of a 10-membered ring (boxed atoms, b), L-Cu_2O. *Source:* Reproduced with permission from [5] / Proceedings of the National Academy of Sciences PNAS.

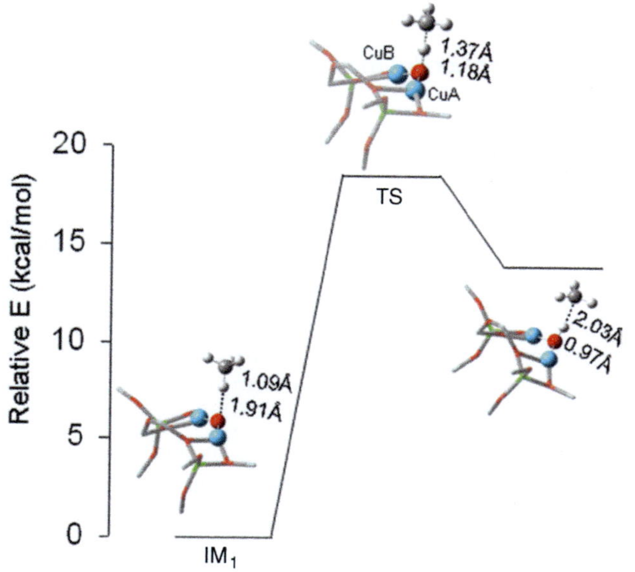

Figure 19.7 DFT-calculated reactivity of L-Cu^{II}_2O with CH_4 and reaction coordinate of H atom abstraction from CH_4 by L-Cu^{II}_2O. *Source:* Reproduced with permission from [5] / Proceedings of the National Academy of Sciences PNAS.

Figure 19.7 presents the intermediate, IM_1 on which a H atom of CH_4 binds to the oxygen atom of the mono-μ(oxo)-di-copper complex, $[Cu–O–Cu]^{2+}$. Considering the proposed intermediate involving the oxidant in Figure 19.7, the elementary step of abstracting a hydrogen atom from CH_4 by an oxygen atom of the intermediate was optimized, revealing a transition state with a barrier of about 16–20 kcal mol^{-1}. Figure 19.7 depicts the energy profile reported in the literature [6]. In this profile, the formed intermediate exhibits a C–H bond distance of 2.03 Å, nearly twice that of the C–H bond in a free-standing CH_4 molecule.

19.2 Cu-MOR

The early investigations of the partial oxidation of CH_4 to CH_3OH on Cu-MOR were reported in 2005 [5]. Cu-MOR with a Cu/Al ratio of 0.43 and a Si/Al ratio of 8.8, was activated at 450 °C in O_2. Subsequently, the appearance of a band at 22 200 cm^{-1} was observed in UV–Vis–NIR DRS spectroscopy during this activation. However, this band disappeared when the activated Cu-MOR was exposed to CH_4 at 200 °C. Clearly, a reactive oxygen species must have formed in Cu-MOR through activation in O_2. The yield of CH_3OH from Cu-MOR is 11.34 $\mu mol\,g^{-1}$ of Cu-MOR. It is worth noting that the intensity of the 22 200 cm^{-1} band was barely observed for Cu-MOR with Si/Al ratios of 141 upon activation at 450 °C in O_2. This disparity between Cu-MOR with a Si/Al ratio of 8.8 and Cu-MOR with Si/Al ratios of 141 suggests that the formation of reactive oxygen species is closely related to the Cu sites introduced through ion exchange with Al^{3+} in Cu-MOR.

19.2.1 Correlation between Pretreatment Condition and Structure of Active Copper Sites

In literature [11], Cu-MOR for the partial oxidation of CH_4 to CH_3OH was synthesized through ion exchange between Cu ions and MOR at room temperature, followed by calcination at 450 °C in air, and reduction in H_2 at 200 °C. The oxidation state of Cu during these pretreatment steps was monitored using XANES. Figure 19.8 depicts the XANES spectra of Cu-MOR. The black line in Figure 19.8 represents the edge feature of Cu atoms after ion exchange, where Cu atoms exhibit an oxidation state of II; The edge feature indicates the formation of hydrated Cu(II) sites, with a weak pre-edge at 8977 eV suggesting the oxidation state of Cu(II) and the major edge at 8997 eV indicating coordination to OH or H_2O [11].

The spectrum in energy space after calcination at 450 °C in air is shown in red dashed lines in Figure 19.8. In comparison to the hydrated Cu(II) species in the black spectrum, the edge at 8997 eV suggests dehydration, and the rising edge at 8986.3 eV indicates the formation of CuO-like Cu sites. The CuO-like species could be oligonuclear Cu(II) species, as no XRD pattern of CuO was observed [11]. Importantly, the EXAFS spectrum derived from the XAS data does not suggest the formation of large CuO clusters, confirming that Cu atoms did not form CuO clusters after calcination in O_2. [11] Conclusive information indicates the oxidation number of Cu atoms after calcination in air is II. According to a subsequent Raman spectroscopy study, the Cu atoms formed mono(μ-oxo)-di-copper species, a structure formed in Cu-ZSM-5 as well.

The blue spectrum in Figure 19.8 was obtained after the reaction with CH_4 at 200 °C. In comparison to the red spectrum collected after calcination in O_2, the blue spectrum exhibits a new feature at 8983.6 eV. Based on the literature [12, 13], this new feature results from the 1s→4p transition of Cu(I). The observation of Cu(I) upon the partial oxidation of CH_4

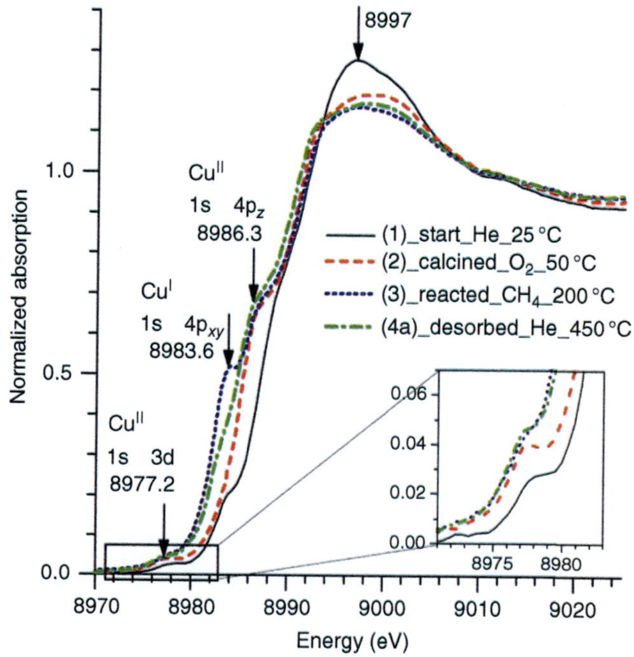

Figure 19.8 XANES spectra of Cu-MOR were recorded at key conditions of the methane conversion reaction, including starting material (1), calcined in oxygen (2), reacted with methane (3), and after heating in helium (4a) to release the adsorbed intermediate. *Source:* Reproduced with permission from [10] / American Chemical Society.

is consistent with the reduction of Cu(II) by CH_4. This assignment was confirmed by the progressive decrease of the 8983.6 eV signal assigned to Cu(I) along with the slight increase in the 8993 eV edge of Cu^{2+} during dry activation with O_2 or wet activation with H_2O, as illustrated in Figure 19.9 [11].

Based on the XANES and EXAFS studies, literature proposed the evolution of the chemical and coordination environments of Cu atoms, as depicted in Figure 19.10 [11]. The scheme suggests that Cu atoms exist in a hydrated copper state with an oxidation state of +2 after aqueous ion exchange. In structure (1) of Figure 19.10, Cu is bonded to four oxygen atoms, which can be oxygen atoms of H_2O, OH, O–Al, or O–Si. These oxygen atoms form strong bonds with H, Al, or Si, rendering the hydrated Cu species incapable of activating the C–H bond of CH_4. The crucial step is the calcination at 450 °C in O_2, which transforms the hydrated Cu species into mono(μ-oxo)di-copper species (step 2 in Figure 19.10). In the mono(μ-oxo)dicopper species, the oxygen atoms situated between two Cu(II) atoms become highly active, capable of activating the C–H bond of CH_4 upon the C–H dissociation, the H and CH_3 species bond to active oxygen atoms of one or two mono(μ-oxo)dicopper species, forming a hydroxyl group bonded between two Cu atoms [one Cu(I) and one Cu(II)] and a methoxy group between two Cu atoms [one Cu(I) and one Cu(II)] (step 3 in Figure 19.10). This process is termed the partial oxidation of CH_4 to CH_3OH. The methoxy species can be hydrolyzed to form CH_3OH in a mixture of H_2O and O_2 at 200 °C (step 4b in Figure 19.10).

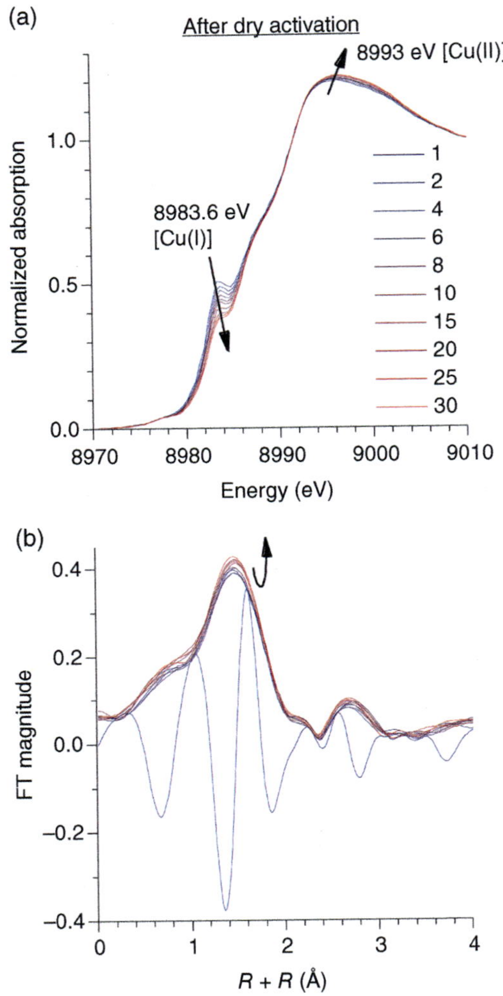

Figure 19.9 Comparison of (a) the XANES and (b) the nonphase corrected k^3 weighted FT EXAFS spectra taken during desorption of methanol after dry activation. For clarity, only the first FT EXAFS spectrum is displayed with an imaginary part. Subsequent FT EXAFS spectra did not display observable changes in their imaginary parts. *Source:* Reproduced with permission from [10] / American Chemical Society.

Based on step 3 in Figure 19.10, half of the Cu atoms bonding to O(–H) or O–(CH_3) are Cu(I) atoms. The observation of Cu(I) by XANES (blue spectrum in Figure 19.8) suggests that the Cu(I)–O(CH_3)–Cu(II) structure formed upon the partial oxidation of CH_4 is stable. Computational studies support the observation of the Cu(I)–O(CH_3)–Cu(I) structure as binding CH_3 and H to [Cu(II)–O–Cu(II)] to form [Cu(I)–O(CH_3)–Cu(II)] and [Cu(I)–O(H)–Cu(II)], respectively, is exothermic by 50 kcal mol^{-1}. The calculated energy profile for the reaction between CH_4 and [Cu(II)–O–Cu(II)] is schematically shown in Figure 19.11 [11]. The computational study also calculated (1) the binding energy of H and CH_3 on the O atom between two Cu atoms, (2) adsorption energy of H on one O atom

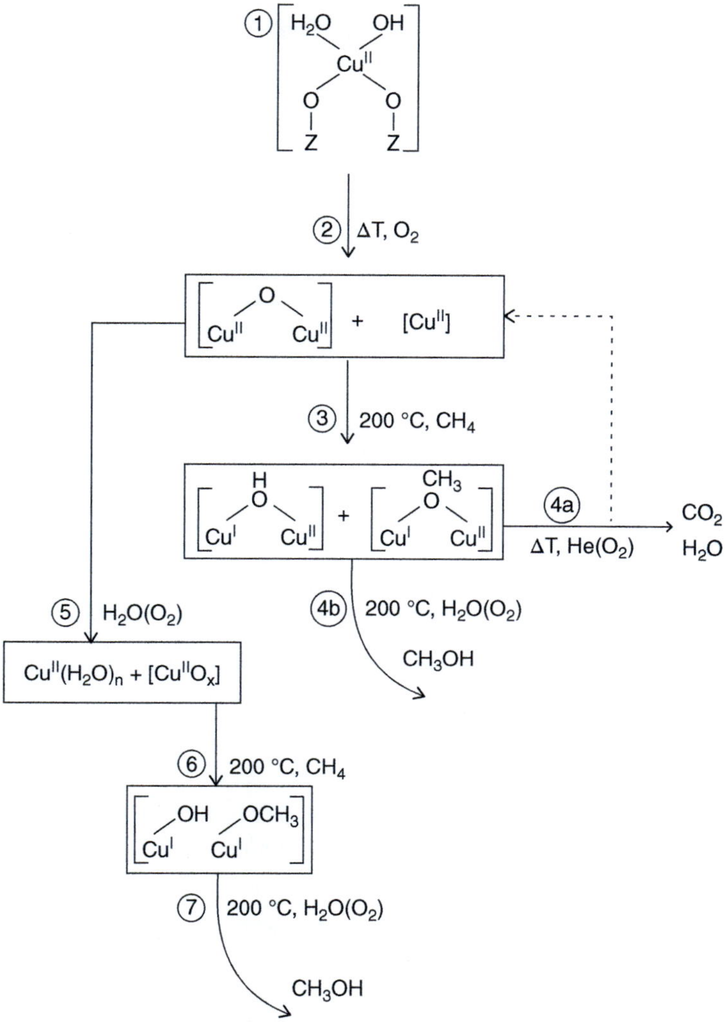

Figure 19.10 Proposed scheme of structural changes of the active Cu species in Cu-mordenite based on former studies, mass spectrometry, UV-visible spectrometry, infrared spectroscopy, and quick XAS data. Cu-MOR after aqueous ion exchange (1) is composed of hydrated Cu^{II} sites. Thermal treatment in (dry) oxygen (2) forms the water-sensitive di-copper active sites. Reaction with methane (3) reduces a significant fraction of the copper sites and forms the sorbed reaction intermediate, which may be removed by heating (4a) as CO_2 or by water-assisted desorption (4b) as MeOH. Treatment of oxygen-activated Cu-MOR with water (5) leaves a water-stable Cu^{II} oxide species that activates methane (6–7). *Source:* Reproduced with permission from [10] / American Chemical Society.

between two Cu atoms, and (3) binding energy of CH_3 on the O atom between the two Cu atoms of another mono(μ-oxo)di-copper species. It was found that the total of binding energies of H and CH_3 on the same O atom of a mono(μ-oxo)di-copper species is lower than the total of adsorption energies of H on one O atom of one mono(μ-oxo)di-copper species and CH_3 on the O atom of another mono(μ-oxo)di-copper species by

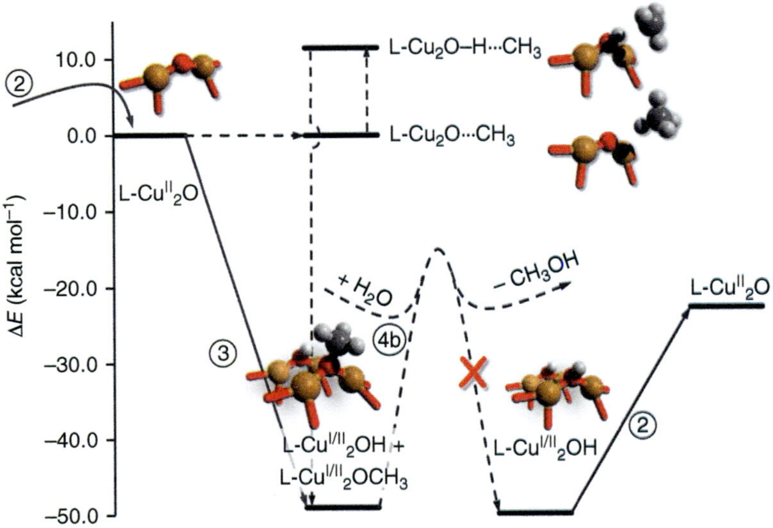

Figure 19.11 B3LYP/6-311+G(d) energy diagram for the di-copper active site L-Cu_2O in the methane to methanol conversion. Solid arrows indicate processes supported by experimental and computational evidence. The presumably short-lived physiosorbed and chemisorbed intermediates are not observed experimentally, and no transition state or final product is reported for Cu_2O-CH_3 hydrolysis. The catalytic cycle is stopped by the irreversible hydration of the active L-Cu_2O site after methanol desorption. *Source:* Reproduced with permission from [10] / American Chemical Society.

38–40 kcal mol^{-1} [11]. In addition, the pathway for the hydrolysis of [Cu(I)–OCH_3–Cu(II)] through a ring-like [Cu_2O···H···OH···CH_3] transition state was simulated.

19.2.2 Formation of Momo(μ-oxo)di-copper Species in Cu-MOR through Activation at 450 °C in O_2

The investigation of bis(μ-oxo) di-copper species and mono(μ-oxo) di-copper species in MOR was conducted through XAS analysis and computational studies [13]. Figure 19.12 schematically illustrates the evolution from hydrated copper sites formed during ion exchange to the formation of mono(μ-oxo) di-copper species in Cu-zeolites [13]. Temperature-dependent XAS indicates that the dehydration step is completed at 200 °C. In the subsequent thermal treatment in O_2, di-copper species were formed at 450 °C. This study delves into how the hydrated copper sites transform into mono(μ-oxo) di-copper species in MOR [13]. The calcination of Cu-MOR during ion exchange leads to the loss of coordinating water or OH ligands. With the removal of H_2O and OH groups, disproportionation and condensation processes take place [13]. According to the literature [14], in disproportionation, an OH· radical is lost, resulting in the formation of Cu(I) species. Simultaneously, another $[Cu(II)(OH)_2]^+$ loses a H_2O molecule to form $[Cu(II)O]^+$. As depicted in Figure 19.12, the condensation process involves two steps. In the first step, two adjacent [Cu(II)OH] species couple to form [Cu(II)–O–Cu(II)] through the elimination of an H_2O molecule. The second step is the auto-reduction of [Cu(II)–O–Cu(II)] to form a

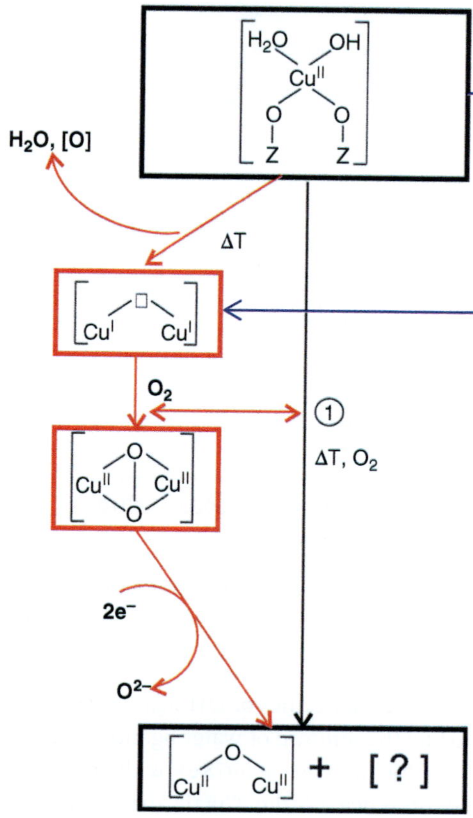

Figure 19.12 Chemical scheme showing the formation of active sites in Cu-MOR. Thermal treatment of Cu-MOR induces dehydration of the copper sites, which is the driving force for autoreduction. The reduced and undercoordinated copper sites cleave molecular oxygen to form the peroxo precursor, followed by elimination of an oxygen atom to form the methane activating site, mono(μ-oxo) di-copper species. *Source:* Reproduced with permission from [13] / Royal Society of Chemistry.

structure with an oxygen vacancy [Cu(I)–□–Cu(I)] by losing an oxygen atom. As shown in Figure 19.12, the Cu atoms in the structure with an oxygen vacancy, [Cu(I)–□–Cu(I)] are undercoordinated. This structure is highly active for cleaving O=O of molecular O_2 to form di-copper species with bridge oxygen atoms, $[Cu–O–Cu]^{2+}$. According to literature [15], studies of UV-visible and resonant Raman spectroscopy on the preparation of Cu-ZSM-5 for the partial oxidation of CH_4 to CH_3OH suggest the formation of a precursor, (μ-η^2:η^2-peroxo)di-copper (II), and then readily restructuring to form mono(μ-oxo)di-copper (II) species, supporting the proposed pathway of formation of active oxygen species, mono(μ-oxo) di-copper species.

Figure 19.13 presents the stacked XANES spectra of Cu K-edge under various thermal treatments in O_2 (Figure 19.13a) and Helium (Figure 19.13b). In Figure 19.13a, a systematic change is evident. With an increase in treatment temperature in O_2, the absorption feature at 9000 eV progressively decreases, indicating a gradual alteration in the chemical

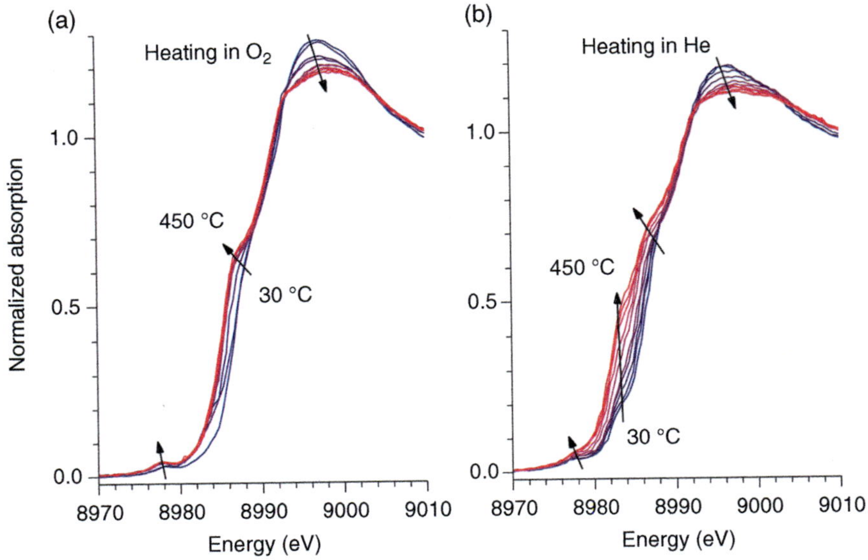

Figure 19.13 XANES spectra were recorded during thermal treatments of Cu-MOR either in (a) oxygen or (b) helium spaced every 50 °C. *Source:* Reproduced with permission from [13] / Royal Society of Chemistry.

environment of Cu atoms. It indicates the loss of water molecules or OH ligands, similar to the decrease in intensity at 9000 eV observed during the removal of water ligands from copper nitrate salt [16]. In addition, the observation of the pre-edge at 8978 eV with an increase in temperature in Figure 19.13 suggests the presence of the Cu(II) oxidation state. Furthermore, the pre-edge increases in intensity and shifts down in energy with the rise in treatment temperature (Figure 19.13a), implying the transformation of copper species from hydrated copper to Cu(II)O-like species [13, 17]. Deconvolution of the collected XANES spectra of Cu K-edge suggests that the fraction of the hydrated Cu(II) component decreases, and the dehydrated Cu(II) component increases to 80% of all copper species along the increase of temperature until 200 °C. In addition, analysis of the r-space spectra of Cu K-edge indicates that the loss of water molecules and OH groups causes the copper atoms to bond with the oxygen atoms of the internal surface of the pore, and annealing to 450 °C in O_2 makes the formation of copper dimers possible.

Notably, the debate on whether the copper dimers are mono(μ-oxo)di-copper, bis(μ-oxo) di-copper, or both remains. This is because the XAS technique cannot distinguish between a mono(μ-oxo)di-copper species and bis(μ-oxo)di-copper species, as XAS measures the average structural parameters of different copper-related species, including mono(μ-oxo) di-copper and bis(μ-oxo)di-copper species. Computational studies were employed to explore the stabilities of mono(μ-oxo)di-copper and bis(μ-oxo)di-copper species encapsulated in MOR [13]. For comparison, the stabilities of mono(μ-oxo)di-copper and bis(μ-oxo)di-copper molecular complexes were studied through DFT calculations. Figure 19.14 presents the optimized structures of mono(μ-oxo)di-copper molecular complex (a), bis(μ-oxo)di-copper molecular complex (b), another mono(μ-oxo)di-copper molecular complex (c), another bis(μ-oxo)di-copper molecular complex (d), di-copper species

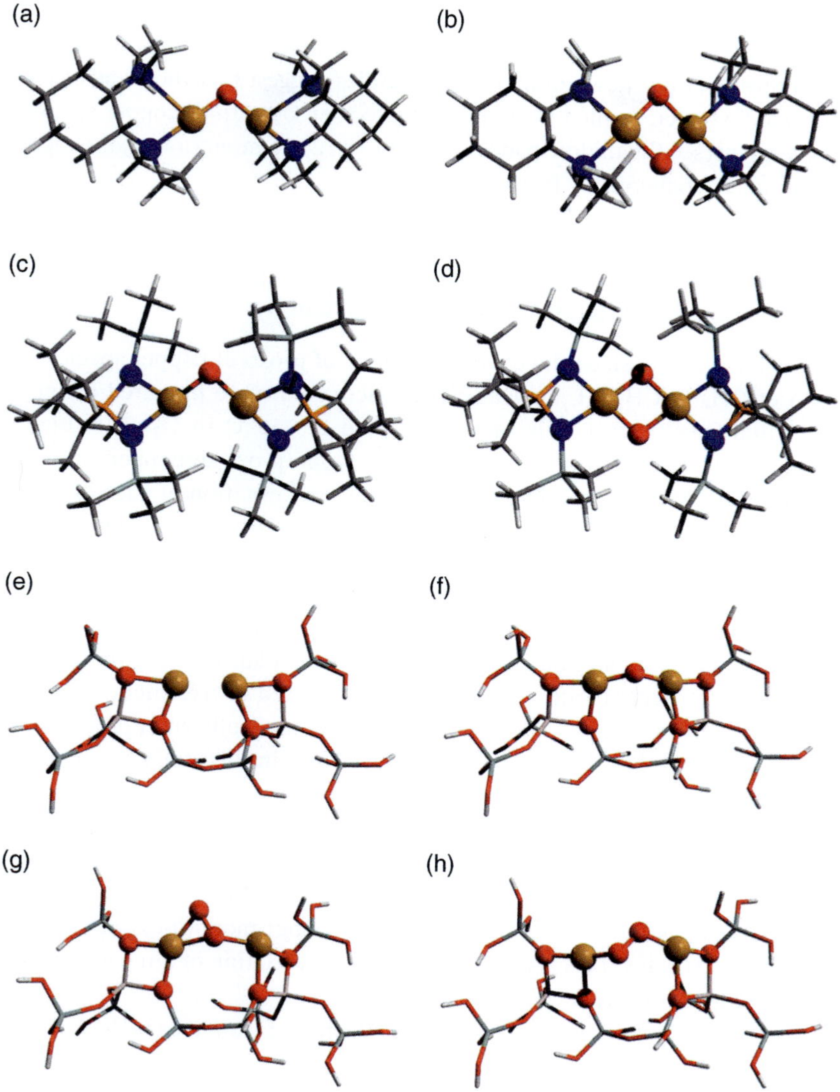

Figure 19.14 Optimized di-copper complexes and LLM structures including mono(μ-oxo)di-copper molecular complex (a), bis(μ-oxo)di-copper molecular complex (b), another mono(μ-oxo)di-copper molecular complex (c), another bis(μ-oxo)di-copper molecular complex (d), di-copper species with oxygen vacancy in MOR (e), mono(μ-oxo)di-copper species anchored in MOR (f), μ-(η^1: η^2)-peroxo-di-copper species anchored in MOR (g), and trans-μ-1,2-peroxo-dicopper species anchored in MOR (h). *Source:* Reproduced with permission from [13] / Royal Society of Chemistry.

with oxygen vacancy in MOR (e), mono(μ-oxo)di-copper species anchored in MOR (f), μ-(η^1: η^2)-peroxo-di-copper species anchored in MOR (g), and trans-μ-1,2-peroxo-dicopper species anchored in MOR (h) [13]. DFT calculations suggest that the transformations of bis(μ-oxo)di-copper species anchored in MOR (g or h) to mono(μ-oxo)di-copper species anchored in MOR (f) are thermodynamically favorable as they are exothermic [13].

In contrast, the transformations from bis(μ-oxo)di-copper molecular complexes (c or d) to mono(μ-oxo)di-copper molecular complexes (a or b) are thermodynamically unfavorable as they are endothermic [13]. These computational studies suggest that the transformation from bis(μ-oxo)di-copper, formed by coupling molecular O_2 with di-copper species with an oxygen vacancy (see the middle panel in Figure 19.12), to mono(μ-oxo)di-copper species in MOR is thermodynamically favorable [13].

19.2.3 Formation of Copper Oxide Clusters Instead of Momo(μ-oxo) di-copper Species in Cu-MOR through Activation at 200 °C in O_2

Numerous reports have emphasized the catalytic activity of μ-oxo di-copper species in partial oxidation of CH_4 to CH_3OH. Typically, these active sites are prepared using a specific protocol that involves a crucial activation step at 450 °C in O_2. This high-temperature activation results in the formation of mono(μ-oxo)di-copper species anchored in the pores of MOR. The evidence supporting the formation of such species is the observation of a distinctive band at 22 000 cm^{-1} in the UV–vis spectrum during activation and the subsequent decay and disappearance of this band upon the partial oxidation of CH_4 to CH_3OH. The role of high-temperature activation of Cu-MOR in O_2 and the structural evolution of Cu atoms in Cu-MOR during this process have been characterized using XAS and simulated through DFT calculations (Figures 19.10 and 19.11). However, in a recent study, Cu-MOR was subjected to treatment in O_2 at 200 °C instead of the conventional 450 °C, aiming to develop an isothermal process for producing CH_3OH from CH_4 [18, 19]. A noteworthy discovery in this study is that activation in O_2 at 200 °C does not lead to the formation of mono(μ-oxo)di-copper species in MOR, even though the resulting Cu-MOR remains active for the partial oxidation of CH_4 to CH_3OH [18]. The absence of any μ-oxo di-copper species in the Cu-MOR activated at 200 °C in O_2 is confirmed by the lack of the band at 22 000–22 700 cm^{-1}. Evidently, the active phase responsible for the oxidation of CH_4 is neither mono(μ-oxo)di-copper nor bis(μ-oxo)di-copper species. This finding suggests that a variation of annealing temperature could create quite different catalytic sites.

The active phase of Cu-MOR upon activation at 200 °C in O_2 is found to be less active than that resulting from activation at 450 °C in O_2. Transmission electron microscopy (TEM) studies of Cu-MOR activated at 200 °C reveal sub-nanometer copper oxide clusters dispersed on or in Cu-MOR. Figure 19.15a clearly illustrates low-contrast dots, attributed to copper oxide clusters with a size of <1 nm [18]. Further evidence for the formation of copper oxide clusters can be observed in HAADF–STEM images in Figure 19.16 [18]. The consistent yield of CH_3OH from the partial oxidation of CH_4 over four cycles suggests the viability of the isothermal process, which includes 200 °C activation in O_2, partial oxidation of CH_4 by the formed copper oxide clusters at 200 °C, and the extraction of CH_3OH with steam at 200 °C [18].

The absence of μ-oxo formation during activation of Cu-MOR at 200 °C in O_2 is not surprising since the precursor [Cu(II)–O–Cu(II)], essential for di-copper species formation with an oxygen vacancy [Cu(I)–□–Cu(I), middle panel in Figure 19.12], can only be generated through a condensation process of hydroxyl groups at temperatures of 300 °C or

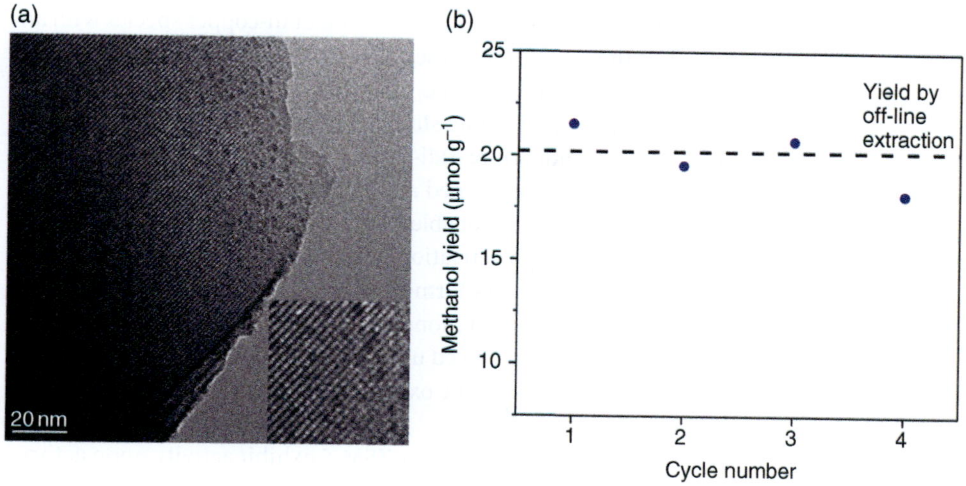

Figure 19.15 Cycling of the isothermal process at 200 °C. (a) TEM image of Cu-MOR activated at 200 °C, showing the intact pore structure of the zeolite and small copper oxide particles. (b) Methanol yields after consecutive cycles. In each cycle, the yield (about 20 μmol g^{-1}) was stable. A cycle consisted of activation for 8 h at 1 bar of oxygen, reaction with methane at 6 bar, and extraction with steam. The liquid was collected by condensation of the reactor effluent. *Source:* Reproduced with permission from [18] / John Wiley & Sons.

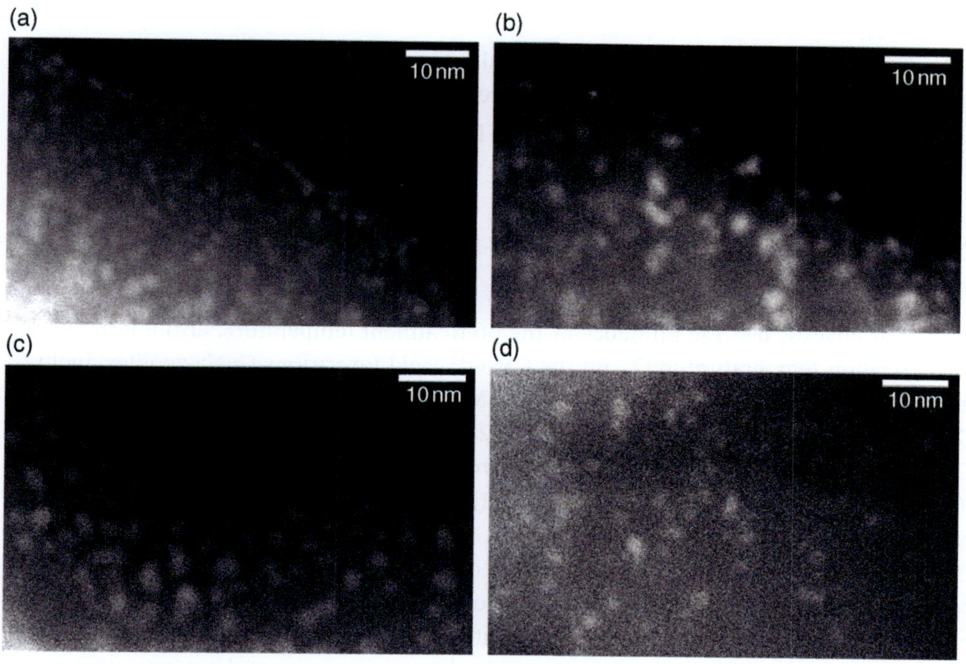

Figure 19.16 HAADF-STEM images of Cu-MOR (a) after activation in oxygen at 200 °C, (b) after one cycle at 200 °C and off-line extraction, (c) after two consecutive 100 nm cycles including on-line extraction and (d) after four consecutive cycles including on-line extraction. *Source:* Reproduced with permission from [18] / John Wiley & Sons.

higher. This precursor acts as an intermediate in the formation of di-copper species with an oxygen vacancy, which serves as the site for the subsequent binding of molecular O_2 to create bis(μ-oxo)di-copper species as an intermediate and then morph to mono(μ-oxo)di-copper species. Due to the lower thermodynamic stability of the intermediate, bis(μ-oxo) di-copper species, it transforms into mono(μ-oxo)di-copper species (Figure 19.10). Given that the activation of Cu-MOR in O_2 was performed at 200 °C, the formation of the critical intermediate [Cu(I)–□–Cu(I)] becomes improbable, and consequently, mono(μ-oxo)di-copper species cannot be generated through activation at 200 °C.

As the sub-nanometer copper oxide clusters formed through activation of Cu-MOR at 200 °C in O_2 exhibit lower activity compared to mono(μ-oxo)di-copper species, the reduction of oxygen atoms of these clusters was achieved using high-pressure CH_4 at 6 bars [18], although 0.05 bar CH_4 is sufficient for reducing the oxygen atoms of mono(μ-oxo)di-copper species anchored in Cu-MOR [13].

Similarly, copper oxide clusters encapsulated in Cu-ZSM-5 exhibit activity upon activation at >300 °C in O_2, as evidenced by the clear observation of the 22 700 cm^{-1} band [2]. Similar to Cu-MOR, when Cu-ZSM-5 was treated in O_2 at 200 °C, no (μ-oxo)di-copper species were formed, as indicated by the absence of the band at 22 700 cm^{-1}. While the Cu-ZSM-5 activated at 200 °C in O_2 exhibits activity CH_4 at 32 bar for the partial oxidation of CH_4 with a yield of 17.7 μmol g^{-1} of catalyst [18], it is not active at 0.01 bar CH_4. Similar to Cu-MOR activated at 200 °C in O_2, the low activity of the sub-nanometer copper oxide cluster in Cu-ZSM-5 necessitates high pressure, such as 32 bar, for the formation of CH_3OH.

As depicted in Figure 19.15b, the isothermal activation of Cu-zeolite at 200 °C, followed by partial oxidation of CH_4 at 200 °C, can be repeated over several cycles without a loss of CH_3OH yield. This suggests that it could be a promising protocol for energy-efficient production of CH_3OH from CH_4 [18].

19.2.4 How Activation Temperature of Cu-MOR in O_2 in 350–550 °C Influence Activity or Reactivity of Cu-MOR

Park et al. investigated the influence of the pretreatment temperature, also referred to as the activation temperature in O_2, on the CH_3OH yield for various CuNa-zeolites, including CuNa-MordeniteH (Cu/Al = 0.38), CuNa-MordeniteH (Cu/Al = 0.08), CuNa-ECR-1H, CuNa-omegaH, CuNa-PST-11H, CuNa-UZM-22H, and CuNa-UZM-4H [25]. In the case of CuNa-MordeniteH (Cu/Al = 0.38), CuNa-MordeniteH (Cu/Al = 0.08), CuNa-MordeniteH (Cu/Al = 0.08), and CuNa-UZM-4H, the yield (μmol mol^{-1} Cu) increased with the elevation of pretreatment temperature in 350–550 °C. However, CuNa-ECR-1H, CuNa-omegaH, CuNa-PST-11H, and CuNa-UZM-4H activated at 550 °C exhibit lower methanol yields compared to samples activated at lower temperatures; In these four Cu-zeolites, a significant amount of copper species may have sintered at the high-activation temperature of 550 °C, resulting in the loss of active sites; This phenomenon occurred because copper-oxo species are the main contributor to the activity in the oxidation of CH_4 to CH_3OH. Although copper oxide clusters are active, their activity is lower. Hence, maintaining a moderate activation temperature is crucial for Cu-zeolites with a substantial copper content. Similar trends have been observed in other metal-zeolites [2, 18].

19.3 Ni-ZSM-5

Recently, Tao et al. reported that the anchored Ni atoms can form mono(μ-oxo)dinickel species in the micropores of ZSM-5 by activating Ni-ZSM-5 in O_2 [20]. The mono(μ-oxo) di-nickel species exhibits activity in the partial oxidation of CH_4 to CH_3OH. The synthesis of Ni-ZSM-5 with mono(μ-oxo)di-nickel species involved several steps. Initially, H-ZSM-5 was prepared from purchased NH_4-ZSM-5 with a Si/Al ratio of 15 at 400 °C. Given the known activity of Fe-ZSM-5 and Cu-ZSM-5 in CH_4 partial oxidation, the concentration of Fe and Cu in the commercial NH_4-ZSM-5 was considered seriously. Inductively coupled plasma (ICP) studies revealed that there is no Cu in ZSM-5, and the concentration of Fe is only 0.0019 wt%. Consequently, the contribution of Cu-based and Fe-based species to the CH_4 partial oxidation was deemed negligible. In the second step, Ni precursor was introduced to H-ZSM-5 via impregnation. The third step involved filtration and subsequent washing with deionized water, with the filtered sample designated as "Ni-ZSM-5 with filtration" and the nonfiltered sample as "Ni-ZSM-5 without filtration." The consequence of filtration is the removal of Ni cations adsorbed on the external surface of ZSM-5 particles as these cations could be readily oxidized into nickel oxide clusters in calcination. These two sample types exhibited distinct surface chemistries. In the fourth step, the samples were dried in a vacuum oven at 60 °C overnight, followed by calcination in air at 550 °C for 3 hours. The fifth step involved the activation of Ni-ZSM-5 in pure O_2 or 5% O_2 balanced with Ar at 450–700 °C. The sixth step is the partial oxidation of CH_4 with Ni-ZSM-5 at 125 –300 °C in a fixed bed flow reactor with a flow rate of 20 mL min^{-1}. The final step was the extraction of formed methanol through a hydrolysis process in an aqueous solution.

Figure 19.17 illustrates that filtration during ion exchange effectively removed Ni^{2+} cations. Thus, no NiO clusters were observed in 2.5 wt% Ni-ZSM-5 upon calcination in air at 550 °C for the sample prepared with filtration in Figure 19.17a and b [20]. Without filtration, NiO clusters formed on the surfaces of ZSM-5 particles, as depicted in Figure 19.17c. Both XRD and TEM confirmed the absence of NiO nanoclusters in a sample (with filtration) upon calcination at 500 °C in air. Figure 19.18a presents Ni 2p spectrum of NiO nanoparticles, 2.5wt%Ni-ZSM-5 before sputtering and 2.5wt%Ni-ZSM-5 after sputtering. The lack

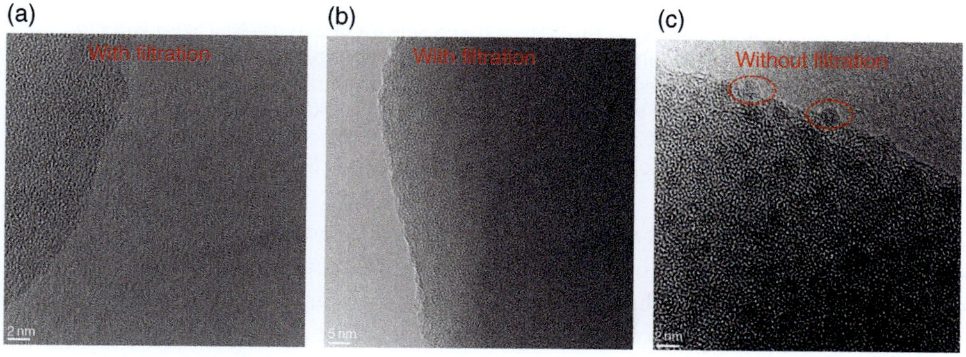

Figure 19.17 TEM images of 2.5 wt% Ni-ZSM5 catalysts. (a, b) 2.5 wt% Ni-ZSM5 catalyst with filtration. (c) 2.5 wt% Ni-ZSM5 catalyst without filtration. *Source:* Reproduced with permission from [20] / American Chemical Society.

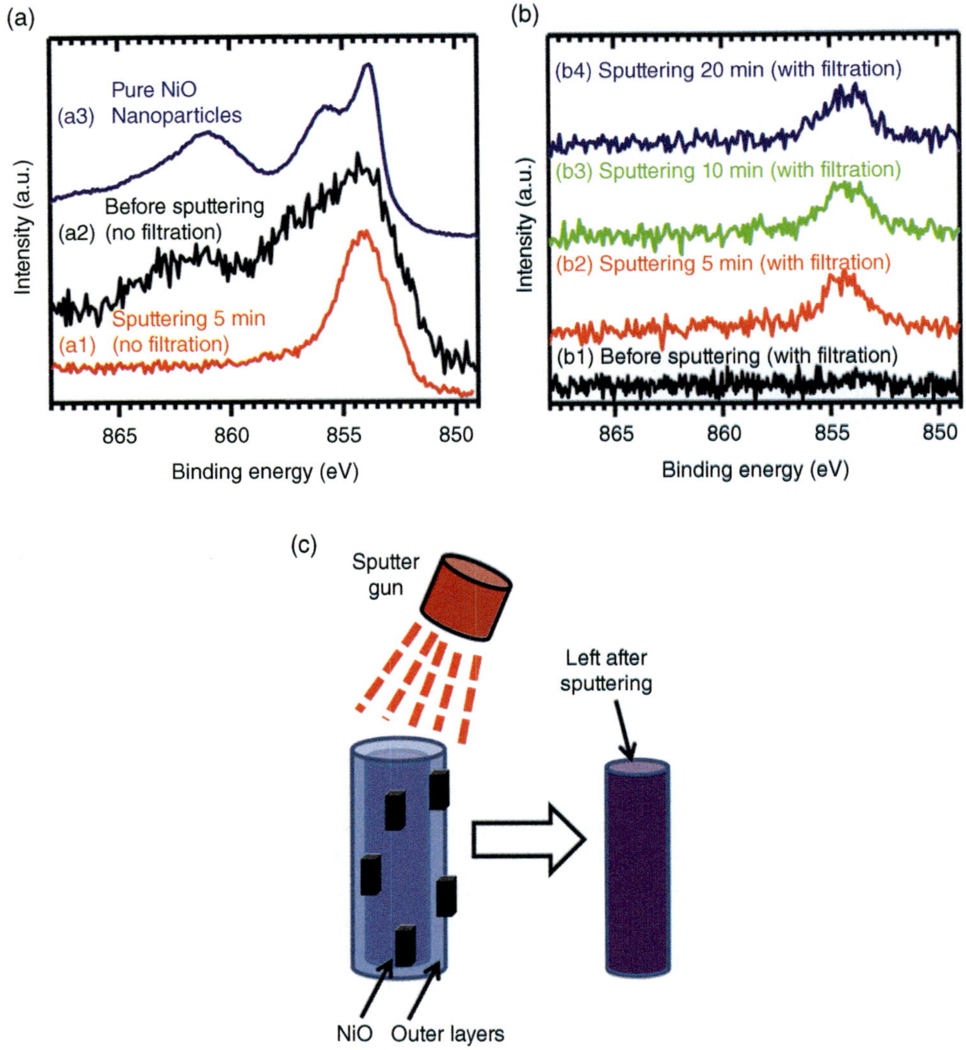

Figure 19.18 XPS studies of 2.5 wt% Ni-ZSM-5. (a) XPS spectra of 2.5 wt% Ni-ZSM-5 without filtration before and after sputtering. For comparison, the XPS spectrum of NiO nanoparticles was included in a3 as well. (b) XPS spectra of 2.5 wt% Ni-ZSM-5 with filtration before and after sputtering for various amounts of time. (c) Schematic showing how XPS spectra were acquired after removing the outer layers of ZSM-5 and the external NiO nanocluster with Ar^+ sputtering. *Source:* Reproduced with permission from [20] / American Chemical Society.

of a satellite peak of Ni $2p_{5/2}$ at 861 eV in the spectrum of 2.5wt%Ni-ZSM-5 after sputtering suggests the lack of formation of NiO nanoparticles. Further evidence is the lack of a Ni 2p signal for a sample prepared through filtration (black spectrum in Figure 19.18b). As XPS is surface-sensitive, the absence of a Ni 2p signal indicates that the Ni cations left on the external surface of the ZSM-5 particles were dissolved in the aqueous solution, and thus, no Ni cations could be found on the external surface of ZSM-5. The successful introduction of Ni atoms into the micropores of ZSM-5 particles was confirmed by the observation of Ni 2p after removing the surface region of the Ni-ZSM-5 particles (red, green, and blue spectra in

Figure 19.18b) by sputtering the surface of nanoparticles with helium ion beam to expose the Ni atoms anchored in the micropores although no Ni signal was observed from the surface of Ni-ZSM-5 before removing the surface region of the Ni-ZSM-5 particle (black spectrum in Figure 19.18b). Figure 19.18c schematically presents the outcome of the sputtering surface atomic layers of ZSM-5 particles. The invisibility of Ni cation before sputtering (black spectrum in Figure 19.18b) together with the observation of Ni cations after sputtering (blue, green, and red spectra in Figure 19.18b) clearly shows that (1) the surface of the "2.5 wt% Ni-ZSM-5 with filtration" does not have Ni atoms and (2) the Ni cations were successfully introduced to the micropores of ZSM-5.

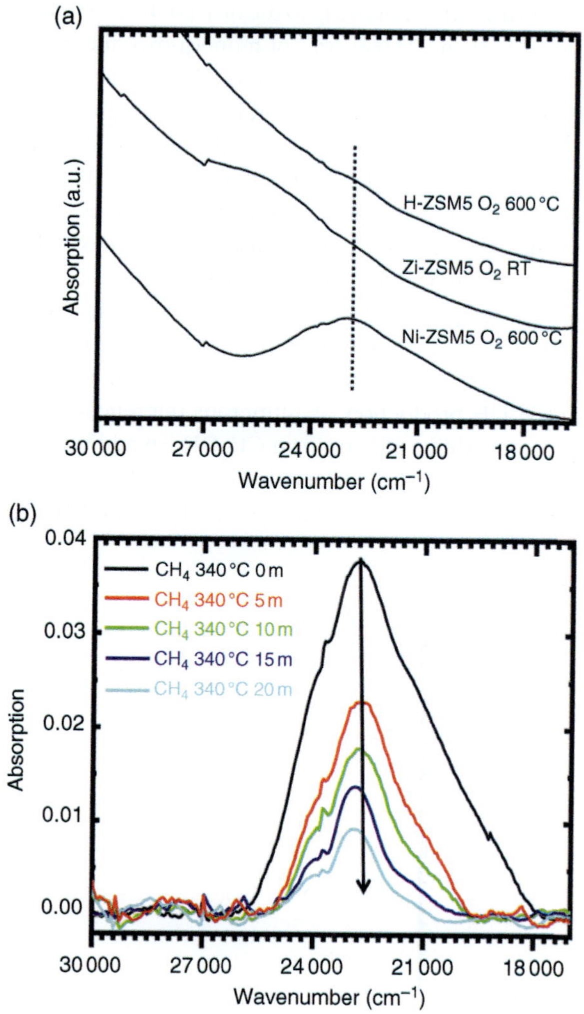

Figure 19.19 In-situ UV–vis spectra studies of 5 wt% Ni-ZSM-5. (a) In-situ UV–Vis spectra of 5 wt% Ni-ZSM-5 with filtration after O_2 treatment at RT or 600 °C. For comparison, the spectrum of H-ZSM-5 after O_2 treatment at 600 °C is included as well. (b) Decrease in the intensity of the absorption band at 22 800 cm^{-1} along oxidation of CH_4 by the oxygen atoms of the activated 2.5 wt% Ni-ZSM-5 at 340 °C for different amount of time as listed. *Source:* Reproduced with permission from [20] / American Chemical Society.

The formation of mono(μ-oxo)di-nickel species in ZSM-5 was indicated by UV–Vis spectroscopy (Figure 19.19). In Figure 19.19a1, a band at 22 800 cm^{-1} was observed from Ni-ZSM-5 upon activation at 600 °C in O_2 (see the spectrum of Ni-ZSM-5 O_2 600 °C in Figure 19.19a). Importantly, the UV–Vis spectra of Ni-ZSM-5 without activation (see the spectrum of Ni-ZSM-5 O_2 RT in Figure 19.19a) did not exhibit this peak, indicating that the band at 22 800 cm^{-1} was formed during activation. In contrast, H-ZSM-5 without Ni did not show the band at 22 800 cm^{-1}, even after activation at 600 °C in O_2 (see the spectrum of H-ZSM-5 O_2 600 °C in Figure 19.19a). Therefore, the band at 22 800 cm^{-1} is attributed to a Ni-based structure formed through activation at 600 °C in O_2. Notably, the literature reports the band of bis(μ-oxo)di-nickel at 25 000 cm^{-1}, which is absent in Figure 19.19. Thus, no bis(μ-oxo)di-nickel structure was formed in Ni-ZSM-5. Referring to the observed 22 700 cm^{-1} of mono(μ-oxo)di-copper species in ZSM-5 [3, 4] and 22,000 cm^{-1} of mono(μ-oxo)di-copper species in MOR [13], it is deduced that the 22 800 cm^{-1} observed from Ni-ZSM-5 activated at 600 °C in O_2 is attributed to mono(μ-oxo)dinickel species [20]. The role of mono(μ-oxo)di-nickel species in the partial oxidation of CH_4 was assessed by monitoring the time-dependent intensity of the band at 22 800 cm^{-1}(Figure 19.19b) [20]. Clearly, as shown in Figure 19.19b, with a longer reaction time, a lesser amount of mono(μ-oxo)dinickel species remained, as evidenced by the decreasing intensity of the band at 22 800 cm^{-1}. Thus, the mono(μ-oxo)di-nickel species provide oxygen atoms for the partial oxidation of CH_4 to CH_3OH.

NMR analysis demonstrated that CH_3OH was produced from Ni-ZSM-5, while the yield of CH_3OH from H-ZSM-5 was negligible. The inactive nature of H-ZSM-5 for the partial oxidation of CH_4 to CH_3OH ruled out the possibility that the trace amounts of Cu or Fe in commercial ZSM-5 could contribute to CH_3OH production, as numerous references have confirmed that Fe-ZSM-5 and Cu-ZSM-5 can oxidize CH_4 to form CH_3OH. In addition to CH_3OH, byproducts HCOOH and $HOCH_2CH_2OH$ were observed. Figure 19.20 presents the yields of methanol generated from 2 wt% Ni-ZSM-5, 5 wt% Ni-ZSM-5, and 10 wt%

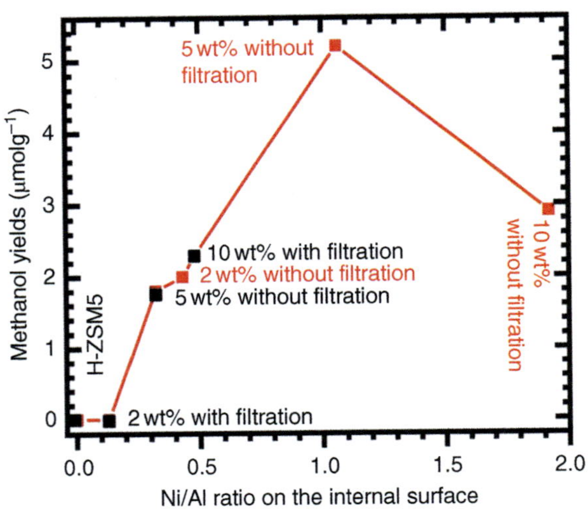

Figure 19.20 Methanol yields as a function of the atomic ratio of Ni/Al on the internal surfaces of micropores for different Ni-ZSM5 catalysts. The catalysts were first activated with pure O_2 at 650 °C for 3 h and then the activated Ni-ZSM-5 was used to oxidize CH_4 at 175 °C for 45 min. *Source:* Reproduced with permission from [20] / American Chemical Society.

Ni-ZSM-5 as a function of the measured Ni/Al ratio. In this figure, the yield is expressed in μmol g^{-1} of catalyst (the mass of ZSM-5 and the introduced Ni) and represents the amount of CH_3OH produced through the oxidation of CH_4 to CH_3OH at 175 °C for 45 minutes. Clearly, the yield of CH_3OH depends significantly on the reaction temperature, time, and CH_4 pressure. Figure 19.20 suggests that a sample subjected to filtration exhibits a lower yield of CH_3OH, as filtration must wash out Ni cations adsorbed in the micropores of ZSM-5 after ion exchange [20]. According to the reference [20, 21], the total yield from 5 wt% Ni-ZSM-5 at 175 °C is approximately 15.0 μmol oxygenates from 1 gram of Ni-ZSM-5. In the case of Cu-ZSM-5 activated with O_2, a yield of 9 μmol CH_3OH was produced per gram of Cu-ZSM-5. Therefore, Ni-ZSM-5 activated by O_2 is even more active than Cu-ZSM-5 activated by O_2 [5]. The advantage of using O_2 as the activating gas for Ni-ZSM-5 or Cu-ZSM-5 is that O_2 is nearly cost-free.

19.4 Zeolite with Small Pore Cu-SSZ-13, Cu-SSZ-16, and Cu-SSZ-39

As the source of oxygen atoms can be provided by N_2O or O_2, the partial oxidation of CH_4 catalyzed by Cu-SSZ-13 was performed with the two different gases. One mixture included CH_4, N_2O, and H_2O vapor, while the other contained CH_4, O_2, and H_2O vapor. In the temperature range of 270–300 °C, when O_2 was used as the oxidant, the selectivity for CH_3OH was lower than 3%, with the majority of products being CO_2. While N_2O was used as the oxidant, the selectivity for CH_3OH was higher, and the main byproduct was CO instead of CO_2. The difference in catalytic selectivity with the use of two different oxidants (N_2O versus O_2) is likely related to two distinct active phases. This is supported by the observations that the sample treated with O_2 appeared dark blue, while the one treated with N_2O appeared pale green. Absorption features at 23 100 and 21 000 cm^{-1} observed in the pale green sample obtained from pretreatment in N_2O indicate the formation of mono-μ-oxo di-copper (II) species. This distinction suggests that different oxidants could form different catalytic sites, even though the zeolite and metal elements are the same.

SSZ-16 and SAPO-34 are active for the oxidation of CH_4 to CH_3OH. A three-step activation process was employed to produce CH_3OH from CH_4. Figure 19.21 presents the concentration of CH_3OH in an effluent stream collected over time. It is observed that the amount of CH_3OH sharply decreases with time, with nearly 250 ppm of CH_3OH initially found in the collected water vapor [22].

The active sites of small-pore Cu-zeolites, including Cu-SSZ-13, and Cu-SSZ-16, for oxidizing CH_4 to CH_3OH are still under debate [22]. However, diffuse reflectance UV–Vis spectroscopy suggests that Cu in these small-pore zeolites exists in the Cu^{2+} state rather than Cu^{+}. The electron configuration of Cu(I) is $3d^{10}$, which does not exhibit d-d transitions. In contrast, the electron configuration of Cu(II) allows d-d transitions. Spectra features of d-d transitions of Cu(II) include bands between 12 000 and 20 000 cm^{-1}, as reported for Cu-SSZ-13 treated with O_2 in the literature [23]. Bands centered at 11 000, 13 600, 16 500, and 19 700 cm^{-1} were observed. The coexistence of multiple bands in the region of 10 000 to 20 000 cm^{-1} results from the John–Teller splitting effect of highly distorted copper (II) species. Figure 19.22 presents the diffuse reflectance UV–vis spectra of Cu-SSZ-39, Cu-SSZ-13, Cu-SSZ-16, Cu-SAPO-34, and Cu-ZSM-5 [22]. Two bands in the energy window of

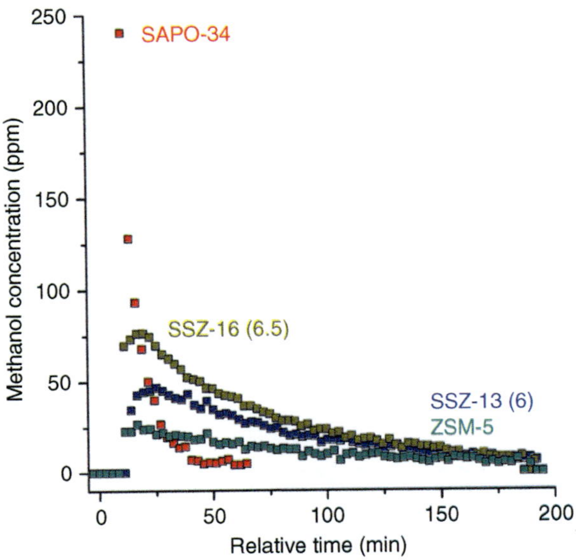

Figure 19.21 The concentration of methanol in the effluent stream is shown versus the time after admission of water vapor (which occurred at relative time = 0 min). The three Cu-zeolites were those exchanged with copper(II) acetate, and Cu-SAPO-34 was exchanged with copper(I) chloride. *Source:* Reproduced with permission from [22] / Royal Society of Chemistry.

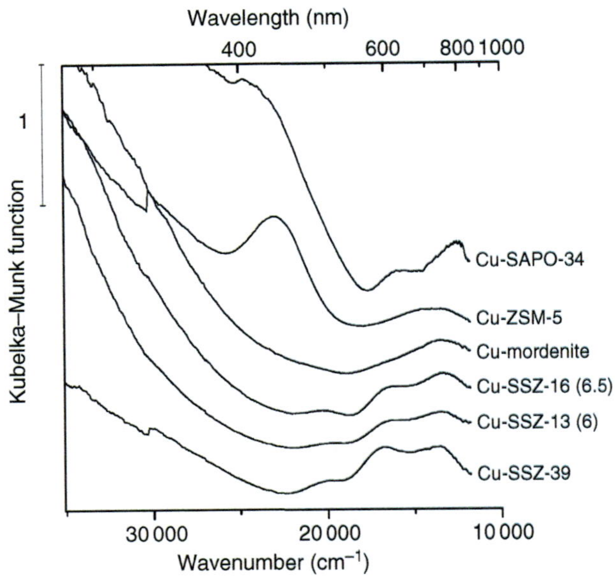

Figure 19.22 Diffuse reflectance UV–vis spectra of Cu-zeolites and zeotypes. Before spectra were collected, the samples were heated to a temperature of 450 °C in flowing O_2, held at that temperature for 2 h, and cooled to room temperature. Samples were exchanged with copper(II) acetate except for SAPO-34, which had been exchanged with copper (I) chloride. The feature at 30 000 cm^{-1} is an artifact caused by a lamp switch in the spectrometer. *Source:* Reproduced with permission from [22] / Royal Society of Chemistry.

10 000–20 000 cm^{-1} for Cu-small-pore zeolites, including Cu-SSZ-39, Cu-SSZ-13, Cu-SSZ-16, Cu-SAPO-34, are observed. This suggests that the copper atoms encapsulated in the small pores are in the Cu(II) state. The lack of an obvious band in this energy window for Cu-ZSM-5 suggests that the copper atoms bonded in the pores of Cu-ZSM-5 are in the Cu(I) state. Thus, the diffuse reflection UV–Vis spectrum of Cu-ZSM-5 is consistent with Cu(I) in the proposed mono-μ-oxo di-copper complex in Cu-ZSM-5.

19.5 Pore Size-dependence on Activity

Partial oxidation of CH_4 to CH_3OH by activated SAPO-34, SSZ-16, SSZ-13, and ZSM-5 was performed in parallel. The oxidant is O_2 [24]. As shown in Figure 19.22, the amount of methanol produced from these Cu-zeolite was presented as a function of time [24]. Compared to ZSM-5, SAPO-34 releases methanol in the first hour, but ZSM-5 continuously releases it in nearly three hours. In addition, the amount of initial release from Cu-SAPO-34 is much higher than Cu-ZSM-5 at the beginning. In other words, the release of CH_3OH from Cu-SAPO-34 is much faster than Cu-ZSM-5. This probably results from a weaker interaction between CH_3OH with SAPO-34 and thus a faster transport of CH_3OH from the catalytic sites in the pores of the SAPO-34 crystal in contrast to CH_3OH in Cu-ZSM-5.

The pore-size effect on activity in producing CH_3OH by oxidization of CH_4 with molecular O_2 was confirmed by systematic studies of oxidation of CH_4 to CH_3OH on over ten Cu-zeolites with different pore sizes [25]. In contrast to Cu-small-pore zeolite, Cu-Y zeolite, an FAU-type with a pore size of 1.0 nm or so, is nearly inactive for producing CH_3OH. This is related to an enhanced confinement effect of molecules in a small-pore zeolite versus a zeolite with a relatively larger pore size. The high activity of small-pore zeolite defined by of the amount of CH_3OH produced from a mol of Cu sites was attributed to a stronger confinement effect in a smaller micropore of a zeolite as discussed in Chapter 15.

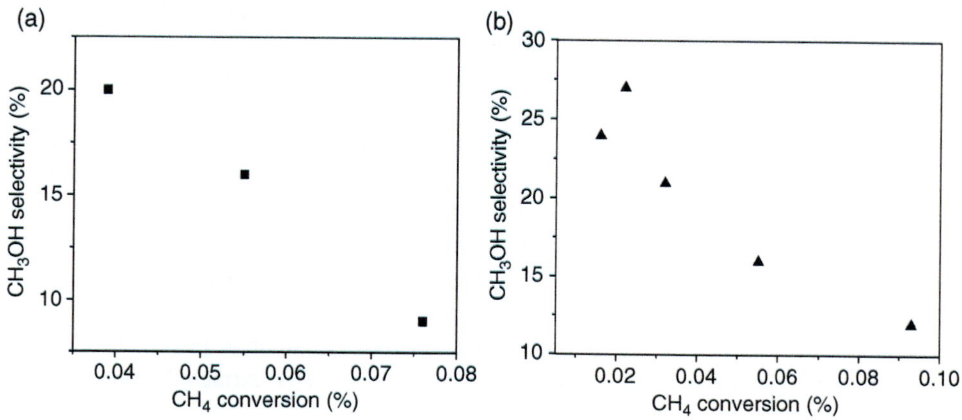

Figure 19.23 CH_4 conversion versus CH_3OH selectivity on Cu-SSZ-13 (a) for CH_4 partial pressures between 20.3 and 40.5 kPa (at 270 °C, 30.4 kPa N_2O, 3.2 kPa H_2O) and (b) for temperatures between 250 and 275 °C (30.4 kPa CH_4, 30.4 kPa N_2O, 3.2 kPa H_2O, 37.3 kPa He). *Source:* Reproduced with permission from [24] / Royal Society of Chemistry.

19.6 Catalytic Oxidation of Methane with O_2 by Cu-zeolite

Production of CH_3OH from CH_4 through partial oxidation by specific oxygen atoms in small-pore zeolites has been well-demonstrated in the literature. Typically, this process involves three steps, as described before, since the formed CH_3OH in the pore must be removed to regenerate the sites fin terms of anchored oxygen atoms/species or the next run of production of CH_3OH. In general, this product is not a catalytic process since an active oxidant must be regenerated with extra effort after a run of production.

The first catalytic process for producing CH_3OH through the partial oxidation of CH_4 with metal-zeolite was achieved using ZSM-5 encapsulated with Fe [26]. In this catalytic oxidation, N_2O serves as the source of oxygen atoms, and thus, CH_4, N_2O, and H_2O were mixed and co-fed to a fixed-bed flow reactor. The catalysis was performed at 275 °C, resulting in a 62% selectivity for CH_3OH with a CH_4 conversion of 0.19% in the catalytic oxidation of CH_4 [26]. Using N_2O as an oxidant, CH_3OH can be catalytically produced by co-feeding CH_4, N_2O, and H_2O to Cu-SSZ-13 at around 210 °C [24]. This has been reviewed before.

O_2 can be used as a source of oxygen atoms for catalytic partial oxidation on the same catalyst, Cu-SSZ-13. At 210 °C, CH_3OH was catalytically produced from Cu-SSZ-13 while co-feeding CH_4, O_2, and H_2O [27]. Literature reports a continuous catalytic production of CH_3OH in the gas phase on Cu-ZSM-5, flowing CH_4 (98.1 kPa), H_2O (3.2 kPa), and O_2 (0.0025 kPa) at 210 °C, reaching ~70% CH_3OH selectivity but with a CH_3OH production rate of 0.88 μmol g cat $^{-1}$ h^{-1} (0.0014% CH_4 conversion) [27]. The capability of performing catalytic oxidation of CH_4 to CH_3OH by co-feeding CH_4, O_2, and H_2O is consistent with the catalytic oxidation of CH_4 to CH_3OH on Cu-SSZ-13 by flowing CH_4, O_2, and H_2O [27]. The catalytic conversion of CH_4 to CH_3OH in Cu-SSZ-13 by co-feeding CH_4, N_2O, and H_2O was reported in the literature [24]. The nature of sites responsible for this catalytic production of CH_3OH has been debated [27]. The observed bands at 20 800, 26 800, and 30 000 cm^{-1} during steady-state CH_3OH production suggest that small copper oxide clusters could have been formed [27]. Similarly, literature proposed the formation of copper oxide in Cu-MOR responsible for the oxidation of CH_4 to CH_3OH [18].

The catalytic oxidation of CH_4 to CH_3OH on Cu-MOR was reported in 2012 [29]. The key feature of this catalytic process is that the formed CH_3OH is not extracted through soaking in aqueous solution at room temperature or purged through flowing steam at a temperature higher than 150 °C, although extraction was a necessary step to obtain methanol after the partial oxidation of CH_4 to CH_3OH in other studies [2, 30, 31]. The Cu-MOR used in the developed catalytic oxidation of CH_4 to CH_3OH was activated by calcinating Cu-MOR at 450 °C in O_2. The activated Cu-MOR exhibits a clear absorption band at 22 700 cm^{-1}, suggesting the formation of mono-(μ-oxo) di-copper center. This formation was further supported by the disappearance of the band at 22 700 cm^{-1} upon exposure to CH_4 at 200 °C. Without the catalytic process, the amount of CH_3OH extracted upon partial oxidation at 200 °C is about 13 μmol g^{-1} of Cu-MOR.

While the activation temperature of Cu-MOR by O_2 to form reactive oxygen species is 450 °C, the reaction temperature between the reactive oxygen species and CH_4 to produce CH_3OH is only 200 °C. Literature reports a well-designed process acting as a complete catalytic reaction system [29]. Cu-MOR can be readily prepared through the aqueous ion

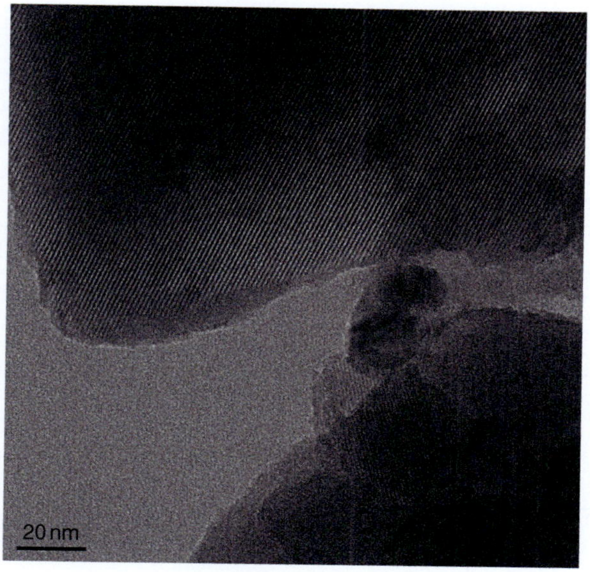

Figure 19.24 TEM image of Cu-MOR after ion exchange and drying. The intact zeolitic channels are visible. *Source:* Reproduced with permission from [29] / Royal Society of Chemistry.

exchange of Na-MOR in the market with Cu(II) acetate. Figure 19.24 shows the TEM image of the prepared Cu-MOR, demonstrating that the aqueous ion exchange and the calcination in O_2 do not destroy the structure of Cu-MOR. A multi-step process and its corresponding reactor system were developed by Alayon et al. for continuously producing CH_3OH [29]. As schematically shown in Figure 19.25, this process includes the oxidation of Cu-MOR in O_2 at 450 °C (red segment in Figure 19.25), the reaction with CH_4 at 200 °C (blue), and the desorption of the formed CH_3OH and its related organic molecules, by flowing a mixture of H_2O and He at 300 °C. So far, a complete reaction cycle is done (first cycle in Figure 19.25). The first step in the second cycle is the regeneration of the active site by calcinating the Cu-MOR in O_2 at 450 °C, forming reactive oxygen species. Upon this activation, the reactive oxygen species react with CH_4 to form CH_3OH. Then, the formed CH_3OH is desorbed from the Cu-MOR at 200 °C in the mixture of He and H_2O. Specific parameters of this process can be found in the literature [29]. Figure 19.26 shows the schematic of the apparatus of the described process. A mass spectrometer was placed at the downstream exit for simultaneously analyzing the composition of the gas generated from the reactor. Figure 19.27a is a typical mass spectrum of the downstream of the reaction system after the first cycle of this process, confirming the on-site extraction of CH_3OH along with the introduction of H_2O at 200 °C. Significantly, the mass spectrum collected in the second cycle (Figure 19.27b) is nearly the same as that of the first cycle. The similarity in the production of CH_3OH in the first and second cycle of production clearly shows that the designed process and reactor system allow for continuously producing CH_3OH [29]. The participation of the reactive oxygen species bonded to Cu atoms in the partial oxidation of CH_4 to CH_3OH was confirmed by the XANES spectra of Cu K-edge, as shown in Figure 19.28. As suggested by the XANES studies, the Cu atoms of Cu-MOR activated at 350 °C in O_2 are at an

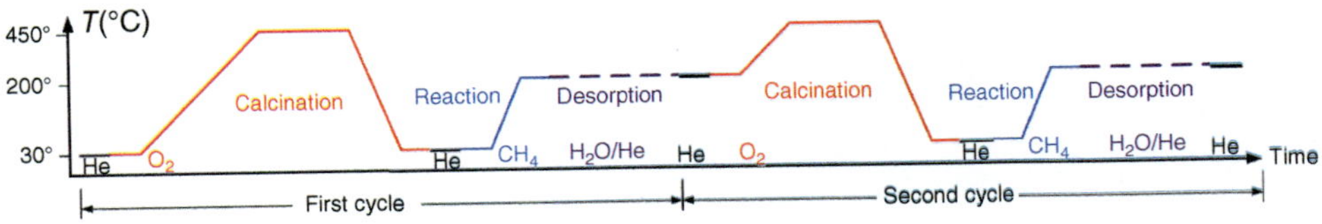

Figure 19.25 Schematic diagram of the cycling process employed for Cu-MOR. *Source:* Reproduced with permission from [29] / Royal Society of Chemistry.

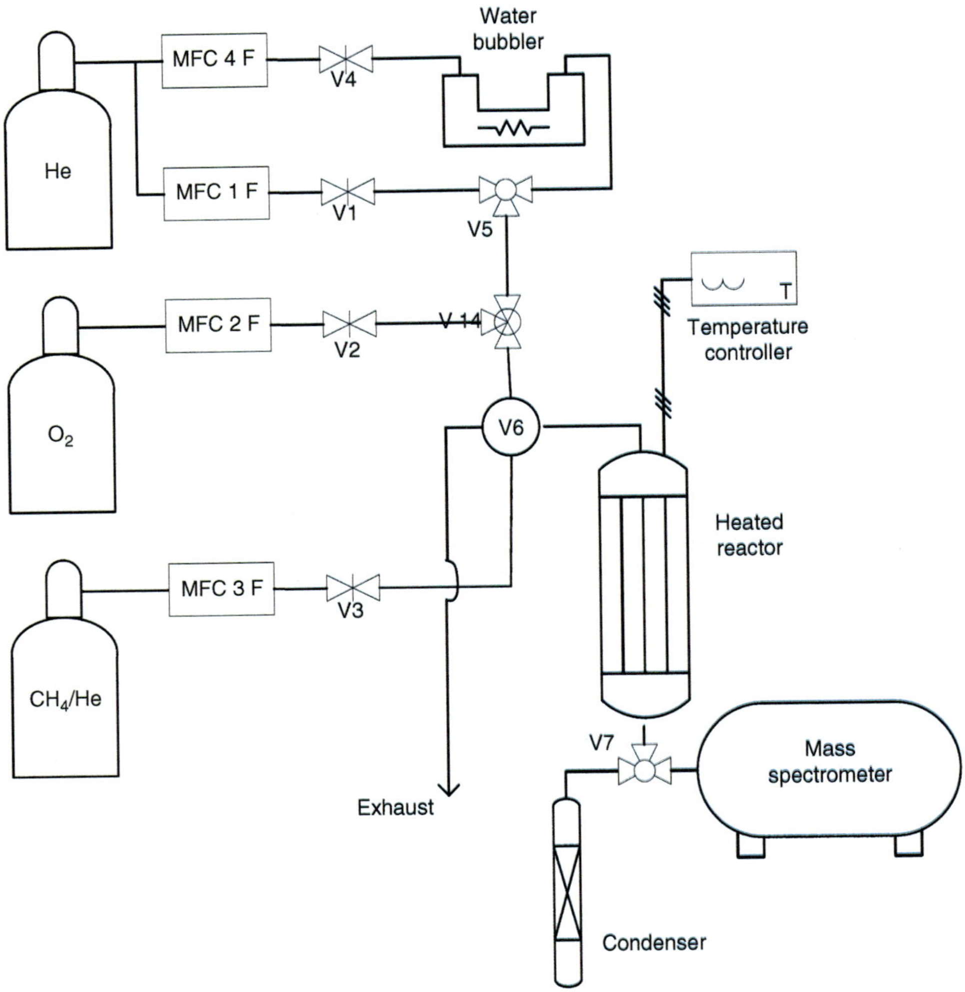

Figure 19.26 Gas mixing scheme and reactor set-up for catalytic experiments. *Source:* Reproduced with permission from [29] / Royal Society of Chemistry.

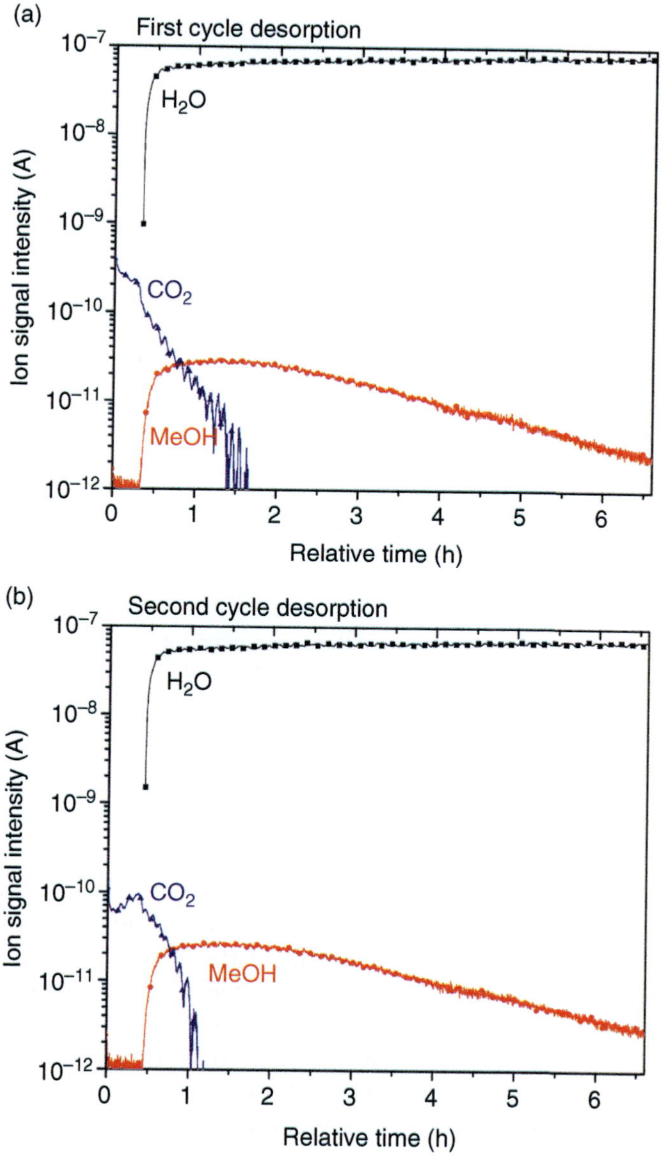

Figure 19.27 Mass spectrometer-detected signals of O_2, H_2O, CO_2 and CH_3OH (a) during the first cycle of desorption by treatment of wet He at 200 °C after methane interaction, and (b) during the second cycle of desorption. *Source:* Reproduced with permission from [29] / Royal Society of Chemistry.

oxidation state of +2, giving an edge position at 8987 eV for Cu^{2+}. This peak at 8987 eV was assigned as 1s→4p of Cu^{2+} (red dash line in Figure 19.28). During the reaction between CH_4 and the reactive oxygen species, Cu^{2+} was reduced to Cu^+ since the O atom of the reactive oxygen species left this special Cu structure to join CH_4 to form CH_3OH (Figure 19.28). As shown in the blue spectrum in Figure 19.28, the edge of the Cu^+ was clearly identified at 8983 eV, attributed to the 1s→4p transition [29].

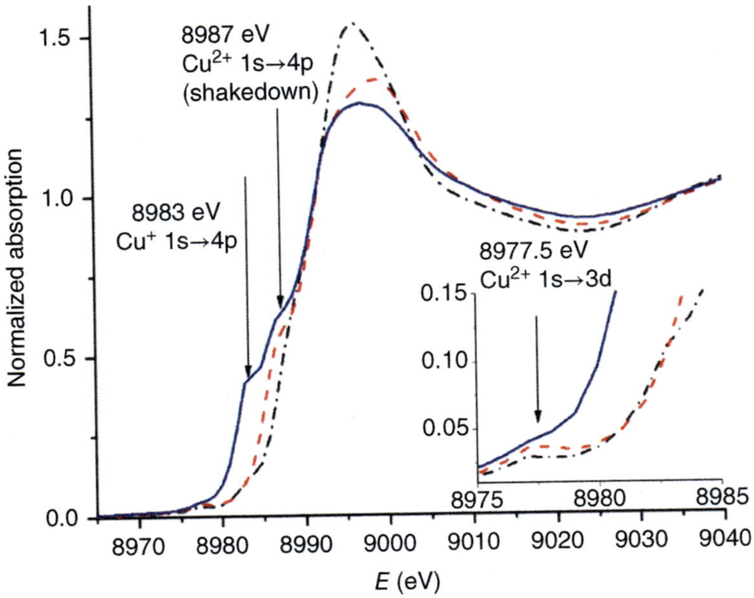

Figure 19.28 In-situ XANES of Cu K edge of Cu–MOR taken of the as-prepared catalyst in He at room temperature (-·-·-·-), during calcination in O_2 at 350 °C (------), and during CH_4 interaction at 200 °C (——). *Source:* Reproduced with permission from [29] / Royal Society of Chemistry.

19.7 Catalytic Coupling between O_2 or HOO· and CH_3· in a Solution with Coexisting O_2 and H_2O_2

Colloidal AuPd nanoparticles, exemplified in Figure 19.29, catalyze the oxidation of CH_3· radicals to form CH_3OO·, and subsequently CH_3OOH, forming CH_3OH. This catalysis occurs at 50 °C or even 23 °C, as illustrated in Figure 19.30a. The reaction involves introducing a mixture of CH_4 and O_2 into an H_2O_2 aqueous solution containing colloidal AuPd nanoparticles encapsulated in a PVP surfactant. H_2O_2 and dissolved CH_4 and O_2 molecules can access the surface of AuPd nanoparticles through spaces between the nanoparticles and the surfactant shell. H_2O_2 is a crucial component in the production of organic oxygenates, including CH_3OOH, CH_3OH, HCOOH, and CO_2. When comparing the reaction system containing CH_4, H_2O_2, and O_2, where the degradation of H_2O_2 is catalyzed by AuPd/TiO_2, to the same reaction system without obvious degradation of H_2O_2, the latter formed substantially more products. This difference suggests that H_2O_2 plays a significant role in the production of CH_3OH for increasing the amount of products and selectivity [32].

The role of O_2 in the partial oxidation of CH_4 is evident in the isotope-labeled experiment presented in Figure 19.30b. When both $H_2{}^{16}O_2$ and $^{16}O_2$ were utilized, $CH_3{}^{16}OH$ was observed as expected (upper panel in Figure 19.30b). More notably, when $^{16}O_2$ was replaced with $^{18}O_2$ under the same conditions, both $CH_3{}^{16}OH$ and $CH_3{}^{18}OH$ were formed. This suggests that both H_2O_2 and O_2 can serve as sources of oxygen atoms for CH_3OH. Even when H_2O_2 and O_2 are both present in the reactants, O_2 significantly contributes to

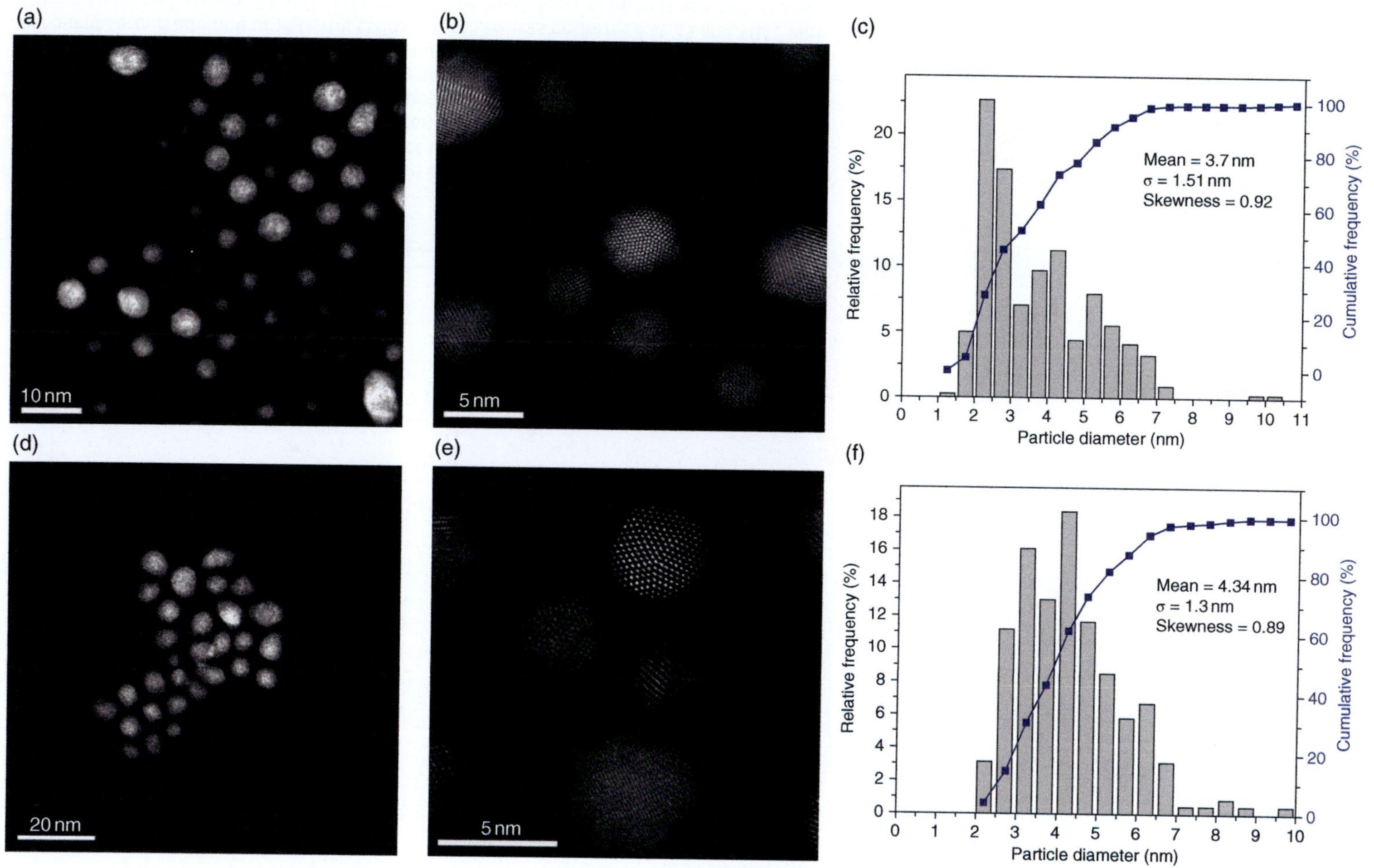

Figure 19.29 Representative HAADF images of unsupported AuPd colloids and particle size distributions for the unsupported AuPd-PVP sol in the fresh (a–c) and after a CH_4 oxidation reaction (d–f). *Source:* [32] /American Association for the Advancement of Science - AAAS / CC BY 4.0.

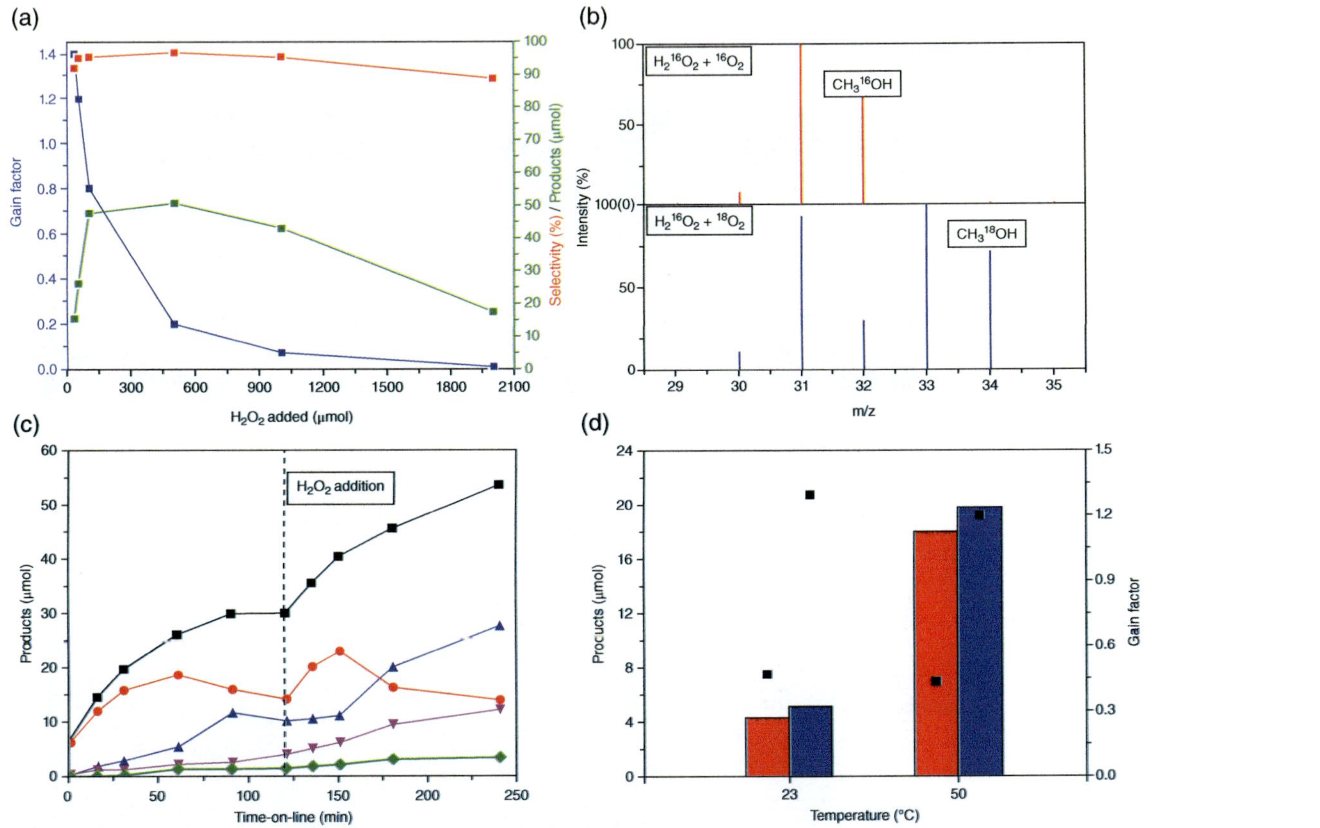

Figure 19.30 Methane oxidation reactions are carried out over unsupported AuPd colloids. (a) Gain factor (blue line), selectivity (red line), and total amount of products (green line) as a function of the different amounts of H_2O_2 used. (b) GC–MS spectra of CH_3OH formed (*m* = 32 and 34 for $CH_3{}^{16}OH$ and $CH_3{}^{18}OH$, respectively) during methane oxidation with an AuPd colloid via $H_2{}^{16}O_2$ + $^{16}O_2$ (*upper panel*) or $H_2{}^{16}O_2$ + $^{18}O_2$ (*lower panel*). For CH_4 oxidation with $^{18}O_2$, > 70% of $^{18}O_2$ molecules were incorporated in the CH_3OH product. (c) Time-on-line and AuPd colloid re-use study for the methane oxidation reaction employing 50 μmol H_2O_2 and 5 bar O_2 showing no induction period. Total products (black line), and individual products [CH_3OOH (red line), CH_3OH (blue line), HCOOH (pink line), and CO_2 (green line)] generated as a function of time-on-line. The black dotted vertical line at 120 min indicates a subsequent second addition of H_2O_2. (d) Oxidation of CH_4 was performed at 23 and 50 °C with 50 μmol H_2O_2. The blue and red bars represent reactions performed with and without 5 bar O_2, respectively. The black squares (■) indicate their respective gain factors. *Source:* [32] /American Association for the Advancement of Science - AAAS / CC BY 4.0.

Table 19.2 List of catalytic activity for synthesis of CH_3OH through AuPd colloids under different catalytic conditions.

Entry	Catalyst	H_2O_2 µmol	O_2 bar	Amount of product (µmol)				Primary oxygenate selectivity / %	Oxygenate productivity/ mol kg_{cat}^{-1} h^{-1}	H_2O_2 consumed / %	Gain factor
				CH_3OOH	CH_3OH	HCOOH	CO_2				
1	1%AuPd/TiO_2	1000	0	0.0	0.4	0.00	1.2	26	0.03	73	2×10^{-3}
2	AuPd colloid	1000	0	11.8	3.3	0.6	1.1	90	29.4	38	3×10^{-2}
3	AuPd colloid	1000	5	17.4	7.6	1.8	1.5	88	53.6	27	9×10^{-2}
4	AuPd colloid	0	5	0	0	0	0.2	0	0	0	–
5	AuPd colloid	50	5	15.7	2.8	1.2	0.3	92	39.4	44	1.2
6	Pd colloid	50	5	0	0	0	0.7	0	0	22	0
7	Au colloid	50	5	0	0	0	0.1	0	0	12	0
8	$PdCl_2$	50	5	0	0	0	0.3	0	0	5	0
9	$HAuCl_4$	50	5	0	0	0	0.2	0	0	10	0

Comparison of catalytic activity of supported and unsupported colloidal Au-only, Pd-only, and AuPd catalysts for the liquid phase oxidation of CH_4 using H_2O_2. *Test conditions:* Reaction time = 0.5 hour; Stirring speed = 1500 rpm; Pressure CH_4 = 30 bar; Reaction temperature = 50 °C with stirred heating ramp rate of 2.25 °C min^{-1}. *Note: Entry 1*: Sol-immobilized solid catalyst, 100 mg (6.6 µmol metal in 10 mL water). *Entries 2–7*: Colloidal catalysts (10 mL, 6.6 µmol of metal). *Entries 8, 9*: Homogeneous metal precursor solutions (6.6 µmol metal in 10 mL water).
Source: [32] /American Association for the Advancement of Science - AAAS / CC BY 4.0.

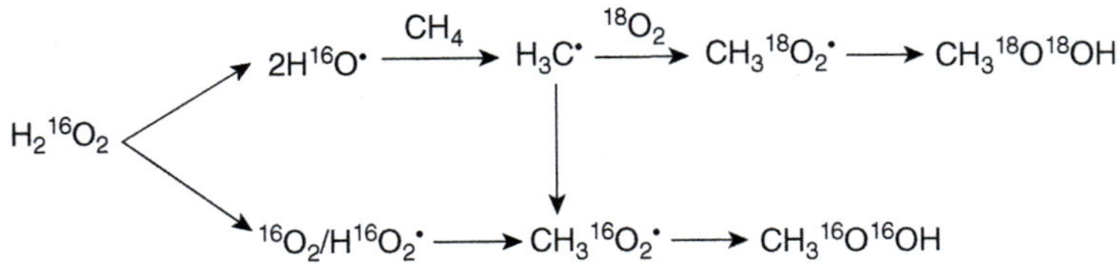

Figure 19.31 Proposed reaction scheme for methane oxidation in the presence of H_2O_2 and molecular O_2. *Source:* [32] /American Association for the Advancement of Science - AAAS / CC BY 4.0.

the formation of CH_3OOH. The role of O_2 was confirmed by two control experiments in entries 2 and 3 in Table 19.2. In comparison to the experiment in Entry 2, the experiment in Entry 3, which involved the introduction of 5 bar O_2 to the reaction system, exhibited nearly double productivity (Table 19.2). Thus, half of the products were contributed by the reaction involving molecular O_2.

A proposed pathway for the partial oxidation of CH_4 involving CH_4, O_2, and H_2O_2 is depicted in Figure 19.31. As illustrated in Figure 19.31 [32], the HO· radical from H_2O_2 breaks a C–H bond in CH_4 through a hydrogen abstraction mechanism, forming H· and CH_3·. The CH_3· combines with molecular O_2, forming an intermediate CH_3OO·. Upon receiving a proton (H^+), CH_3OOH is formed and subsequently transforms to CH_3OH. In the absence of molecular O_2 in the reaction system, as shown in entry 2 in Table 19.2, the rate of producing CH_3OH is 29.4 mol kg (catalyst) $^{-1}$ h^{-1} was achieved. Without molecular O_2, CH_3· can couple with either HOO·, generated through the decomposition of H_2O_2, forming CH_3OOH. In other words, without molecular O_2, H_2O_2 plays dual roles in two subsequent steps, involving hydrogen abstraction to break the C–H bond by HO· to form CH_3· and then the coupling of CH_3· with HOO· or O_2 obtained from decomposition of another H_2O_2 molecule. While external O_2 coexists with H_2O_2, H_2O_2 is primarily responsible for providing HO· for the hydrogen absorption mechanism to form CH_3· and H_2O [32]. Thus, the coexisting H_2O_2 and O_2 play a synergistic role. H_2O_2 can form OH· radical through dissociation of H_2O_2, and then OH absorbs an H atom from a CH_4 molecule to form CH_3·. The coexisting O_2 can readily bond to CH_3· to form CH_3OOH, the intermediate leading to form the ideal product CH_3OH.

References

1 Elliott, S. J., Zhu, M., Tso, L., Nguyen, H. H. T., Yip, J. H. K., Chan, S. I. 1997. "Regio- and stereoselectivity of particulate methane monooxygenase from methylococcus capsulatus (bath)." *J. Am. Chem. Soc.* 119, 9949–9955.

2 Smeets, P. J., Groothaert, M. H., Schoonheydt, R. A. 2005. "Cu based zeolites: A UV–vis study of the active site in the selective methane oxidation at low temperatures." *Catal. Today* 110, 303–309.

3 Groothaert, M. H., van Bokhoven, J. A., Battiston, A. A., Weckhuysen, B. M., Schoonheydt, R. A. 2003. "Bis(μ-oxo)dicopper in Cu-ZSM-5 and its role in the decomposition of NO: a combined in situ XAFS, UV–Vis–near-IR, and kinetic study." *J. Am. Chem. Soc.* 125, 7629–7640.

4 Groothaert, M. H., Smeets, P. J., Sels, B. F., Jacobs, P. A., Schoonheydt, R. A. 2005. "Selective oxidation of methane by the bis(mu-oxo)dicopper core stabilized on ZSM-5 and mordenite zeolites." *J. Am. Chem. Soc.*127,1394–1395.

5 Woertink, J. S., Smeets, P. J., Groothaert, M. H., Solomon, E. I. 2009. "A $[Cu_2O]^{2+}$ core in Cu-ZSM-5, the active site in the oxidation of methane to methanol." *PNAS* 106, 18908–18913.

6 Baldwin, M. J., Root, D. E., Pate, J. E., Fujisawa, K., Kitajima, N., Solomon, E. I. 1992. "Spectroscopic studies of side-on peroxide-bridged binuclear copper(II) model complexes of relevance to oxyhemocyanin and oxytyrosinase." *J. Am. Chem. Soc.* 114, 10421–10431.

7 Maiti, D., Fry, H. C., Woertink, J. S., Vance, M. A., Solomon, E. I., Karlin, K. D. 2007. "A 1:1 copper-dioxygen adduct is an end-on bound superoxol copper(II) complex which undergoes oxygenation reactions with phenols." *J. Am. Chem. Soc.* 129, 264–265.

8 Chen, P., Root, D. E., Campochiaro, C., Fujisawa, K., Solomon, E. I. 2003. "Spectroscopic and electronic structure studies of the diamagnetic side-on Cu-II-superoxo complex $Cu(O_2)[HB(3\text{-}R\text{-}5\text{-}^iPrpz)_3]$: Antiferromagnetic coupling versus covalent delocalization." *J. Am. Chem. Soc.* 125, 466–474.

9 Yoon, J., Fujii, S., Solomon, E. I. 2009. "Geometric and electronic structure differences between the type 3 copper sites of the multicopper oxidases and hemocyanin/tyrosinase." *PNAS* 106, 6585–6590.

10 Alayon, E. M. C., Nachtegaal, M., Bodi, A., van Bokhoven, J. A. 2014. "Reaction conditions of methane-to-methanol conversion affect the structure of active copper sites." *ACS Catal.* 4, 16–22.

11 Prestipino, C., Berlier, G., Xamena, F. X. L. I., Spoto, G., Bordiga, S., Zecchina, A., Palomino, G. T., Yamamoto, T., Lamberti, C. 2002. "An in situ temperature dependent IR, EPR and high resolution XANES study on the NO/Cu^+–ZSM-5 interaction." *Chem. Phys. Lett.* 363, 389–396.

12 Kau, L. S., Spirasolomon, D. J., Pennerhahn, J. E., Hodgson, K. O., Solomon, E. I. 1987. "X-ray absorption edge determination of the oxidation state and coordination number of copper. Application to the type 3 site in Rhus vernicifera laccase and its reaction with oxygen." *J. Am. Chem. Soc.* 109, 6433–6442.

13 Alayon, E. M. C., Nachtegaal, M., Bodi, A., Ranocchiari, M., van Bokhoven, J. A. 2015. "Bis(μ-oxo) versus mono(μ-oxo)dicopper cores in a zeolite for converting methane to methanol: An in situ XAS and DFT investigation." *Phys. Chem. Chem. Phys.* 17, 7681–7693

14 Larsen, S. C., Aylor, A., Bell, A. T., Reimer, J. A. 1994. "Electron paramagnetic resonance studies of copper ion-exchanged ZSM-5." *J. Phys. Chem.* 98, 11533–11540.

15 Smeets, P. J., Hadt, R. G., Woertink, J. S., Vanelderen, P., Schoonheydt, R. A., Sels, B. F., Solomon, E. I. 2010. "Oxygen precursor to the reactive intermediate in methanol synthesis by Cu-ZSM-5." *J. Am. Chem. Soc.* 132, 14736–14738.

16 Cheng, N., Wei, Y. L., Yang, Y. W., Lee, J. F. 2015. "Bis(μ-oxo) versus mono(μ-oxo)dicopper cores in a zeolite for converting methane to methanol: An in situ XAS and DFT investigation." *Phys. Scr.* T115, 907–908.

17 Forsyth, J. B., Hull, S. 1991. "The effect of hydrostatic pressure on the ambient temperature structure of CuO." *J. Phys.: Condens. Matter*, 3, 5257.

18 Tomkins, P., Mansouri, A., Bozbag, S. E., Krumeich, F., Park, M. B., Alayon, E. M. C., Ranocchiari, M., van Bokhoven, J. A. 2016. "Isothermal cyclic conversion of methane into methanol over copper-exchanged zeolite at low temperature." *Angew. Chem. Int. Ed.* 128, 5557–5561.

19 Tomkins, P., Mansouri, A., Sushkevich, V. L., van der Wal, L. I., Bozbag, S. E., Krumeich, F., Ranocchiari, M., van Bokhoven, J. A. 2019. "Increasing the activity of copper exchanged mordenite in the direct isothermal conversion of methane to methanol by Pt and Pd doping." *Chem. Sci.* 10, 167–171.

20 Shan, J., Haung, W., Nguyen, L., Yu, Y., Zhang, S., Li, Y., Frenkel, A. I. 2014. "Conversion of methane to methanol with a bent mono(μ-oxo)dinickel anchored on the internal surfaces of micropore." *Langmuir* 30, 8558–8569.

21 Starokon, E. V., Parfenov, M. V., Pirutko, L. V., Abornev, S. I., Panov, G. I. 2011. "Room-temperature oxidation of methane by alphaoxygen and extraction of products from the FeZSM-5 surface." *J. Phys. Chem. C* 115, 2155–2161.

22 Wulfers, M. J., Teketel, S., Ipek, B., Lopo, R. F. 2015. "Conversion of methane to methanol on copper-containing small-pore zeolites and zeotypes." *Chem. Commun.* 51, 4447–4450.

23 Giordanino, F., Vennestrom, P. N. R., Lundegaard, L. F., Stappen, F. N., Mossin, S., Beato, P., Bordiga, S., Lamberti, C. 2013. "Characterization of Cu-exchanged SSZ-13: A comparative FTIR, UV-Vis, and EPR study with Cu-ZSM-5 and Cu-β with similar Si/Al and Cu/Al ratios." *Dalton Trans.* 42, 12741–12761.

24 Ipek, B., Lobo, R. F. 2016. "Catalytic conversion of methane to methanol on Cu-Ssz-13 using N_2O as oxidant." *Chem. Commun.* 52, 13401–13404.

25 Park, M. B., Ahn, S. H., Mansouri, A., Ranocchiari, M., van Bokhoven, J. A. 2017. "Comparative study of diverse copper zeolites for the conversion of methane into methanol." *ChemCatChem* 9, 3705–3713.

26 Parfenov, M. V., Starokon, E. V., Pirutko, L. V., Panov, G. I. 2014. "Quasicatalytic and catalytic oxidation of methane to methanol by nitrous oxide over FeZSM-5 zeolite." *J. Catal.* 318, 14–21.

27 Narsimhan, K., Iyoki, K., Dinh, K., Roman-Leshkov, Y. 2016. "Catalytic oxidation of methane into methanol over copper exchanged zeolites with oxygen at low temperature." *ACS Cent. Sci.* 2, 424–429.

28 Vanelderen, P., Hadt, R. G., Smeets, P. J., Solomon, E. I., Schoonheydt, R. A., Sels, B. F. 2011. "Cu-ZSM-5: A biomimetic inorganic model for methane oxidation." *J. Catal.* 284, 157–164.

29 Alayon, E. M., Nachtegaal, M., Ranocchiari, M., van Bokhoven, J.A. 2012. "Catalytic conversion of methane to methanol over Cu–mordenit." *Chem. Commun.* 48, 404–406.

30 Dubkov, K. A., Sobolev, V. I., Panov, G. I. 1998. "Low-temperature oxidation of methane to methanol on FeZSM-5 zeolite." *Kinet. Catal.* 39, 72–79.

31 Beznis, N. V., Weckhuysen, B. M., Bitter, J. H. 2010. "Cu-ZSM-5 zeolites for the formation of methanol from methane and oxygen: Probing the active sites and spectator species." *Catal. Lett.* 138, 14–22.

32 Agarwal, N., Freakley, S. J., McVicker, R. U., Althahban, S. M., Dimitratos, N., He, Q., Morga, D. J., Jenkins, R. L., Willock, D. J., Hutchings, G. J. 2017. "Aqueous Au-Pd colloids catalyze selective CH_4 oxidation to CH_3OH with O_2 under mild conditions." *Science* 358, 223–227.

20

Oxidation of CH_4 and CO with O_2 through M–O Clusters Anchored in Zeolite in Liquid at Low Temperature

In addition to the extensive investigations of the partial oxidation of CH_4 for the formation of CH_3OH, a novel reaction involving the co-oxidation of CH_4 and CO to produce acetic acid has been reported. Among the heterogeneous catalysts explored for this purpose, Rh_1-ZSM-5 has emerged as an efficient catalyst [1]. Notably, this reaction, although conducted in a batch reactor, is a genuinely catalytic process. Moreover, it represents a carbon-addition reaction, providing an alternative pathway for utilizing CH_4 to generate feedstocks for chemical industries. Acetic acid, a key industrial intermediate, is produced in this process, with methanol being a byproduct although the synthesis of acetic acid does not go through methanol carbonylation.

The catalyst was prepared through the ion exchange of NH4-ZSM-5 with a Si/Al ratio of 11.5. Various Rh/ZSM-5 samples with loadings of 0.01 wt% Rh, 0.05 wt% Rh, 0.10 wt% Rh, and 0.50 wt% Rh were synthesized. Rhodium (III) nitrate hydrate served as the precursor of rhodium. The introduction of Rh precursor occurred through ion exchange after purging the air from the pores of H-ZSM-5 using a vacuum pump. Subsequently, the wet powder was dried in an oven at 80 °C and calcinated at 550 °C in air for 3 hours. Figure 20.1a-f schematically illustrates the preparation of an Rh_1-ZSM-5 catalyst. As depicted in Figure 20.1g, the chemical and coordination environments of Rh_1 atoms in ZSM-5 are evidently distinct from rhodium oxide nanoparticles.

As depicted in Figure 20.2a and b, the amount of loading plays a crucial role in preventing the formation of rhodium oxide clusters. The absence of any rhodium oxide clusters in the red spectrum in Figure 20.2c suggests that Rh atoms are dispersed in the micropores of ZSM-5 at the loading of 0.10wt%Rh. This is further supported by Extended X-ray Absorption Fine Structure (EXASF) studies of the 0.10 wt% Rh/ZSM-5, with fitting parameters listed in Figure 20.2f. Approximately 4–5 oxygen atoms are directly bonded to Rh_1 atoms. The single dispersion of Rh atoms is confirmed by the absence of a Rh–O–Rh peak of rhodium oxide clusters in the r-space spectrum of Rh K-edge (Figure 20.2e), which is clearly observed in the r-space spectrum of Rh_2O_3 supported on Al_2O_3 (Figure 20.2h).

Catalysis was conducted in a Para reactor capable of maintaining gas pressure up to 200 bar and heating the catalyst to temperatures as high as 200 °C or even 300 °C. A 28 mg catalyst (0.10 wt% Rh/ZSM-5) was added to an empty reactor, followed by the addition of 10 mg deionized water. A mixture of CH_4, CO, and CH_4 with specific partial pressures was

Low-temperature Activation and Catalytic Transformation of Methane to Non-CO_2 Products, First Edition. Franklin Tao.

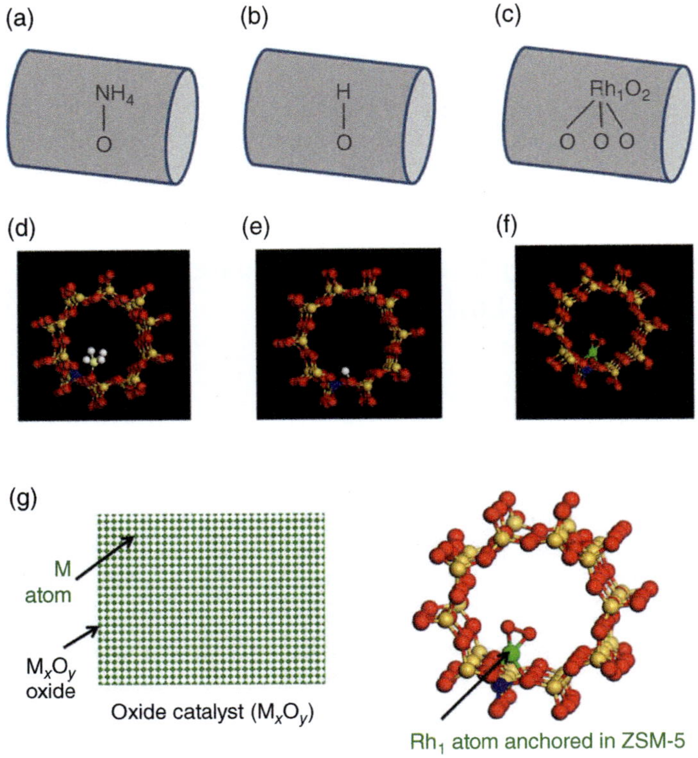

Figure 20.1 Supplementary Figure 1. Schematic showing synthesis and structural feature of isolated Rh atoms in ZSM-5. Schematic (a–c) and pictures (d–f) showing the preparation of 0.10 wt% Rh/ZSM-5 catalyst in which Rh_1O_5 sites were formed in the micropores of aluminosilicate. (a) NH4-ZSM-5 as purchased. (b) Formation of H-ZSM-5 upon calcination of NH4-ZSM-5 at 450 °C for 3 hours. (c) Formation of ZSM-5 with anchored Rh_1O_5 sites through integrated vacuum pumping and incipient wetness impregnation (IWI) of Rh^{3+} to pores through ion exchange in aqueous solution; after ion exchange, catalyst precursor was centrifuged; the precipitant was dried in an oven at 80 °C for 3 hours and calcined in air at 550 °C for 3 hours. (d–f) are ball-stick model figures of the corresponding schematics (a–c). (g) Schematic showing the difference between continuously packed cationic sites on M_xO_y oxide nanoparticles and separately anchored Rh cations in ZSM-5. *Source:* [1] / Springer Nature / CC BY 4.0.

then introduced into the Para reactor. During catalysis, the solid powder was continuously stirred (600 rpm) to ensure homogeneous distribution in the aqueous solution. Reactant molecules access the catalytic sites encapsulated in the micropores of ZSM-5 by dissolving in water and subsequently diffusing to the catalytic sites in the micropores of Rh-ZSM-5. High pressure of reactants was necessary to increase their solubility. Temperature, being a key parameter, influenced the reaction rate and yield at a given temperature. The catalysis was carried out at 150 °C. As illustrated in Figure 20.3, for a reaction time of 2 hours, approximately 80 μmol of formic acid and 30 μmol of acetic acid were produced at a CH_4 pressure of 10 bar. The yields of formic acid and acetic acid increased to 190 and 160 μmol, respectively, when the CH_4 pressure was increased to 50 bar. Figure 20.3 suggests that a longer reaction time results in larger amounts of products.

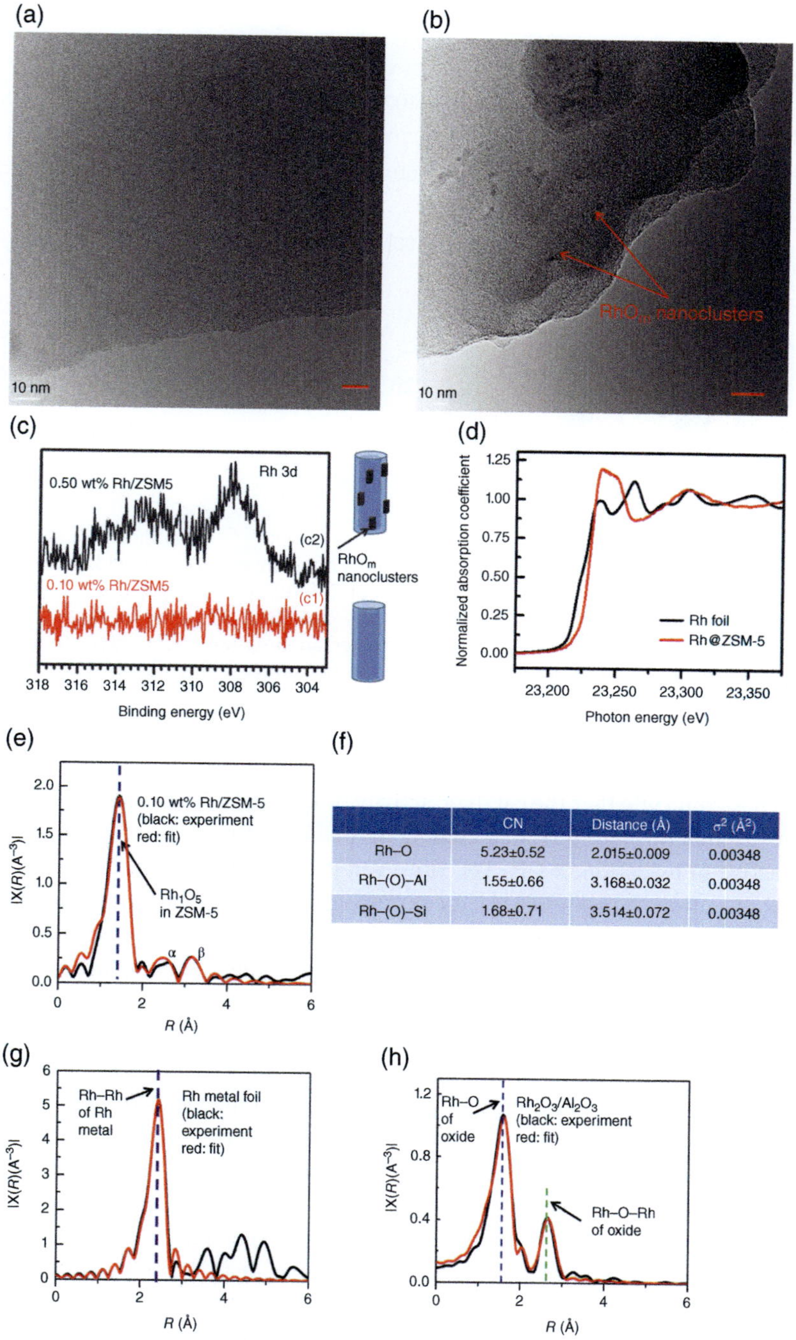

	CN	Distance (Å)	σ^2 (Å^2)
Rh–O	5.23±0.52	2.015±0.009	0.00348
Rh–(O)–Al	1.55±0.66	3.168±0.032	0.00348
Rh–(O)–Si	1.68±0.71	3.514±0.072	0.00348

Figure 20.2 Structural characterization of isolated rhodium atoms in ZSM-5. (a) TEM image of particles of 0.10 wt% Rh/ZSM-5; scale bar: 10 nm. (b) TEM image of particles of 0.50 wt% Rh/ZSM-5; scale bar: 10 nm. (c) Rh 3d XPS peak of 0.10 wt% Rh/ZSM-5 and 0.50 wt% Rh/ZSM-5. (d) Energy space of Rh K-edge of 0.10 wt% Rh/ZSM-5 and Rh foil (reference sample) of X-ray absorption near edge spectra (XANES). (e) r-space of Rh K-edge of experimental (black) and calculated (red) data of the k^2-weighted Rh K-edge EXAFS spectra of used 0.10 wt% Rh/ZSM-5. (f) Coordination number and bond length on average of the used 0.10 wt% Rh/ZSM-5. (g) r-space of Rh K-edge of experimental (black) and calculated (red) data of the k^2-weighted Rh K-edge EXAFS spectra of Rh metal foil. (h) r-space of Rh K-edge of experimental (black) and calculated (red) data of the k^2-weighted Rh K-edge EXAFS spectra of Rh_2O_3 nanoparticles supported on Al_2O_3. *Source:* [1] / Springer Nature / CC BY 4.0.

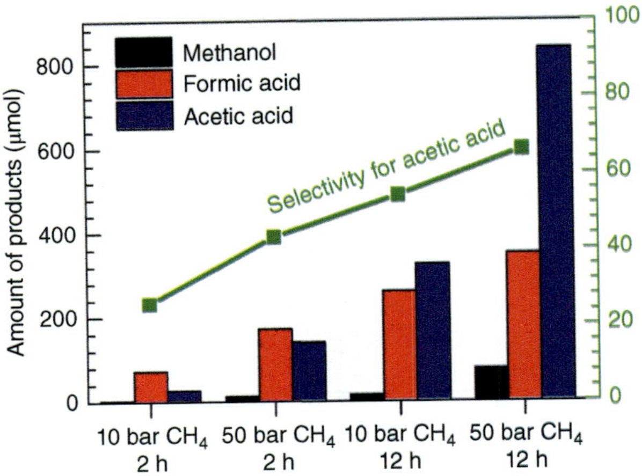

Figure 20.3 Catalytic performance of 0.10 wt% Rh/ZSM-5. Yields of acetic acid, formic acid, and methanol, as well as the selectivity to acetic acid as a function of different pressures of CH_4 (10 or 50 bar) and different reaction times (2 or 12 h). 28 mg of 0.10 wt% Rh/ZSM-5, was used for each catalysis test. Each catalysis test used 10 bar CO, 8 bar O_2, and a certain pressure of CH_4 as noted on the X-axis (10 or 50 bar). The catalysis temperatures of all studies here were 150 °C. *Source:* [1] / Springer Nature / CC BY 4.0.

A concern arises regarding whether the catalytic activity of Rh_1/ZSM-5 results from free-standing Rh^{3+} ions formed through potential etching of anchored Rh atoms in ZSM-5 or not since Rh^{3+} is active in oxidation of CH_4 [2]. However, it has been confirmed that the amount of etched Rh^{3+} cations is negligible [1]. A contrasting experiment indicates that the turnover rate (TOR) of each free-standing Rh^{3+} is much lower than that of an anchored Rh_1O_5 site by four magnitudes. Thus, even if all the rhodium atoms anchored in ZSM-5 were etched into the solution, their yield would still be much lower than the observed yield, as shown in Table 20.1. Clearly, the catalytic sites of 0.10 wt% Rh/ZSM-5 are Rh_1O_5 instead of free-standing Rh^{3+}.

The partial pressures of O_2 and CO exhibit distinct influences on the yields of products. As illustrated in Figure 20.4, the influence of O_2 pressure follows a volcano shape. At 16 bar of O_2, the yields of CH_3COOH and HCOOH are nearly the same as at 2 bar, but 8 bar of O_2 result in yields higher by tens of times. A similar volcano evolution was observed in Figure 20.4b with increasing CO pressure. At 15 bar of CO, the yields for CH_3COOH and HCOOH decrease noticeably, and they are nearly zero at 30 bar of CO. The two similar volcano evolutions suggest that the Rh_1O_5 sites catalyze CO oxidation. Once the pressure of CO or O_2 is too high, it likely consumes most O_2 or CO in the batch reactor, and the synthesis of CH_3COOH cannot be processed. In comparison to the volcano evolutions of CO and O_2, the influence of CH_4 pressure on yield is different. As shown in Figure 20.4c, yields of CH_3COOH and HCOOH continuously increase with the increasing CH_4 pressure. This is simply because more CH_4 dissolved in H_2O at a higher pressure of CH_4 forms more CH_3COOH and HCOOH.

Table 20.1 Comparison of TOR for the formation of acetic acid or oxygenates, including acetic acid, formic acid, and methanol on 0.10 wt% Rh/ZSM-5.

Entry	Catalyst	Catalytic temperature	TOR of acetic acid (molecule per site per second)	TOR of organic oxygenated (number of molecules per site per second)	Selectivity for the production of acetic acid
1	0.10 wt% Rh/ZSM-5[a]	150 °C	0.040[b]	0.099[b]	40.0%
2	0.10 wt% Rh/ZSM-5[a]	150 °C	0.070[c]	0.10[c]	70.1%
3	$Rh(NO_3)_3$[a]	150 °C	6.3×10^{-6}[d]	2.4×10^{-5}[d]	26.3%

[a] Here, the organic oxygenates include acetic acid, formic acid, and methanol (CO_2 was not included). Calculations of TORs of these catalysts were described in Supplementary Methods.
[b] The catalysis condition of 0.10 wt% Rh/ZSM-5: a mixture of 50 bar CH_4, 10 bar CO, 8 bar O_2, 2 h.
[c] The catalysis condition of 0.10 wt% Rh/ZSM-5: mixture of 50 bar CH_4, 10 bar CO, 8 bar O_2, 12 h.
[d] In total, 5 mL of 0.01 mol L^{-1} $Rh(NO_3)_3$ were added to the Parr reactor, and 50 bar CH_4, 10 bar CO, and 8 bar O_2 were introduced to the Parr reactor, and then the Parr reaction was sealed; the reaction was performed at 150 °C for about 90 h. This measurement was done for comparison with 0.10 wt% Rh/ZSM-5 in which Rh_1 cations were anchored in micropores. *Source:* [1] / Springer Nature / CC BY 4.0.

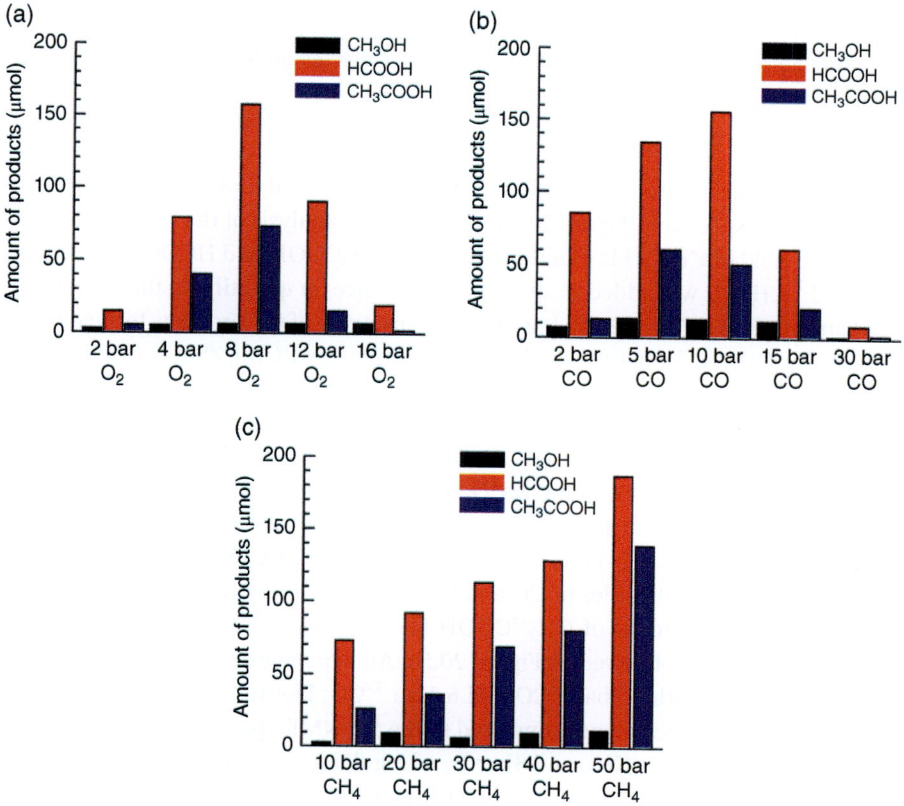

Figure 20.4 Influence of partial pressure O_2, CO, and CH_4 on catalytic performance. Yields of methanol (black), formic acid (blue), and acetic acid (red) in the chemical transformation of CH_4 at 150 °C in aqueous solutions at different pressures of O_2, CO, and CH_4. (a) 35 bar CH_4, 10 bar CO, and different pressure of O_2 at 150 °C for 2 h. (b) 50 bar CH_4, 8 bar O_2, and different pressure of CO at 150 °C for 1.5 h. (c) 10 bar CO, 8 bar O_2 and different pressure of CH_4 at 150 °C for 2 h. *Source:* [1] / Springer Nature / CC BY 4.0.

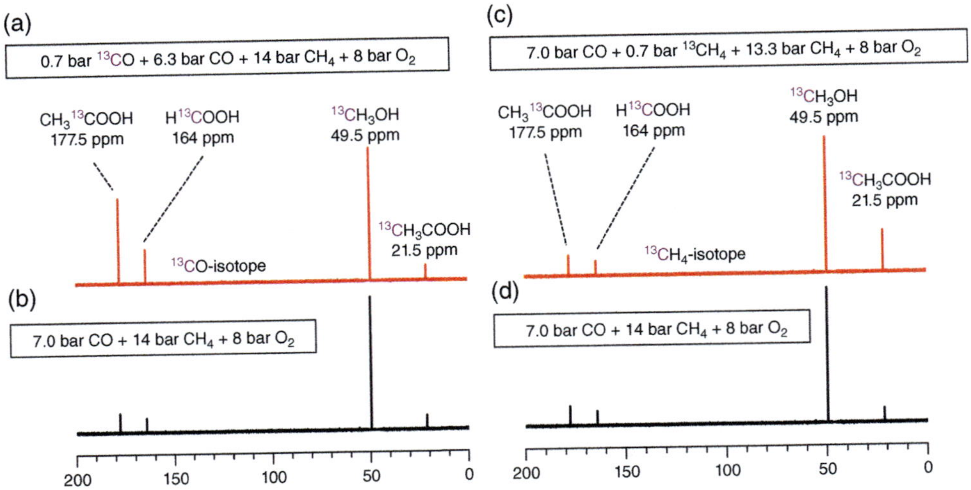

Figure 20.5 ^{13}C NMR studies of reaction using ^{13}CO or $^{13}CH_4$. ^{13}C NMR spectra of products of acetic acid, formic acid, and methanol on 28 mg 0.10 wt% Rh/ZSM-5 at 170 °C for 10 h in gas of (a) mixture of 0.7 bar ^{13}CO, 6.3 bar CO, 14 bar CH_4, and 8 bar O_2, (b) mixture of 7 bar CO, 14 bar CH_4, and 8 bar O_2, (c) mixture of 7 bar CO, 0.7 bar $^{13}CH_4$, 13.3 bar CH_4, and 8 bar O_2, and (d) mixture of 7 bar CO, 14 bar CH_4, and 8 bar O_2. (a) and (b) are isotope experiments; (c) and (d) are their corresponding contrast experiments. (b) and (d) were collected under the same condition. *Source:* [1] / Springer Nature / CC BY 4.0.

Isotope-labeled experiments confirmed that sources of CH_3 and C=O of CH_3COOH are CH_4 and CO gases, respectively. Figure 20.5 is the ^{13}C NMR analysis of the CH_3COOH and HCOOH. As the yield of CH_3OH is much smaller than CH_3COOH and HCOOH, 1.0 mmol isotope-labeled $^{13}CH_3OH$ was added as an internal reference for quantifying the amount of CH_3COOH and HCOOH. Figure 20.5b is a ^{13}C NMR spectrum of the products formed from 7.0 bar CO, 14 bar CH_4, and 8 bar O_2 catalyzed by 28 mg 0.10 wt% Rh/ZSM-5. The highest intensity in the ^{13}C NMR spectrum of Figure 20.5b is $^{13}CH_3OH$, which was added as an internal reference for other potential products containing ^{13}C, such as $^{13}CH_3COOH$ at 21.5 ppm, $H^{13}COOH$ at 164 ppm, and $CH_3{}^{13}COOH$ at 177.5 ppm. In principle, Figure 20.5a should have only $^{13}CH_3OH$ at 49.5 ppm as CO and CH_4 do not contain ^{13}C. In fact, commercial CO and CH_4 still contain a small amount of ^{13}CO and $^{13}CH_4$, respectively, although the majority of carbon atoms of CO and CH_4 are ^{12}C. Since there are small amounts of ^{13}CO and $^{13}CH_4$ in the commercial bottles of CO and CH_4, the reaction of regular CH_4, CO, and O_2 gases formed a small amount of $CH_3{}^{13}COOH$ (177.5 ppm), $H^{13}COOH$ (164 ppm), and $^{13}CH_3COOH$ (21.5 ppm) as observed in Figure 20.5b. An isotope experiment was done by replacing the 7 bar ^{12}CO with 0.7 bar ^{13}CO and 6.3 bar ^{12}CO; Figure 20.5a is the ^{13}C NMR spectrum of this isotope experiment. Compared to the ^{13}C NMR spectrum of the products formed from 7.0 bar CO, 14 bar CH_4, and 8 bar O_2 of regular gases (Figure 20.5b), the ratio of $CH_3{}^{13}COOH$ to the internal reference $^{13}CH_3OH$ increased a few times, but the ratios of $H^{13}COOH$ and $^{13}CH_3COOH$ to $^{13}CH_3OH$ remained constant upon the replacement of 0.7 bar CO with 0.7 bar ^{13}CO (Figure 20.5a). Thus, the isotope experiment of Figure 20.5a confirmed that CO is the source of carbonyl C=O of CH_3COOH. A similar isotope experiment was done in Figure 20.5c, where 0.7 bar CH_4 was replaced with 0.7 bar $^{13}CH_4$. ^{13}C

NMR spectrum of the products of 7.0 bar CO, 0.7 bar $^{13}CH_4$, 13.3 bar CH_4, and 8 bar O_2 was presented in Figure 20.5c. Upon this replacement, the ratio of $^{13}CH_3COOH$ to the internal reference $^{13}CH_3OH$ largely increased while the ratios of $CH_3{}^{13}COOH$ and $H^{13}COOH$ to $^{13}CH_3OH$ remained constant (Figure 20.5c); the isotope experiment of Figure 20.5c clearly shows the source of C of CH_3 of CH_3COOH is CH_4.

The reaction pathway for the synthesis of CH_3COOH from CH_4, CO, and O_2 was simulated. Figure 20.6 presents 26 elementary steps of this synthesis. Rh_1O_5 is the active site identified with *in situ* EXAFS studies. A C–H bond of CH_4 couples with Rh–O, forming a four-membered ring of C, H, O, and Rh as the first transition state (c3). Upon the C–H bond breaking, CH_3 is formed, and it bonds to the Rh atom, while H from CH_4 bonds to O to form an H–O (c4). Immediately, the intermediate c4 transforms to c5 by cleaving the O–O bond. Then, CO is inserted into Rh–O of c5, forming a carboxylic group on c6. In the intermediate c6, Rh is fully coordinated with six atoms or groups, including $HOOC^-$, O^-, H_3C, O, O, and O. The full coordination makes C of CH_3 and C of $HOOC^-$ couple readily through the second transition state (c7). This coupling breaks the C–Rh bond between CH_3 and the Rh atom, forming an intermediate c8 having a weakly bound CH_3COOH. CH_3COOH desorbs, and thus, a reactive Rh=O structure (c9) is formed. A second CH_4 molecule couples with Rh=O (c10), forming the third transition state (c11), where a four-membered ring of C, H, O, and Rh atoms is formed. Cleave of the C–H bond of CH_4 in the transition state (c11), CH_3 and OH groups are formed and bond to the Rh atom. Different from the insertion of the first CO into c5 to form c6, a second CO can bind to the Rh atom of c12 directly instead of insertion since only five atoms bond to the Rh atom in c12. In other words, the undercoordination makes CO directly bind to the Rh atom of c12, forming c13. The crowding packing of the three groups (CH_3, CO, and OH) in the hemisphere space in c13 makes the CH_3 and CO couple readily in the fourth TS (c14), forming the CH_3CO group in c15. Then, the electron-rich O atom of OH performs a nucleophilic attack on the C atom of the carbonyl of CH_3CO, forming CH_3COOH in c17. This is the second CH_3COOH molecule. Desorption of this CH_3OH creates the active site, Rh_1O_3 (c18). The c18 bonds a molecular O_2, forming the initial active site Rh_1O_5 (c1), which is the catalytic site for performing the first step of the next catalytic cycle. In summary, in the proposed catalytic cycle, two CH_4, two CO, and one O_2 participate in a catalytic cycle, forming two CH_3COOH molecules. Figure 20.6b is the energy profile of this synthesis of acetic acid catalyzed by Rh_1O_5 anchored in the pore of ZSM-5.

Other than the above synthesis of acetic acid through the CH_4 carbonylation involving molecular O_2, acetic acid can be synthesized through selective oxidation of ethane by H_2O_2 in a new catalytic process. This process is an integration of photocatalysis and heterogeneous catalysis. Pd atoms are introduced to the pores of the graphitic carbon nitride (g-C_3N_4). They are singly dispersed on g-C_3N_4. The 4d orbitals of anchored Pd_1 single atoms are the main contributor of the valence band of the Pd_1/g-C_3N_4. The C 2p of g-C_3N_4 is the main contributor of the conduction band. Upon photoirradiation, electrons in Pd 4d are excited to the C 2p of g-C_3N_4, making Pd_1 atoms positively charged. The electrons on g-C_3N_4 reduces Fe^{3+}, a product of Fenton reaction, to Fe_{2+}. The Fe^{2+} can drive the Fenton reaction to generate hydroxide radicals which oxidize C_2H_6 at 273K. Thus, this process synthesizes acetic acid from ethane and H_2O_2 with a selectivity of 97% for acetic acid at low temperature through the integrated photocatalysis and heterogeneous catalysis. More details can be found in reference [3].

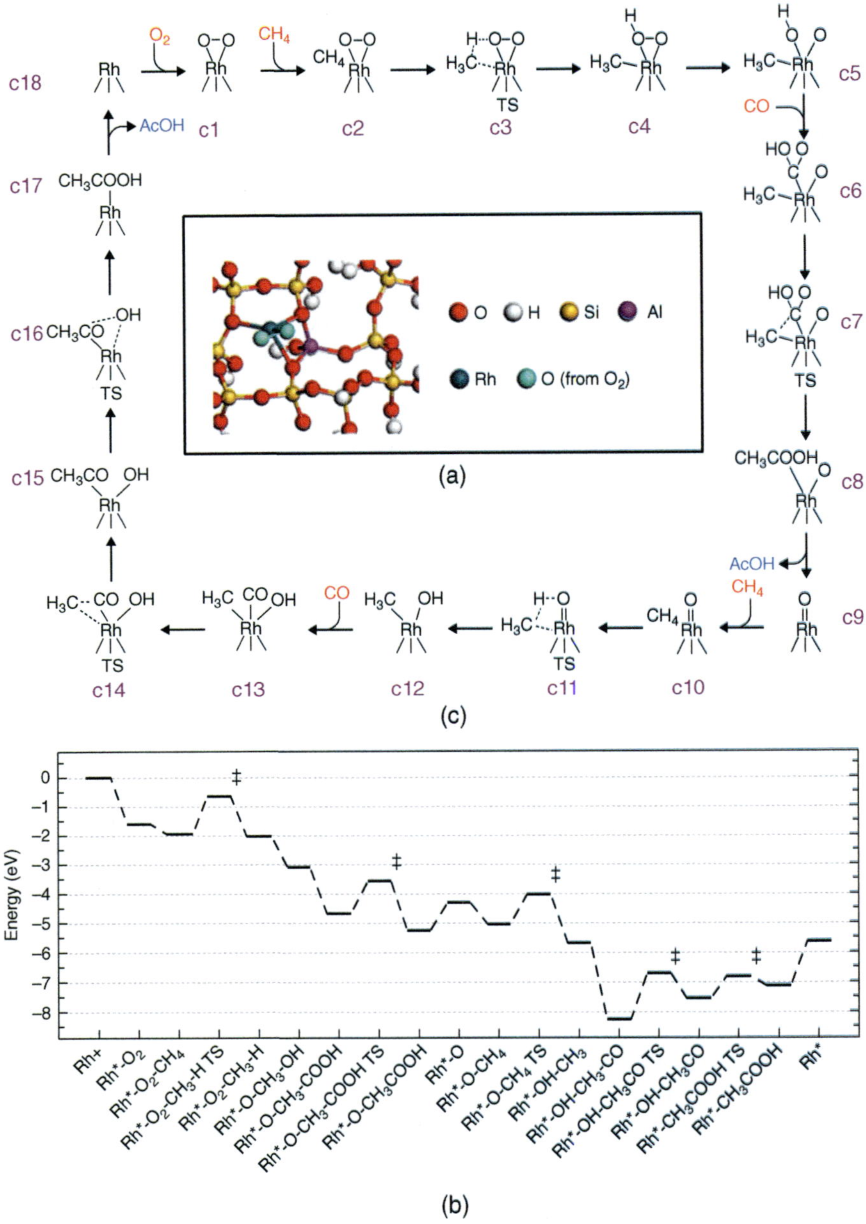

Figure 20.6 Computational studies of reaction pathway. Minimum energy paths and reaction schematic for the formation of acetic acid from CH_4, CO, and O_2 on Rh_1O_5/ZSM-5. The formation of acetic acid is illustrated in a catalytic cycle starting with the singly dispersed Rh_1O_5 site. The balanced reaction cycle consumes one O_2, two CH_4, and two CO to make two CH_3COOH molecules ($2CH_4 + 2CO + O_2 = 2CH_3COOH$). (a) The optimized catalytic sites, Rh_1O_5, were anchored on the Brønsted site in the microspore of ZSM-5. (b) Energy profile for the pathway of transforming CH_4, CO, and O_2 to CH_3COOH. (c) Intermediates and transition states for a complete catalytic cycle, starting with Rh_1O_5 (c1). Transition states are highlighted with the double dagger symbols. *Source:* [1] / Springer Nature / CC BY 4.0.

References

1 Tang, Y., Li, Y., Fung, V., Jiang, D., Huang, W., Zhang, S., Iwasawa, Y., Sakata, T., Nguyen, L., Zhang, X., Frenkel, A. I., Tao, F. 2018. "Single rhodium atoms anchored in micropores for efficient transformation of methane under mild conditions." *Nat. Commun.* 9, 1231.

2 Periana, R. A., Mironov, O., Taube, D., Bhalla, G., Jones, C. J. 2003. "Catalytic, oxidative condensation of CH4 to CH3COOH in one step via CH activation." *Science* 301, 814–818.

3 Zong, X., Chu, Y. C., Tang, Y., Li, Y., Wu, X. P., Sun, Z., Tao, F. 2024. "Highly selective photocatalytic synthesis of acetic acid at 0-25C". *Angew. Chem. Int. Ed.* 136, e202404598.

21

Challenges and Prospect

21.1 Challenge in Achieving High Selectivity for a Specific Product

Achieving high selectivity in catalytic reactions for the formation of an ideal product is imperative. A catalytic chemical transformation with high selectivity holds significant cost-saving potential, as the separation of the desired product from byproducts is typically expensive and, in some cases, technically unfeasible. Developing a chemical route for effectively separating a product from the solution requires considerable effort and resources. In addition, high selectivity in catalytic transformations has environmental advantages. A catalytic chemical transformation with low selectivity produces unwanted byproducts, necessitating additional inputs of time, cost, and technology to mitigate their negative impact on the environment. However, attaining high selectivity poses a challenge in the low-temperature catalytic transformation of CH_4 to value-added chemicals through partial oxidation. In this context, "partial oxidation" refers to the oxidation of CH_4 to various products except CO_2. The primary obstacle lies in the inert nature of CH_4. Most intermediates resulting from the activation of the first C–H bond, as well as intermediate compounds produced from the partial oxidation of CH_4, exhibit activity greater than CH_4 itself. For example, CH_4 can undergo catalytic oxidation with H_2O_2 to form the initial product CH_3OOH, which then transforms into CH_3OH – a desired product. Unfortunately, compared to CH_4, CH_3OH is considerably more active in oxidation by H_2O_2. In general, the ideal product, CH_3OH, is susceptible to further oxidation, leading to the formation of formaldehyde, formic acid, and even carbon dioxide, depending on the specific catalytic condition. The inertness of CH_4 poses a significant hurdle in achieving high selectivity for a product in these catalytic processes.

21.2 Challenge in Achieving High Conversion of CH_4

The majority of activation and catalytic transformations described above occur at a solid–liquid interface, where the gas-phase CH_4 remains above the liquid phase in high-pressure reactor such as the Parr reactor. Typically, the liquid phase comprises an

Low-temperature Activation and Catalytic Transformation of Methane to Non-CO_2 Products, First Edition. Franklin Tao.

aqueous solution containing solid catalyst particles. Given that CH_4 is in the gas phase, CH_4 molecules can only access catalytic or activation sites through first dissolution and then diffusion in the liquid phase. However, the low solubility of CH_4 in an aqueous solution results in a limited conversion of gas-phase CH_4. The conversion of CH_4 catalyzed by the zeolite-based catalyst is contingent upon the solubility of CH_4 in the solvent, dictated by Henry's law. Higher solubility enables more CH_4 molecules to access the catalyst, resulting in a more significant transformation of CH_4 into product molecules. To address this limitation, there is a need to identify a solvent with substantially higher solubility for CH_4 than water. However, a considerable challenge arises as most solvents exhibit higher reactivity than CH_4 in oxidation processes. Consequently, an ideal solvent should exhibit a low activity in oxidation by the oxidant, such as H_2O_2, in addition to having high solubility for CH_4. Furthermore, the ideal solvent must meet the following criteria. **Size Compatibility:** To prevent competitive adsorption of solvent molecules on catalytic sites within the micropores of zeolite, an appropriate solvent molecule having a size larger than the diameter of the pore should be used, avoiding undesired adsorption of solvent molecules on the catalytic sites in the micropores. **Affinity for Product Molecules:** To prevent the accumulation of product molecules in the local environment of catalytic sites, a solvent should exhibit a reasonably high affinity for product molecules. For an efficient catalyst system in methane low-temperature conversion, the selection of a well-designed solvent becomes crucial. A well-designed solvent system is a key factor for creating a favorable environment, leading to a high conversion of CH_4 and a high yield of the ideal product.

21.3 Challenge in Finding a New Reaction

Finding new reactions that are thermodynamically favorable in the low-temperature regime (25–200 °C) presents a challenge. The partial oxidation of CH_4 to CH_3OH with O_2 is only slightly exothermic, and the use of this reaction is hindered by the low activity of the oxidant, O_2. In comparison, N_2O and H_2O_2 exhibit higher activity in oxidizing CH_4, making them more attractive. Another alternative oxidant worth considering is SO_2. It is widespreadly used in oxidative dehydrogenation processes, such as transforming isobutane to isobutylene and catalyzing the conversion of butene into butadiene with the support of Cu catalyst [1, 2]. SO_2 has also been utilized in various transformations involving isobutane, butane, hexane, and 3-methylpentane [3]. The advantages of SO_2 include its lower oxidation temperature compared to O_2 for the same reductant, suggesting higher activity in oxidizing CH_4, and its cost-effectiveness as it is a byproduct of chemical industries. In addition to SO_2, literature references indicate the use of carbonyl sulfide (OCS, O=C=S) [4–6]. Compared to O_2 and N_2O, OCS stands out for its high selectivity for alkene in oxidative dehydrogenation processes involving light alkanes [7].

Designing more reactions that are thermodynamically favorable in 25–200 °C is crucial for low-temperature activation and catalytic conversion. Currently, there are fewer than ten reported reactions for the low-temperature transformations of CH_4 into value-added chemicals. Given the limited reported reactions, it is anticipated that there are a multitude of thermodynamically favorable reactions yet to be explored. For example, thermodynamic calculations suggest that the transformation of CH_4 to HCHO is thermodynamically favorable ($CH_4 + O_2 \rightarrow HCHO + H_2O$, $\Delta G^o{}_{100\,°C} = -281.9\,kJ\,mol^{-1}$). However, there is a scarcity of reports on catalysts actively facilitating the transformation of CH_4 to HCHO at low temperatures. It highlights the significance of further exploration and innovation in this area.

21.4 Challenge in Reproducible Preparation of Metallozeolite with Homogeneous Catalytic Sites

A hurdle in the partial oxidation of CH_4 to CH_3OH lies in achieving the reproducible preparation of metallozeolite with uniformly distributed catalytic sites. Various methods are known for preparing metallozeolites, such as Co-ZSM-5, including incipient wetness impregnation [8, 9], wet ion exchange [10–12], solid-state ion exchange [13, 14], and sublimation [14]. The common challenge is to devise a method for consistently creating metallozeolites with active sites tailored for partial oxidation of CH_4 to specific oxides, such as methanol or acetic acid.

Reproducibility is another challenge in the preparation of metallozeolites. Lack of reproducibility often arises due to unidentified factors, referred to as unknown factors, contributing to the structure formation. These factors may not be explicitly detailed in experimental protocols, as they are assumed to be common knowledge or routine procedures. Depending on the awareness of the experimentalists, these unknown factors or unmentioned details can significantly or marginally influence the formation of an ideal catalyst structure. Unfortunately, since these factors are not recognized as critical for synthesis, experimentalists may not adhere to specific guidelines, resulting in varied outcomes. For example, a catalyst synthesis step involving pH adjustment to 9.5 may be mentioned without specifying the base used, leading to variations in catalytic performance as the authors were not aware that the sodium of the base, NaOH, is the significant promoter to their observed catalytic performance. Some experimentalists may use ammonia solution for pH adjustment, inadvertently creating a sodium-free catalyst with lower catalytic performance, thereby compromising reproducibility.

To address the challenge of reproducibility, it is essential to provide comprehensive details, particularly in the preparation of catalyst samples, and correctly assign the actual factor dominating the catalytic performance. In addition, inaccuracies in synthesis parameters and undisclosed ordering of experimental steps contribute to uncertainties that can yield different results among experimentalists. One solution to enhance reproducibility is to use robotic systems in exploring new metallozeolite synthesis methods since robots has already employed for synthesizing nanoparticles and offered precision and consistency in the experimental process [15].

21.5 Challenge in Characterizing the Actual Catalyst during Catalysis

Literature reveals that the active phase of Cu atoms in Cu-MOR, prepared by aqueous ion exchange from copper acetate and the sodium form of mordenite, initially exists as dehydrated Cu (II) oxide species, specifically in the form of mono-(μ-oxo) and bi-(μ-oxo)dicopper species [16]. However, it has been observed that the mono(μ-oxo)di-copper species restructure to a hydrated Cu(II) species due to the introduction of small amounts of moisture. This highlights the sensitivity of the active phase structure to moisture, emphasizing the necessity of in-situ/operando studies. Unfortunately, conducting in-situ or operando studies of the catalysis of partial oxidation of CH_4 at the solid–liquid–gas interface has been quite challenging.

In the case of low-temperature transformations at the solid–liquid interface, where the gas phase must be maintained at high pressure, catalysis essentially occurs at the solid–liquid interface under a high-pressure gas phase of CH_4. Consequently, in-situ and operando studies face considerable obstacles. Several factors contribute to the difficulty of these studies. For example, the encapsulation of catalytic sites in micropores of zeolite particles makes surface-sensitive techniques inaccessible. Moreover, most surface analytical techniques struggle to access the surface of catalyst particles dispersed in a liquid phase under a high-pressure gas phase. While XPS can study the surface of catalyst nanoparticles in a liquid solution by loading it into a miniature cell separated from the high vacuum environment of a traditional XPS system [17–22], this method faces challenges when studying catalyst nanoparticles in an aqueous solution under a high-pressure gas phase, due to the presence of high-pressure methane.

X-ray absorption spectroscopy (XAS), including X-ray absorption near edge structure (XANES) and extended X-ray absorption fine structure (EXAFS), offers valuable but limited information. XAS is a photon-in and photon-out analytical technique. Unlike supported metal or oxide catalysts in a gas phase, the catalytic reaction of partial oxidation of CH_4 to an ideal product occurs on catalyst particles dispersed in a liquid under a high-pressure gas phase (10 bars to several tens of bars). Literature has demonstrated the feasibility of characterizing metal atoms on the walls of the micropores of zeolites dispersed in a liquid under a gas phase at high pressure during catalysis through the XAS technique by using a well-designed reactor [23–25]. Furthermore, quick XAS has been shown to track the evolution of chemical and coordination environments of metal atoms as a function of time [26].

A technical challenge of using XAS is the characterization of catalysts during catalysis at high temperatures or relatively high temperatures. Most reported characterizations of catalysts for CH_4 oxidation and catalysis using XAS are performed at or near room temperature, even though the activation of metallozeolites such as Cu-MOR is conducted at 450 °C in O_2, or the following catalysis is at 100–200 °C [25, 26]. This is due to the temperature-dependent Debye–Waller factor (DWF) contributing to the width of a peak in the radial distribution function. At higher temperatures, the DWF is larger, resulting in a wider peak, causing

EXAFS to lose its spatial resolution. To achieve better resolution for meaningful insights into the coordination environment of a chosen element, most data for XAS analyses are actually collected at room temperature or near room temperature in the mixture of reactants of a reaction, although still referred to as in-situ or operando studies in the literature. It is well known that structural information deduced from XAS data collected at room temperature or near room temperature may not represent the actual phase during catalysis at high temperatures or relatively high temperatures. This limitation poses a challenge in accessing the actual phase of a catalyst during pretreatment at high temperatures or during catalysis at relatively high temperatures through the XAS technique.

Using XPS to generate data from metal sites in micropores of zeolite faces challenges. While XANES can provide insights into the oxidation station of metal atoms, XPS provides more straightforward information on oxidation states and readily distinguishes the oxidation states of coexisting species. Another significant difference is that XPS can quantify elemental compositions and atomic fractions of species with different oxidation states of the same element but XANES cannot. However, the short mean free path of photoelectrons from atoms of a specific element in micropores excited by a soft X-ray, Al Kα, or even tender X-ray makes an XPS study of the elements in the micropores at the solid–liquid or solid–liquid–gas interface extremely challenging. A promising solution is to use a hard X-ray source to excite high-energy electrons from these atoms, as the longer mean free path of these high-energy electrons may allow XPS to access those metal atoms encapsulated in micropores near to the surface of a zeolite particle [28].

21.6 Challenge in the Fundamental Understanding of the Catalytic Mechanism

Most low-temperature activation and catalytic transformations take place on catalytic sites either anchored on the surface of a catalyst nanoparticle or encapsulated in a micropore. These catalytic sites are dispersed in a liquid solution that contains liquid reactants, solvents, and dissolved molecules of gaseous reactants, all under a high-pressure gas phase of reactants such as a mixture of CH_4, CO, and O_2. The complexity of the reaction system, arising from unknown solvent effects, inhomogeneous sites, and etching of catalytic sites in a liquid environment, can significantly obscure the intrinsic correlation between the actual catalyst structure during catalysis and the corresponding catalytic performance. Thus, achieving a fundamental understanding of the catalytic mechanism at the atomic or molecular level for a reaction occurring at a solid–liquid or even a solid–liquid–gas interface is more challenging when compared to catalysis performed at a solid–gas interface.

21.7 Challenge in Transforming Low-concentration Methane of Waste Gas

Waste gas typically arises from organic or biomass containing waste materials and often contains methane (CH_4) at low concentrations. As discussed in preceding chapters, the low-temperature oxidation of CH_4 molecules is typically conducted at the solid–liquid interface, with oxidation events taking place on or in catalyst particles dispersed in a liquid solution. The CH_4 molecules involved in the reaction are those dissolved in the liquid phase. Achieving a relatively high concentration of CH_4 in the liquid requires high pressure, often in the range of several tens of bars.

Given the inherently low concentration of CH_4 in the waste gas, it becomes necessary to pressurize the waste gas to several hundred bars to attain a reasonable concentration of CH_4 dissolved in the solvent. The high-pressure working condition increases footprint cost and rises safety concern. Another significant challenge lies in the potential interference of other components present in the waste gas on catalytic performance of the catalyst, which may either readily react with or poison the catalytic sites. For instance, gases such as pentane could exhibit higher activity than CH_4 in partial oxidation, complicating the generation of value-added chemicals from waste gas. Despite these challenges, this task holds environmental significance by contributing to the mitigation of CH_4 emissions into the atmosphere.

References

1 Rosen, R. 1938. *U. S. Pat.* 2,126,817.

2 Stone, C. M., Korpi, K. J., 1947. *U. S. Pat.* 2,423,418.

3 Adams, C. R. 1968. "Catalytic oxidations with sulfur dioxide: I. Exploratory studies." *J. Catal.* 11, 96–112.

4 Boswell, D. E. 1968. *U. S. Pat.* 3,399,243.

5 Haag, W. O., Miale, J. N. 1974. *U. S. Pat.* 3,787,517 (1974); 1974. *U. S. Pat.* 3,281,278; 1975. *U. S. Pat.* 3,875,252.

6 Haag, W. O., Miale, J. N. 1976. In G. C. Bond, P. B. Wells, & F. C. Tompkins (Eds.), *Proceedings of the Sixth International Congress on Catalysis, Imperial College, London, 12–16 July, 1976*. Chemical Society.

7 Ward, M. B. 1977. "The oxidative dehydrogenation of ethane by nitrous oxide over molybdenum oxide supported on silica gel." *J. Catal.* 50, 306–318.

8 Pierella, L. B., Saux, C., Caglieri, S. C., Bertorello, H. R., Bercoff, P. G. 2008. "Catalytic activity and magnetic properties of Co–ZSM-5 zeolites prepared by different methods." *Appl. Catal. A Gen.* 347, 55–61.

9 Wang, X., Chen, H., Sachtler, W. M. H. 2000. "Catalytic reduction of NOx by hydrocarbons over Co/ZSM-5 catalysts prepared with different methods." *Appl. Catal. B, Environ.* 2000, 26, L227–L239.

10 Dědeček, J., Kaucký, D., Wichterlová, B. 2000. "Co2+ ion siting in pentasil-containing zeolites, part 3.: Co2+ ion sites and their occupation in ZSM-5: a VIS diffuse reflectance spectroscopy study." *Micropor. Mesopor. Mater.* 2000, 35–36, 483–494.

11 Mhamdi, M. B., Khaddar-Zine, S., Ghorbel, A. 2008. "Influence of the Co/Al ratio and the temperature of thermal treatment on cobalt speciation and catalytic properties of Co-ZSM-5 prepared by solid-state ion exchange." *Appl. Catal. A. Gen.* 337, 39–47.

12 Montanari, T., Bevilacqua, M., Resini, C., Busca, G., Pirone, R., Ruoppolo, G. 2007. "A spectroscopic study of the nature and accessibility of protonic and cationic sites in H- and partially exchanged Cu- and Co-MFI zeolites." *J Porous Mater* 14, 291–297.

13 da Cruz, R. S., Macarenhas, A. J. S., Andrade, H. M. C. 1998. “Co-ZSM-5 catalysts for N2O decomposition.” *Appl. Catal. B Environ.* 18, 223–231.

14 Peng, A., Kung, M. C., Ross, M. O., Hoffman, B. M., Kung, H. H. 2020. “The role of Co-ZSM-5 catalysts in aerobic oxidation of ethylbenzene.” *Top. Catal.* 63, 1708–1716.

15 Jiang, Y., Salley, D., Sharma, A., Keenan, G., Mullin, M. 2022. “An artificial intelligence-enabled chemical synthesis robot for exploration and optimization of nanomaterials.” *Sci. Adv.* 8, eabo2626.

16 Alayon, E. M. C., Nachtegaal, M., Bodi, A., van Bokhoven, J. A. 2014. “Reaction conditions of methane to methanol conversion affect the structure of active copper sites.” *ACS Catal.* 4, 16–22.

17 Kolmakov, A., Dikin, D. A., Cote, L. J., Huang, J., Abyaneh, M. K., Amati, M., Gregoratti, L., Günther, S., Kiskinova, M. 2011. “Graphene oxide windows for in situ environmental cell photoelectron spectroscopy.” *Nat. Nanotech* **6**, 651–657.

18 Velasco-Velez, J. J., Pfeifer, V., Hävecker, M., Weatherup, R. S., Arrigo, R., Chuang, C. H., Stotz, E., Weinberg, G., Salmeron, M., Schlögl, R., Knop-Gericke, A. 2015. “Photoelectron spectroscopy at the graphene–liquid interface reveals the electronic structure of an electrodeposited cobalt/graphene electrocatalyst.” *Angew. Chem. Int. Ed.* 54, 14554–14558.

19 Falling, L. J. R., Mom, V., Diaz, L. E. S., Nakhaie, S., Stotz, E., Ivanov, D., Hävecker, M., Lunkenbein, T., Knop-Gericke, A., Schlögl, R., Velasco-Vélez, J. J. 2020. “Graphene-capped liquid thin films for electrochemical operando X-ray spectroscopy and scanning electron microscopy.” *ACS Appl. Mater. Interfaces* 12, 37680–37692.

20 Nguyen, L., Tao, P., Liu, H., Al-Hada, M., Amati, M., Sezen, H., Gregoratti, L., Tang, Y., House, S. D., Tao, F. 2018. “X-ray photoelectron spectroscopy studies of nanoparticles dispersed in static liquid.” *Langmuir* 34, 9606–9616.

21 Nguyen, L., Tao, P., Liu, H., Al-Hada, M., Amati, M., Sezen, H., Tang, Y., Gregoratti, L., Tao, F. (2018) “Studies of the surface of metal nanoparticles in a flowing liquid with XPS,” *Chem. Commun.* 54, 9981–9984.

22 Tao, F. 2025. “Photoelectron spectroscopic operando studies of surface of nanoparticles in liquid or gas at bar pressure with an X-ray photoelectron spectrometer.” *Nat. Protocol20,* https://doi.org/10.1038/s41596-024-01092-x

23 Nguyen, L., Tao, F. 2018. “Reactor for tracking catalyst nanoparticles in liquid at high temperature under a high-pressure gas phase with X-ray absorption spectroscopy.” *Rev. Sci. Instrument.* 89, 024102.

24 Tang, Y., Nguyen, L., Li, Y., Tao, F. 2023. “In situ and operando study of catalysts during high-temperature high-pressure catalysis in a fixed-bed plug flow reactor with x-ray absorption spectroscopy.” *Rev. Sci. Instrument.* 94, 054102.

25 Nguyen, L., Tang, Y., Li, Y., Zhang, X., Wang, D., Tao, F. 2018. “Dual reactor for in situ/operando fluorescent mode XAS studies of sample containing low-concentration 3d or 5d metal elements.” *Rev. Sci. Instrument.* 89, 054103.

26 Alayon, E., M., C., Nachtegaal, M., Bodi, A., Ranocchiari, M., van Bokhoven, J. A. 2015. “Bis(μ-oxo) versus mono(μ-oxo)dicopper cores in a zeolite for converting methane to methanol: an in situ XAS and DFT investigation.” *Phys. Chem. Chem. Phys.* 17, 7681–7693.

27 Shan, J., Huang, W., Nguyen, L., Yu, Y., Zhang, S., Li, Y., Frenkel, A. I., Tao, F. 2013. "Conversion of methane to methanol with a bent mono (μ-oxo) dinickel anchored on the internal surfaces of micropores." *Langmuir* 30, 8558–8569.

28 Tao, F. 2023. *Application of Ambient Pressure X-ray Photoelectron Spectroscopy to Heterogeneous Catalysis*. Wiley

Index

Low-temperature Activation and Catalytic Transformation of Methane to Non-CO_2 Products, First Edition. Franklin Tao.

i

k

l

m

n

o

p

q

r

s

t

Printed and bound by CPI Group (UK) Ltd, Croydon, CR0 4YY
17/09/2026
14964269-0002